Artificial Cilia

RSC Nanoscience & Nanotechnology

Series Editors:
Professor Paul O'Brien, *University of Manchester, UK*
Professor Sir Harry Kroto FRS, *University of Sussex, UK*
Professor Ralph Nuzzo, *University of Illinois at Urbana-Champaign, USA*

Titles in the Series:
1: Nanotubes and Nanowires
2: Fullerenes: Principles and Applications
3: Nanocharacterisation
4: Atom Resolved Surface Reactions: Nanocatalysis
5: Biomimetic Nanoceramics in Clinical Use: From Materials to Applications
6: Nanofluidics: Nanoscience and Nanotechnology
7: Bionanodesign: Following Nature's Touch
8: Nano-Society: Pushing the Boundaries of Technology
9: Polymer-based Nanostructures: Medical Applications
10: Metallic and Molecular Interactions in Nanometer Layers, Pores and Particles: New Findings at the Yoctolitre Level
11: Nanocasting: A Versatile Strategy for Creating Nanostructured Porous Materials
12: Titanate and Titania Nanotubes: Synthesis, Properties and Applications
13: Raman Spectroscopy, Fullerenes and Nanotechnology
14: Nanotechnologies in Food
15: Unravelling Single Cell Genomics: Micro and Nanotools
16: Polymer Nanocomposites by Emulsion and Suspension
17: Phage Nanobiotechnology
18: Nanotubes and Nanowires: 2nd Edition
19: Nanostructured Catalysts: Transition Metal Oxides
20: Fullerenes: Principles and Applications, 2nd Edition
21: Biological Interactions with Surface Charge Biomaterials
22: Nanoporous Gold: From an Ancient Technology to a High-Tech Material
23: Nanoparticles in Anti-Microbial Materials: Use and Characterisation
24: Manipulation of Nanoscale Materials: An Introduction to Nanoarchitectonics
25: Towards Efficient Designing of Safe Nanomaterials: Innovative Merge of Computational Approaches and Experimental Techniques
26: Polymer–Graphene Nanocomposites
27: Carbon Nanotube-Polymer Composites
28: Nanoscience for the Conservation of Works of Art
29: Polymer Nanofibers: Building Blocks for Nanotechnology
30: Artificial Cilia

How to obtain future titles on publication:
A standing order plan is available for this series. A standing order will bring delivery of each new volume immediately on publication.

For further information please contact:
Book Sales Department, Royal Society of Chemistry, Thomas Graham House, Science Park, Milton Road, Cambridge, CB4 0WF, UK
Telephone: +44 (0)1223 420066, Fax: +44 (0)1223 420247
Email: booksales@rsc.org
Visit our website at www.rsc.org/books

Artificial Cilia

Edited by

Jaap M J den Toonder
Philips Research, High Tech Campus 7, Eindhoven, The Netherlands
Email: jaap.den.toonder@philips.com

Patrick R Onck
University of Groningen, Zernike Institute for Advanced Materials, The Netherlands
Email: p.r.onck@rug.nl

RSC Publishing

RSC Nanoscience & Nanotechnology No. 30

ISBN: 978-1-84973-597-1
ISSN: 1757-7136

A catalogue record for this book is available from the British Library

Published by The Royal Society of Chemistry,
Thomas Graham House, Science Park, Milton Road,
Cambridge CB4 0WF, UK

Registered Charity Number 207890

For further information see our web site at www.rsc.org

Printed in the United Kingdom by CPI Group (UK) Ltd, Croydon, CR0 4YY

Preface

The idea for this book was born in December 2010 during the 2nd European Conference on Microfluidics (μFlu'10). During that conference we organized a special session on cilia-driven flow, with beautiful contributions by many groups showing a variety of approaches to flow generation inspired by biological cilia. It became clear to us that this exciting field would deserve to be brought to the attention of a wider audience because of its direct relevance for practical applications in micro-fluidics and lab-on-a-chip devices, because of its inherent multi-disciplinary nature (combining fields such as fluid mechanics, materials science, biology, numerical modelling and experimental fluid characterization), but also simply because this field exemplifies how biology can inspire scientific and technological developments.

Most of the presenters in that special session have contributed to this book, and we believe it gives a good overview of the status of the field of artificial cilia, ranging from design and actuation approaches, fabrication methods, characterization and modelling, to micro-fluidics applications.

We thank the European Committee for funding the project 'Artic', which made it possible for us to play a role in the initial stages of this emerging field. We thank the μFlu'10 committee for the opportunity to organize the session dealing with cilia-driven flow. We thank the Royal Society of Chemistry for the possibility to publish the book and for all the practical support. And, we thank all the authors for their excellent contributions!

We hope that readers will be inspired by this book, just as we have been inspired by these intriguing little hairs called cilia!

Jaap den Toonder and Patrick Onck

RSC Nanoscience & Nanotechnology No. 30
Artificial Cilia
Edited by Jaap M J den Toonder and Patrick R Onck

Published by the Royal Society of Chemistry, www.rsc.org

Contents

RSC Nanoscience & Nanotechnology No. 30
Artificial Cilia
Edited by Jaap M J den Toonder and Patrick R Onck

Published by the Royal Society of Chemistry, www.rsc.org

Section 2 Experimental Approaches to Create Artificial Cilia

CHAPTER 1

Introduction

JAAP DEN TOONDER*[a,b] AND PATRICK ONCK[c]

[a] Philips Research, High Tech Campus 34, 5656 AE Eindhoven, The Netherlands; [b] Eindhoven University of Technology, Postbox 513, 5600 MB Eindhoven, The Netherlands; [c] University of Groningen, Zernike Institute for Advanced Materials, Micromechanics of Materials, Nijenborgh 4, 9747AG Groningen, The Netherlands
*Email: j.m.j.d.toonder@tue.nl

1.1 Natural Cilia

Nature has devised many different ways of creating fluid flow, most of them for animal propulsion, that is, for flying or swimming. At larger scales, examples are the flapping wings of birds, and the waving tails of fishes. Flapping wings are also found at smaller scales in insects. At really small scales, typically for sub-millimetre sizes, a fluid manipulation mechanism used by nature is that by cilia or flagella.

Cilia can be viewed as small hairs or flexible rods, with a typical length between 2 and 15 μm. They cover the outer surface of micro-organisms, such as Paramecia, shown in Figure 1.1a. The length of Paramecia is about 100 μm, and its surface contains over 4000 cilia. These cilia move back and forth in a concerted manner, and are very effective in generating flow: the swimming speed of Paramecium, for example, can be approximately 1 mm s^{-1} (*i.e.* it can travel a distance of 10 times its own body length in a second).

An individual cilium moves in a particular, asymmetric manner, as illustrated in Figure 1.1b. It has a so-called effective stroke, during which the cilium is more or less straight, and its effect on the fluid is maximized. During the recovery stroke, its effect on the surrounding fluid is minimized since the cilium

RSC Nanoscience & Nanotechnology No. 30
Artificial Cilia
Edited by Jaap M J den Toonder and Patrick R Onck

Published by the Royal Society of Chemistry, www.rsc.org

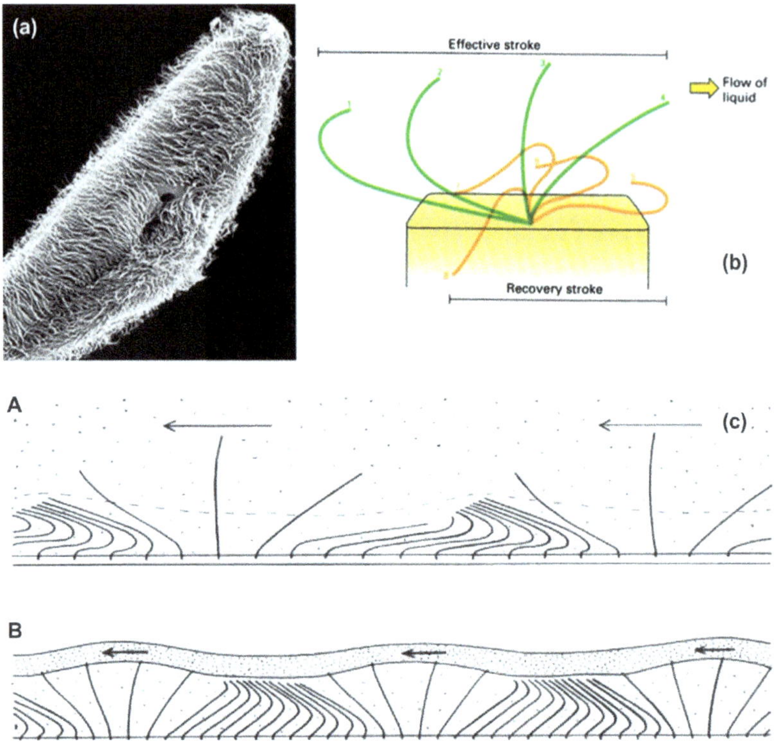

Figure 1.1 (a) Paramecium, covered with thousands of cilia. The size of the Paramecium is about 100 mm. (Reproduced from den Toonder *et al.*[1]) (b) The movement of an individual cilium. The motion is asymmetric: there is an effective stroke and a recovery stroke. (c) The collective wave-like motion of many cilia on a surface. This is called a 'metachronic wave'. If the wave travels in the direction of the effective stroke, it is called 'symplectic'; if it travels against it, it is called 'antiplectic'. (Reproduced from Sleigh.[2])

has a more curved shape. The micro-organism propulsion is in the direction opposite to the effective stroke. The movement of the cilium is always in a plane perpendicular to the surface during the effective stroke. The recovery stroke movement may lie in the same plane, but also in a plane perpendicular to the effective-stroke plane, so that the movement of a cilium may be truly three-dimensional (the latter is, in fact, the case for Paramecium). The beating frequency of the cilia, typically, is tens of hertz.

The collective movement of the cilia seems to occur in a concerted fashion. Neighbouring cilia move somewhat out of phase, so that a collective wave-like motion, going over the micro-organism's surface, takes place. It is interesting that this wave may travel either in the same direction as the swimming direction of the micro-organism (but opposite to the effective stroke; this is called an antiplectic metachronic wave, and occurs for Paramecium) or in the opposite direction (called symplectic metachronic wave behaviour). This is illustrated in

Figure 1.1c. The origin, as well as the physiological reason, for this metachronic co-ordination is not yet completely understood.

Flagella have the same internal structure as cilia, but they are usually longer, typically between 20 and 100 μm, and they do not cover surfaces in large quantities, such as cilia, but as a single flagellum or with a few. Also, their movement is usually different from those of cilia. Instead of having an effective and a recovery stroke, a flagellum mostly makes a helical (cork-skrew) or wave-like motion. The best known example is the flagellum of spermatozoa. A micro-organism that is classified as a flagellate by most authors, but as a ciliate by others, is Chlamydomonas (or green algae), shown in Figure 1.2. The length of its body is 10 μm, and it has two flagellates with a length of 10–15 μm. The flagella make an effective recovery stroke, beating with a frequency of about 50 Hz, giving the organism a swimming speed of 100 $\mu m\ s^{-1}$. Interestingly, the beating pattern in Chlamydomonas can change from a typical ciliary beating to flagellar beating upon an external trigger such as intense light or Ca^{2+}.

There are many different micro-organisms that make use of ciliary propulsion or fluid manipulation. Figure 1.3 depicts some of those that can be found in lakes and rivers. Note the scale bar on the right, representing a length of 1 mm.

Next to active cilia, which are used to induce movement, cilia are also used for sensing, for example of fluid flow, but also for detecting other physical and chemical signals. Both motile and non-motile cilia and flagella are present in the human body, at various locations and with various functions, as shown in Figure 1.4. For example, there are cilia in the cochlear, the inner ear, that contribute in detection of vibration caused by sound. As already mentioned, each spermatozoon swims by beating a flagellum. Also, the Fallopian tubes of females are covered with cilia that move the fertilized ovum from the ovary to

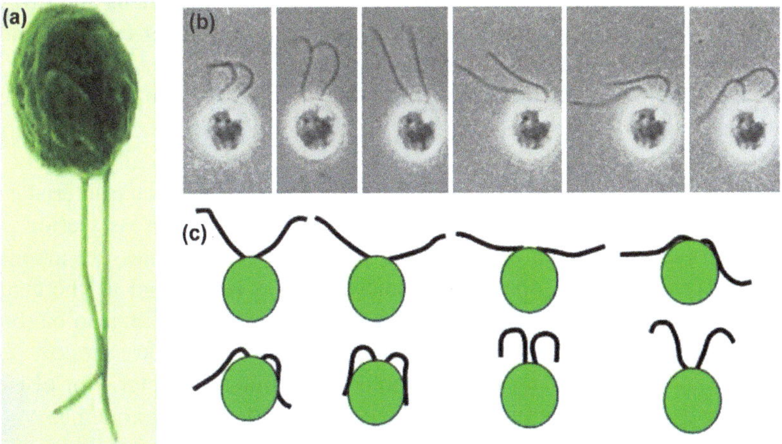

Figure 1.2 Chlamydomonas (green algae), a micro-organism with two flagella. It swims by an asymmetric beating of the two flagella. The size of the body of Chlamydomonas is about 10 mm, and its flagella are about 12 mm long. (a) and (c) are reproduced from Guirao;[3] (b) is © 1984 Rockefeller University Press; originally published as Hoops *et al.*[4]

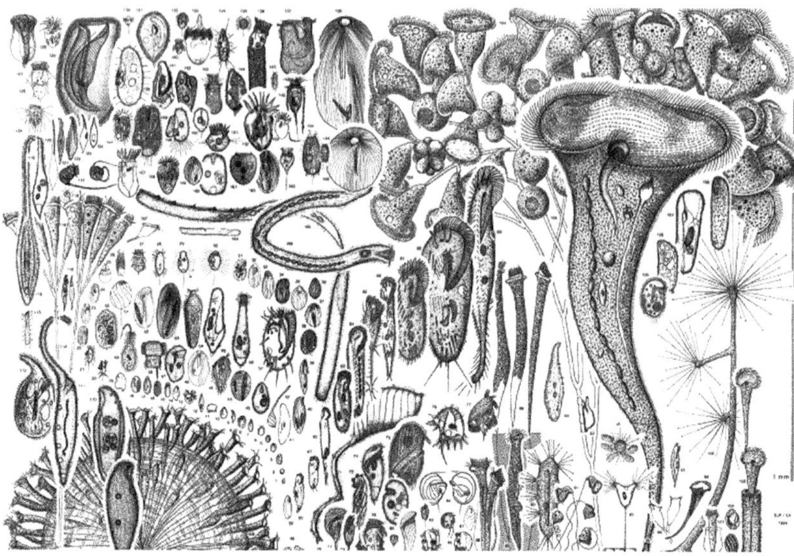

Figure 1.3 Ciliated micro-organisms living in lakes and rivers. Those placed towards the top left are typically found in the open water of lakes; those close to the centre at the base are all anaerobic. The remainder are generally found in sediments and detritus, and attached to submerged surfaces (*e.g.* aquatic animals and plants). All are drawn to scale (see scale bar showing 1 mm at the right hand side). The Ciliate Diversity Chart, reproduced with friendly permission of B. J. Finlay and G. F. Esteban, Institute of Freshwater Ecology, Windermere Laboratory, UK.

the uterus, where the ovum attaches itself. Motile cilia are also present in the lining of human lungs and the windpipe (trachea), to sweep mucus and dirt out of the airways in order to avoid infections. More examples of cilia within the human body can be found in Ibanez-Tallon *et al.*[5]

Motile cilia and flagella both have the same characteristic internal structure. The skeleton of a cilium (or flagella) is made up by a flexible cylindrical structure that is called the axoneme. Figure 1.5a schematically depicts the constituents of the axoneme, showing the structure that is so characteristic for all motile cilia and flagella. An electron micrograph of the cross section of a cilium can be seen in Figure 1.5b. Nine pairs of micro-tubules are arranged along the periphery, and one pair of micro-tubules is situated at the centre. Micro-tubules are biopolymer filaments, which, for example, can also be found in the cytoskeleton (the internal skeleton of cells). They are hollow rods approximately 25 nm in outer diameter and 14 nm inner diameter. For obvious reasons, the axoneme structure shown is known as the $9+2$ axoneme.

The nine outer pairs of micro-tubules are connected by nexin links. Also, each of the outer pairs is linked to the central pair by a radial spoke. A closer inspection of the outer micro-tubules shows that they have side arms of proteins, called dyneins. These proteins are actually mechanical motors, and they are instrumental in the movement of the cilia.[9]

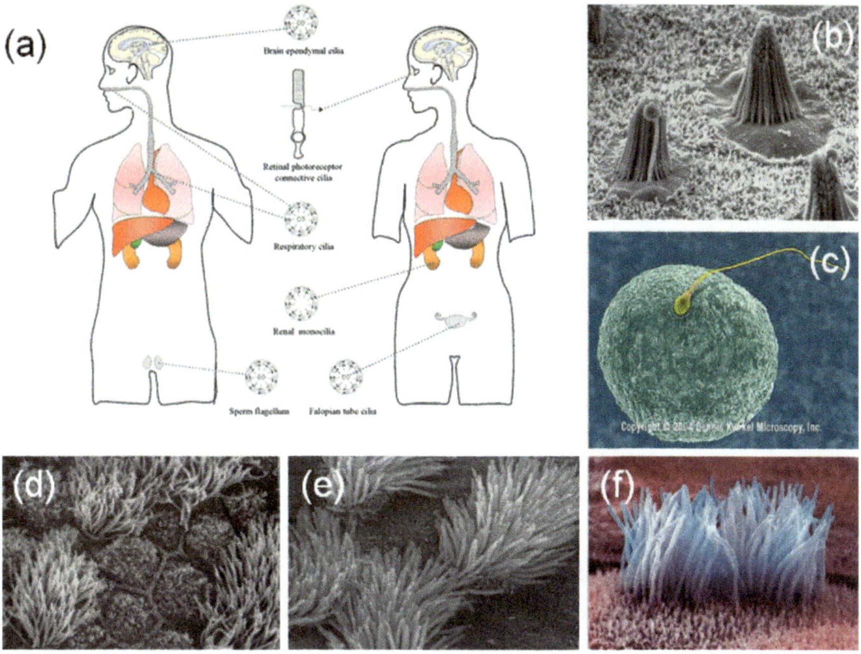

Figure 1.4 Cilia and flagella in the human body. (a) Several locations at which cilia can be found. (Reproduced from Ibanez-Tallon *et al.*[5]) (b) Hair bundles of three sensing hair cells from the frog inner ear. Each bundle is composed of 50–60 actin-based stereocilia, and a single micro-tubule-based kinocilium.[6] (c) Spermatozoon and ovum, from Guirao.[3] (d) Cilia in the fallopian tube, from Kelly *et al.*[7] (e) Cilia in the trachea (windpipe), from Daghlian;[8] and (f) in the lungs, from Guirao.[3]

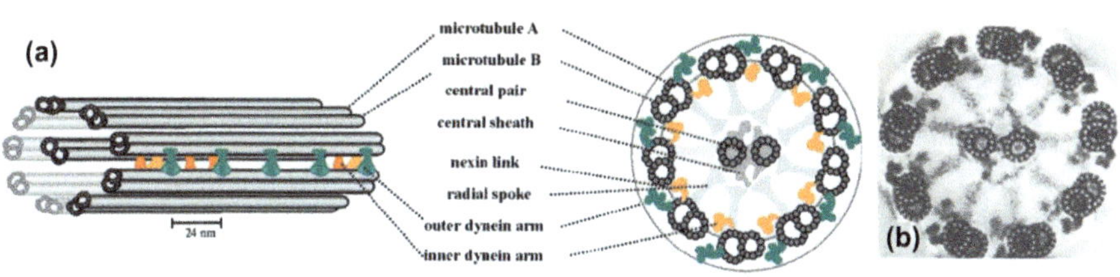

Figure 1.5 The internal structure of motile cilia or flagella: the 'axoneme'. (a) A schematic, showing the 9 + 2 arrangement of micro-tubules, and the various other constituents: nexin links between the outer micro-tubules, radial spokes connecting the inner and outer tubules, and the dynein arms attached to the outer micro-tubules, from Ibanez-Tallon *et al.*[5] (b) Electron micrograph of a cross section through a cilium, from Afzelium *et al.*[10]

Dyneins convert energy, which is released in the body by hydrolysis of ATP (adenosine triphosphate), into mechanical work by changing their molecular conformation. These successive conformational changes allow the molecule to take 'steps' along the micro-tubules in a certain direction. The outer dynein arms attached to the micro-tubule pairs are different in molecular structure than the inner dynein arms. The molecular structures are shown in Figure 1.6.

In the absence of ATP, the dyneins fixed to each peripheral micro-tubule pair are also attached to the adjacent micro-tubule pair. The presence and hydrolysis of ATP provokes the detachment, movement and re-attachment of the dyneins in such a way that the pairs of neighbouring micro-tubule pairs slide along each other, see Figure 1.6 (the dyneins indeed take steps along the micro-tubules). Since the micro-tubules are connected, through the nexin links and radial spokes, this gliding motion results in an overall bending of the cilium.

The details of the control of the beating of cilia are still not completely understood, and this is a topic of ongoing research. It is the combination of the properties of the molecular motors, their interaction with the elastic properties of the axoneme, as well as the hydrodynamic coupling with the surrounding fluid, that determines the typical ciliary movement characterized by their asymmetry in motion, their beat frequency, and their metachronic behaviour.

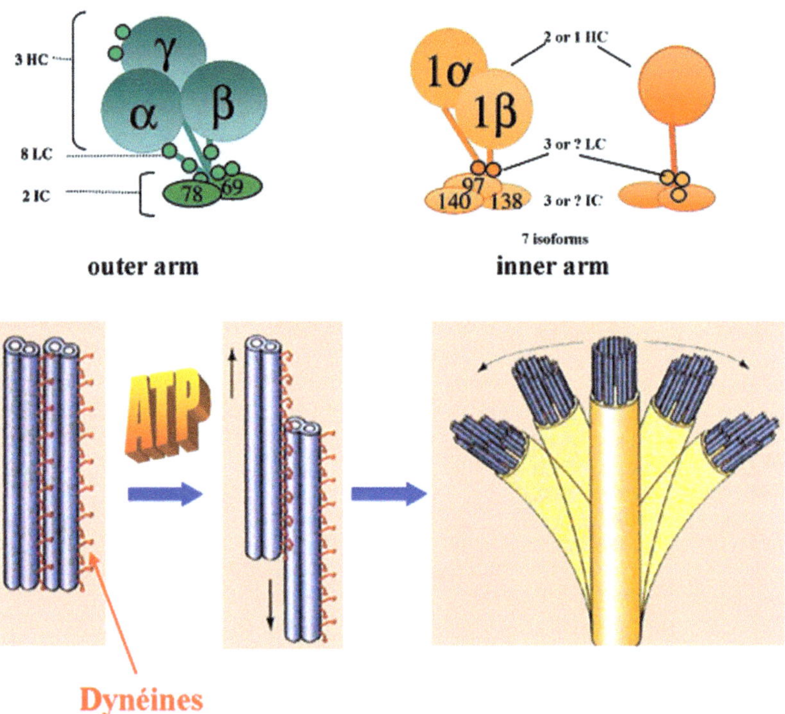

Figure 1.6 The mechanism of cilia movement. The micro-tubules have side arms consisting of dyneins. The molecular structure of the dyeins is different for the inner and the outer arms; both are shown here (see also Figure 1.5). The dyneins change molecular conformation by converting energy released by the hydrolysis of ATP. The change in conformation leads to a relative change in position of the pairs of micro-tubule, resulting in a bending motion of the axoneme. Top images reproduced from Ibanez-Tallon *et al.*[5] Bottom images reproduced from Guirao.[3]

1.2 Low Reynolds Number Flows

The *Reynolds number* is the most important dimensionless number in fluid mechanics. It represents the ratio of the inertial forces to viscous forces in the flow, and is defined by:

$$Re = \frac{\rho U L}{\mu} \tag{1.1}$$

in which U is a characteristic velocity scale, L is a characteristic length scale, ρ is the density of the fluid and μ is its dynamic viscosity. The characteristic velocity and length scales are different for different problems. For a relatively simple and well defined flow, such as the flow through a cylindrical tube, the characteristic scales are easily defined: U is the mean flow velocity in the pipe and L is the pipe diameter. For more complex problems, the definition of the characteristic scales may be more difficult, and sometimes, the problem even cannot be described by just one single Reynolds number.

'Inertia' is the property of an object to remain at a constant velocity, unless an outside force acts on it. An object with a large inertia will resist strongly to a change in velocity, or in other words, it is difficult to start or stop its movement. An object with a small inertia, on the other hand, will almost instantaneously start or stop when acted upon by some external or internally generated force. Inertia of fluid flows is caused by nonlinear interactions within the flow field. These nonlinearities may cause instabilities in the flow to grow, and therefore the flow can become turbulent when inertial effects are dominant, that is, for large Reynolds numbers. For small Reynolds numbers, on the other hand, the flow will always be laminar. For pipe flow, the critical Reynolds number above which turbulence may exist is about 2000.

'Viscosity' is the resistance of a fluid to flow under the influence of an applied external force. It is the source of drag on objects moving through the fluid. For such an object, inertia strives to keep the object going, whereas viscosity tries to stop it.

Some characteristic Reynolds numbers for self-propelled organisms from nature are shown in Table 1.1. The characteristic length scale is the size of the

Table 1.1 A spectrum of Reynolds numbers for self-propelled organisms (after Vogel[11]).

Organism	*Reynolds number*
A large whale swimming at 10 m s^{-1}	300 000 000
A tuna swimming at 10 m s^{-1}	30 000 000
A duck flying at 20 m s^{-1}	300 000
A large dragon fly moving at 7 m s^{-1}	30 000
A copepod in a speed burst of 0.2 m s^{-1}	300
Flapping wings of the smallest flying insects	30
An invertebrate larva, 0.3 mm long, at 1 mm s^{-1}	0.3
A sea urchin sperm advancing the species at 0.2 mm s^{-1}	0.03
A bacterium, swimming at 0.01 mm s^{-1}	0.00001

organism, the characteristic velocity is its swimming speed. The density and viscosity are those of water, *i.e.* $\rho = 1000\,\text{kg}\,\text{m}^{-3}$ and $\mu = 1\,\text{mPa s}$ or air, *i.e.* $\rho = 1.2\,\text{kg}\,\text{m}^{-3}$ and $\mu = 0.018\,\text{mPa s}$. We can see that the Reynolds numbers range from very large values, for a swimming whale, to extremely small values for swimming bacteria. That means that for a whale, inertial effects dominate, and thus, after stopping to swim, the whale will continue to move further, or 'coast' for a substantial distance and time. For a bacterium, on the other hand, inertial effects will not be important at all and viscous effects dominate, so that a bacterium will stop almost instantaneously.

Due to their small sizes, cilia in nature operate under low Reynolds number conditions, which are also called the Stokes flow regimes. This means that inertial effects are not important and the flow is dominated by viscous effects. Consequently, to generate any net flow, the motion of the cilia must be asymmetric, *i.e.* having a different forward trajectory compared to the backward trajectory during one beat cycle. This is the reason for the presence of an effective and a recovery stroke in natural cilia, as shown in Figure 1.1. If the motion was symmetric, fluid would just be displaced back and forth without any net flow, even in the case of temporal asymmetry (moving faster forwards than backwards).

1.3 Artificial Cilia: Micro-fluidics Applications

Micro-fluidics is the science and technology of manipulating and analysing fluid flow in structures of sub-millimetre dimensions.[12–14] This field is particularly relevant for the development of lab-on-chip devices, which can be pictured as credit-card-sized fluidic systems containing tiny channels and chambers in which processes such as mixing and routing of the liquids, and separation, reaction, and detection of individual components (DNA, proteins, cells) present in these liquids are integrated. In this way, a conventional large-scale biomedical analysis laboratory is miniaturized and combined on a single chip. The typical cross-sectional dimensions of the micro-fluidic channels and structures are between 10 and 100 µm. An artist's impression of a lab-on-chip device is depicted in Figure 1.7.

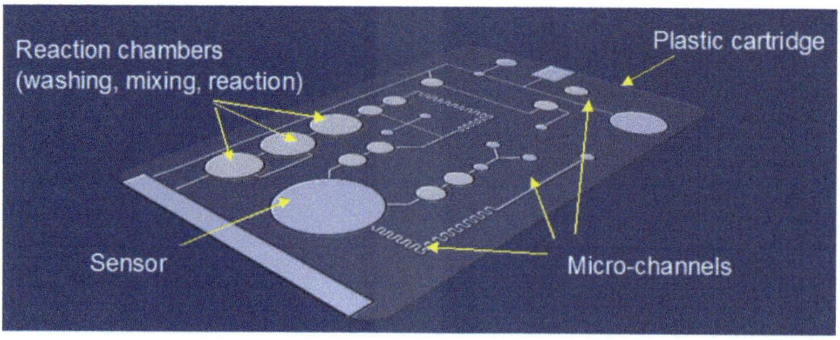

Figure 1.7 Artist's impression of a disposable cartridge, a key component of a lab-on-a-chip device, containing micro-fluidic structures.

Many different physical phenomena can be used to manipulate fluids on sub-millimetre scales. Small droplets can be manipulated by actively changing their surface tension, for instance using electrical potentials.[15] Fluid can be transported through micro-channels by electro-osmosis, in which a spontaneously occurring charged surface layer is set into motion by an applied electrical field.[16] Other physical principles that can be exploited in micro-fluidics are acoustic streaming,[17] optical manipulation,[18] dielectrophoresis,[19] magneto-phoresis[20] and thermophoresis.[21] Use has also been made of micro-fabricated valves to control flow in micro-fluidic channels.[22]

A special challenge in micro-fluidic systems is to create efficient mixing flows. Due to the small channel sizes, the Reynolds number is generally small and flows are non-turbulent. On the other hand, the channel size is often too large for molecular diffusion to be effective in mixing within a reasonable time. To obtain efficient mixing, special strategies must therefore be followed. An approach is to create repeatedly stretching and folding flow patterns, leading to so-called chaotic advection that causes effective mixing.[23] Nguyen and Wu[24] have reviewed and classified the relatively large number of micro-mixers proposed in the literature. The existing micro-mixers can be divided into two general classes, namely passive and active micro-mixers. Passive micro-mixers do not require external energy, and the mixing process relies entirely on chaotic advection or diffusion. The effect is often achieved by special geometrical features like channel shape or corrugations on the channel walls. A well-known and elegant concept uses specially designed grooves or ridges on the channel walls, as described by Stroock *et al.*[25] Active micro-mixers use the disturbance generated by an external field for the mixing process, and thus require external energy. Examples are the use of micro-machined magnetic-bar mixers,[26] sinu-soidal pressure pulses,[27] electro-hydrodynamic forcing,[28] AC electro-osmotic flow,[29] and acoustic streaming.[30]

A recent development, inspired by nature, is the use of *artificial cilia* to create pumping and/or mixing in micro-fluidic devices. Microscopic actuators resembling cilia, actuated to move under the influence of a number of different stimuli such as electric fields, magnetic fields, and even light, have been developed by a number of groups and shown to be capable of generating flow and mixing in micro-fluidic environments. The research on artificial cilia started about a decade ago, and is rapidly expanding. Next to being relevant for potential application in lab-on-a-chip devices, the work on artificial cilia forms a beautiful example of how a biological system can form the successful basis for both scientific research and for technological applications.

1.4 This Book: An Overview of Artificial Cilia Research and Technologies

This book will give an overview of the research field of artificial cilia, a novel technology for controlling and sensing fluid flow at microscopic scales. The various chapters are clustered in three sections.

Section 1, '*Theoretical and numerical descriptions of artificial cilia*', contains three chapters that describe various approaches to model artificial cilia and their interaction with the surrounding fluid. In each of the chapters, particular questions are addressed using the models. In Chapter 2, 'Numerical modelling for artificial cilia' by Holger Stark, a numerical model is introduced for artificial cilia based on a bead-spring chain with bending rigidity whose beads interact hydrodynamically. This is used to describe a super-paramagnetic filament actuated by an external magnetic field. The influence of two-dimensional versus three-dimensional motion of the cilium is investigated, as well as the effect of metachrony on the pumping performance of an array of artificial cilia.

Another approach is taken in Chapter 3, 'Computational design of magnetic artificial cilia' by Syed Khaderi, Jaap den Toonder and Patrick Onck. Using a coupled magneto-mechanical solid–fluid numerical model, artificial cilia are designed that can be realized using thin films consisting of a polymer matrix filled with magnetic nanoparticles, so that they can be actuated using an external magnetic field. The model is used to establish under what conditions (geometrical/mechanical/magnetic design, magnetic field control) a magnetic film will mimic the asymmetric motion of natural cilia and how the flow rates can be maximized using fluid inertia and out-of-phase beating.

Chapter 4, 'Modelling the interaction of active cilia with species in solution: from chemical reagents to microscopic particles' by Pratyush Dayal, Olga Kuksenok, Amitabh Bhattacharya, Gavin A. Buxton, O. Berk Usta and Anna C. Balazs, focuses on a different topic. A computational model is introduced that is used to study cilia formed from chemo-responsive gels, and assess their sensing and communication behaviour. As a second example, the transport of a microscopic particle is modelled via a regular array of beating elastic cilia, whose tips experience an adhesive interaction with the particle's surface. Conditions under which the particle can be 'released', 'propelled' or 'trapped' by the cilia array are investigated.

Section 2, '*Experimental approaches to create artificial cilia*', shows that a variety of different approaches can be used to create artificial cilia, which have different shapes, dimensions, and working principles. In the first contribution, Chapter 5, 'Electrostatic artificial cilia' by Jaap den Toonder, a microsystems technology approach is used to make artificial cilia that can be actuated electrostatically. When actuated, these cilia can generate substantial fluid velocities. In addition, very efficient mixing is obtained using specially designed geometrical cilia configurations in a micro-channel. A numerical model shows that, even though the cilia are small, inertial effects do play a role in the working mechanism of the electrostatic cilia.

Chapter 6, 'Ferromagnetic cilia' by Francis Fahrni and Leo van Ijzendoorn, demonstrates the fabrication of ferromagnetic cilia consisting of polymer films with ferromagnetic particles. The ferromagnetic cilia offer the possibility to tune the remanent magnetization and thereby control asymmetric or selective movements while actuated through an externally applied magnetic field.

Another actuation trigger, namely light, is introduced in Chapter 7, 'Light-actuated artificial cilia based on liquid crystal networks' by Caspar van Oosten,

Kees Bastiaansen and Dick Broer. The optically driven cilia are made by inkjet printing and consist of liquid crystal networks. The cilia have subunits that can be selectively driven by changing the wavelength of the light trigger.

The artificial cilia described in Chapter 8, Hydrogel-actuated integrated responsive systems (HAIRS): Creating cilia-like "hairy" surfaces' by Alison Grinthal and Joana Aizenberg, are chemomechanically driven nanohairs, actuated by a muscle-like responsive hydrogel layer that translates temperature, pH, light, and other stimuli into directional mechanical forces. The elastic hairs can reversibly bend, twist and assemble into bundles with diverse shapes, sizes and complexity.

Chapter 9, 'Microwalkers' by Charles Singh and Alfredo Alexander-Katz, explores the motion of a class of microscopic walkers (microwalkers) made of a chain of colloids self-assembled by magnetic forces. They can be used for a multitude of tasks such as fully controlled two-dimensional walking, creating controlled flows at the micron scale, or for the manipulation of nanoscopic objects.

Artificial flagella are described in Chapter 10, 'Artificial flagellar microswimmers' by Rémi Dreyfus, Jean Baudry and Jérôme Bibette. The artificial flagella are made from magnetic microbeads that self-assemble into filaments using a magnetic field, and linked together using DNA molecules to obtain a flexible filament structure. Using an external magnetic field, the artificial flagella can move in a viscous fluid and even carry cargo.

Section 3, '*The potential of artificial cilia: experimental evidence*', presents two examples of advanced experiments of the fluid flow actuation of artificial cilia, indicating the potential in micro-fluidic applications. Chapter 11, 'Fluid manipulation by artificial cilia' by Jeanette Hussong, Wim-Paul Breugem and Jerry Westerweel, introduces magnetically actuated artificial cilia that are thin polymer films containing dispersed super-paramagnetic nanoparticles. The cilia, integrated in a micro-fluidic cartridge, are actuated using a rotating permanent magnet. The flow generated is characterized quantitatively using micro-particle image velocimetry. Average fluid flow velocities of up to hundreds of micrometres per second are found.

The final contribution, Chapter 12, 'Measurement of fluid flow generated by artificial cilia', by A. Vilfan, D. Babič, B. Kavčič, G. Kokot, N. Osterman, I. Poberaj and M. Vilfa, presents biomimetic cilia constructed as long chains of spherical super-paramagnetic particles, which self-assemble parallel to the external magnetic field. The cilia are actuated by a rotating magnet such that they execute a so-called 'tilted conical beat'. Flow visualization was used to quantify the fluid flow induces by this motion: a net fluid flow was obtained with a maximum average velocity of about $8\,\mu m\,s^{-1}$.

1.5 Perspectives: Looking at the Future

The nature and appearance of the artificial cilia is diverse, and various actuation principles are being applied, from electrostatic, magnetic, optical, to pH controlled. Sizes of artificial cilia range from micrometres to millimetres.

Fabrication processes are very different for the various approaches. The variety of approaches shows that inspiration by nature is not synonymous with literally copying; the working mechanism of the artificial cilia may be completely different from those in nature, but still break the configurational beating symmetry that is essential to create fluid flow or mixing.[31]

Magnetic actuation is most often used for artificial cilia (Chapters 6, 9–11 and 12). This has the obvious advantages of being compatible with biological fluids, and the possibility to actuate the artificial cilia within micro-fluidic channels through external means: (rotating) permanent magnets or electromagnets.

The fluid velocities induced by artificial cilia in experiments range from micrometres per second (*e.g.* Chapters 9 and 12) to hundreds of micrometres per second (*e.g.* Chapters 5, 6 and 11). In terms of potential flow rates, the latter range is certainly interesting for real micro-fluidic applications. However, the artificial cilia are most often not capable of generating substantial pressure gradients, *i.e.* of working against a pressure head. The artificial cilia are therefore most suitable for closed-loop pumping, or for creating local flows *e.g.* for micro-mixing.

Modelling of artificial cilia motion and the induced fluid flow has proven to be very valuable for understanding the mechanisms behind the observed phenomena, but also for designing the artificial cilia for practical application, in terms of properties, size and actuation procedure, see Chapters 2, 3 and 4. The models developed for artificial cilia will be also particularly useful for modelling and understanding complex fluid–solid interaction in general.

The field of artificial cilia is still young, and implementation in commercial products is still remote. To bring artificial cilia to a real application in lab-on-a-chip devices, it is essential that cost-effective and robust fabrication methods are developed instead of the intricate and expensive approaches used in most studies so far. However, if these barriers are overcome, the fluid manipulation effectiveness, the ability to locally control flows, and the elegance of potential full integration into the micro-fluidic channels, make them interesting candidates for future implementation: nature has proven that it works!

References

1. J. den Toonder, F. Bos, D. Broer, L. Filippini, M. Gillies, J. de Goede, T. Mol, M. Reijme, W. Talen, H. Wilderbeek, V. Khatavkar and P. Anderson, *Lab Chip*, 2008, **8**, 533.
2. M. A.Sleigh , *Cilia and Flagella*, Academic Press, London, 1974.
3. Boris Guirao, Propriétés physiques des cellules ciliées, PhD thesis l'Université Pierre et Marie Curie – Paris VI, 2007.
4. H. Hoops, R. Wright, J. Jarvik and G. Witman, *J. Cell Biol.*, 1984, **98**, 818–824.
5. I. Ibanez-Tallon, N. Heintz and H. Omran, *Hum. Mol. Genet.*, 2003, **12**, R27–R35.
6. Courtesy of D. P. Corey, Harvard Medical School.

7. D. E. Kelly, R. L.Wood and A. C.Enders, *Bailey's Textbook of Microscopic Anatomy*. Williams & Wilkins, Baltimore, 18th edition, 1983; http://www.cgu.edu.tw/Anatomy/His-picture-female1.htm.
8. C. Daghlian, SEM Image, http://remf.dartmouth.edu/imagesindex.html.
9. There are three large families of molecular motors: kinesines, myosines, and dyneins.
10. B. A. Afzelius, R. Dallai and S. Lanzavecchia, *Tissue Cell*, 1995, **27**, 241–247.
11. S.Vogel, *Life in Moving Fluids*, Princeton University Press, Princeton, 1994.
12. T. M. Squires and S. R. Quake, *Rev. Mod. Phys.*, 2005, **77**, 977–1026.
13. N.-T.Nguyen and S. T.Wereley, *Fundamentals and Applications of Microfluidics*, Artech House, Boston, 2002.
14. G. M. Whitesides, *Nature*, 2006, **442**, 368–373.
15. F. Mugele and S. Herminghaus, *Appl. Phys. Lett.*, 2002, **81**, 2303–2305.
16. H. Ramos, N. G. Morgan, A. Green and A. Castellanos, *J. Phys. D*; *Appl. Phys.*, 1998, **31**, 2338–2353.
17. P. Marmottant and S. Hilgenfeldt, *Nature*, 2003, **423**, 153–156.
18. D. G. Grier, *Nature*, 2003, **424**, 810–816.
19. J. Voldman, *Annu. Rev. Biomed. Eng.*, 2006, **8**, 425–454.
20. H. Watarai, M. Suwa and Y. Iiguni, *Anal. Bioanal. Chem.*, 2003, **387**, 1693–1699.
21. F. Zheng, *Adv. Colloid Interface Sci.*, 2002, **97**, 255–278.
22. M. A. Unger, H. P. Chou, T. Thorsen, A. Scherer and S. R. Quake, *Science*, 2000, **288**, 113–116.
23. H. Aref, *J. Fluid Mech.*, 1984, **143**, 1–21.
24. N.-T. Nguyen and Z. Wu, *J. Micromech. Microeng.*, 2005, **15**, R1–R16.
25. A. D. Stroock, S. K. W. Dertinger, A. Ajdari, I. Mezic, H. A. Stone and G. M. Whitesides, *Science*, 2002, **295**, 647–651.
26. L. H. Lu, K. S. Ryu and C. Lu, *J. Microelectromech. Syst.*, 2002, **11**, 462–469.
27. I. Glasgow and N. Aubry, *Lab Chip*, 2003, **3**, 114–120.
28. A. O. El Moctar, N. Aubry and J. Batton, *Lab Chip*, 2003, **3**, 273–280.
29. N. Sasaki, T. Kitamori and H. B. Kim, *Lab Chip*, 2006, **6**, 550–554.
30. K. Sritharan, C. J. Strobl, M. F. Schneider, A. Wixforth and Z. Guttenberg, *Appl. Phys. Lett.*, 2006, **88**, 054102.
31. J. M. J. den Toonder and P. R. Onck, *Trends Biotechnol.*, 2013, **31,** Published online 12 December 2012. DOI: http://dx.doi.org/10.1016/j.tibtech.2012.11.005.

Section 1
Theoretical and Numerical Descriptions of Artificial Cilia

CHAPTER 2

Numerical Model for Artificial Cilia

HOLGER STARK

Technische Universität Berlin, Institut für Theoretische Physik,
Hardenbergstr. 36, 10623 Berlin, Germany
Email: Holger.Stark@tu-berlin.de

2.1 Introduction

In the microscopic world inertia is negligible and friction dominates in a fluid environment. At such low Reynolds numbers, one always has to actively stir fluid to keep it in motion. Otherwise the fluid will stop "immediately" since it cannot just drift by inertia. As often, nature offers strategies to deal with this situation. A sperm cell uses a long elastic filament called flagellum for its own locomotion.[1–3] Internal molecular motors create a bending wave that travels from the head to the tail and pushes the sperm cell forward using solely friction with the surrounding fluid.[4] The wave realizes a non-reciprocal motion meaning the time-reversed motion is not the same as the original one. As Purcell pointed out, this is necessary for transporting fluid or equivalently for moving a microorganism forward.[4,5] Some unicellular microorganisms, such as the protozoans paramecium or opalina, are covered by a carpet of shorter flagella now called cilia. Each cilium performs a characteristic often three-dimensional stroke which is coordinated with the strokes of neighboring cilia. As a result, so-called metachronal waves run along the surfaces of the protozoans and propel them through the fluid.[1–3,6–8] Arrays of collectively beating cilia also transport mucus in the respiratory tract[9–13] or fluid in the

RSC Nanoscience & Nanotechnology No. 30
Artificial Cilia
Edited by Jaap M J den Toonder and Patrick R Onck

Published by the Royal Society of Chemistry, www.rsc.org

brain.[14] During the early stage of a developing embryo, rotating cilia are responsible for establishing the left-right asymmetry in the placement of organs.[15] Genetic defects that produce defective cilia then lead to the so-called *situs invertus* where organs are placed on the wrong side of the animal or human body.

Microfluidics deals with pumping and mixing fluids on the microscopic level and with the transport of solutes. These tasks are at the center of a successful lab-on-chip technology.[16,17] Therefore, in recent years there have been growing experimental efforts to copy nature's successful concept of moving fluid around by developing biomimetic or artificial cilia that are actuated by external fields.[17–21] In this chapter, I review one method to model artificial cilia in their viscous environment using a bead-spring chain with bending elasticity.[22–30] To couple the model cilium to its viscous fluid environment, we introduce the method of mobilities and thereby implement hydrodynamic interactions between the beads. We have used the model cilium to investigate in detail a superparamagnetic filament[31–33] that has been introduced recently to construct an artificial microswimmer actuated by an oscillating magnetic field.[18] We have modeled the microswimmer,[22] but also explored basic strategies for pumping fluid with the surface-attached filament.[23,24] I review our work on planar[23] and three-dimensional strokes[24] which have been implemented in experiments.[21,34,35] In addition, I demonstrate that the superparamagnetic filament is useful for studying properties of metachronal waves.

A ciliary beating pattern consists of a transport stroke which pumps fluid and a recovery stroke where the filament returns to its initial position with the goal to keep the amount of fluid moved opposite to the pumping direction as small as possible. An ideal transport stroke keeps the filament away from the surface and moves it perpendicular to its axis where friction with the surrounding fluid is large and therefore also the amount of pumped fluid. On the other hand, the recovery stroke occurs preferentially close to the surface where fluid cannot be moved due to the no-slip boundary condition. In addition, it is performed along the filament axis where hydrodynamic friction is smaller.

In the following I will demonstrate for the superparamagnetic filament how this concept is implemented in planar and three-dimensional strokes. The superparamagnetic filament consists of superparamagnetic colloids linked together by polymer or DNA strands.[31–33] Its beating patterns under an actuating magnetic field were studied in experimens,[21,34,35] also in chains without chemical linkers where, in addition, the initiated flow was determined.[36–39] An attractive alternative approach for magnetic cilia uses magnetic nanoparticles embedded in a polymer film[17,20,40–42] which is complemented by numerical modeling using a magneto-mechanical solid-fluid model.[43–47] Even light-driven artificial cilia based on a liquid-crystal network exist.[48] Finally, recent numerical work also addresses propulsion and trapping of particles in a ciliary array.[49,50]

We now start with introducing the bead-spring chain to model an artificial cilium.

2.2 Bead-spring Model for Artificial Cilia

We model artificial cilia as bead-spring chains that resist bending, like a worm-like chain.[51] So, each bead experiences stretching and bending forces and additional forces due to the actuation of the filament. Figure 2.1 illustrates the geometry. The bead-spring chain is attached to a planar surface with the help of two virtual beads that fix its position and give it an orientation orthogonal to the surface.[23] An alternative approach introduces only one anchor bead and the filament can rotate freely about the anchor point.[24]

In the following we describe the model in detail. We first explain how we couple the bead-spring chain to the viscous fluid.

2.2.1 Method of Mobilities

At low Reynolds number fluid inertia can be neglected on time scales larger than the time $\tau = d^2/(6\eta/\rho)$ fluid momentum needs to diffuse a characteristic distance d, where we have introduced the viscosity η and the mass density ρ of the fluid. The flow field $\boldsymbol{u}(\boldsymbol{r},t)$ in the viscous fluid is then described by the Stokes equations supplemented by the incompressibility condition:

$$\mathbf{0} = -\boldsymbol{\nabla} p + \eta \boldsymbol{\nabla}^2 \boldsymbol{u} \quad \text{and} \quad \text{div}\boldsymbol{v} = 0, \tag{2.1}$$

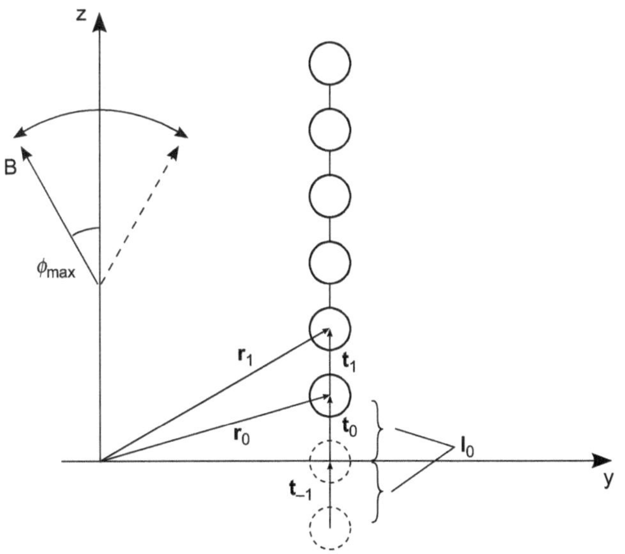

Figure 2.1 The artificial cilium modelled as a bead-spring chain is attached to a bounding wall. The virtual beads drawn as dotted lines are kept at fixed positions to anchor the filament perpendicular to the xy plane through the implemented bending rigidity.[23] Alternatively, one just uses one fixed bead and does not constrain the orientation of the cilium.[24] Superparamagnetic filament: in the two-dimensional stroke the actuating magnetic field oscillates around the z axis. (Figure reprinted from Gauger *et al.*,[23] with kind permission of Springer Science + Business Media.)

where p is pressure. Typically, on all bounding surfaces of the fluid including the beads of the bead-spring chain one assumes no-slip boundary conditions.

Each bead in the bead-spring chain moves under the force $\boldsymbol{F}_j$ which is the sum of stretching, bending, and actuating forces. The bead generates a flow field that influences the motion of other beads. These so-called hydrodynamic interactions are long-ranged. Since the Stokes equations are linear in the flow field, the velocities $\boldsymbol{v}_i$ of the beads are proportional to the forces $\boldsymbol{F}_j$ and the beads obey the following equations of motion,[52]

$$\boldsymbol{v}_i = \sum_j \boldsymbol{\mu}_{ij}\boldsymbol{F}_j. \tag{2.2}$$

In general, the mobilities $\boldsymbol{\mu}_{ij}$ depend on all the coordinates $\boldsymbol{r}_i$ of the beads. A good approximation is to include in the cross-mobilities ($i \neq j$) only two-body interactions and to take for the self-mobilities ($i = j$) in the unbounded fluid the Stokes mobility $1/(6\pi\eta a)$, where a is the bead radius. If the mobilities are known, eqn (2.2) can numerically be integrated, for example, by the simple Euler method.

At distances much larger than the bead radius a, the force $\boldsymbol{F}_0$ applied to the bead at position $\boldsymbol{r}'$ acts as a point force on the fluid and creates a flow field called stokeslet,

$$\boldsymbol{u}(\boldsymbol{r}) = \boldsymbol{O}(\boldsymbol{r} - \boldsymbol{r}')\boldsymbol{F}_0, \tag{2.3}$$

where the Oseen tensor

$$\boldsymbol{O}(\boldsymbol{r}) = \frac{1}{8\pi\eta r}\left(\mathbf{1} + \frac{\boldsymbol{r} \otimes \boldsymbol{r}}{r^2}\right) \tag{2.4}$$

is the Green function of the Stokes equations in an unbounded fluid. For a general body force $\boldsymbol{b}(\boldsymbol{r}')$ acting on the fluid, the flow field is $\boldsymbol{u}(\boldsymbol{r}) = \int \boldsymbol{O}(\boldsymbol{r} - \boldsymbol{r}')\boldsymbol{b}(\boldsymbol{r}')d^3r'$. If the stokeslet is created by bead j at position $\boldsymbol{r}_j$, then in leading order the fluid velocity is taken over by bead i at position $\boldsymbol{r}_i$, $\boldsymbol{v}_i = \boldsymbol{O}(\boldsymbol{r}_i - \boldsymbol{r}_j)\boldsymbol{F}_j$. Hence, the cross-mobility for pointlike beads is $\boldsymbol{\mu}_{ij} = \boldsymbol{O}(\boldsymbol{r}_i - \boldsymbol{r}_j)$. It goes as the inverse distance of the two beads and therefore demonstrates the long-range nature of hydrodynamic interactions.

To transport fluid, cilia are attached to a surface and the initiated flow occurs in a bounded fluid. Near a planar surface with no-slip boundary condition, the velocity and pressure fields of a point force were first derived by Lorentz more than 100 years ago.[53] In eqn (2.3) the Oseen tensor is replaced by what we now call the Blake tensor which was incroduced by Blake in Ref. 54 (see also Ref. 55). To satisfy the no-slip boundary condition at an infinitely extended plane, one introduces appropriate mirror images, similar to the image charges used in electrostatics. However, in contrast to the scalar Poisson equation in electrostatics, where it suffices to simply mirror the charge distribution, the hydrodynamic image system is more complicated due to the vectorial nature of the Stokes equations and the incompressibility condition. In addition to the stokeslet of the mirrored point force (also called anti-stokeslet), so-called stokes-doublet and source-doublet contributions are needed to formulate Blake's tensor,

$$\begin{aligned} \boldsymbol{G}^{\mathrm{Blake}}(\boldsymbol{r},\boldsymbol{r}') &= \boldsymbol{O}(\boldsymbol{r}-\boldsymbol{r}') + \boldsymbol{G}^{im}(\boldsymbol{r},\bar{\boldsymbol{r}}') \\ &= \boldsymbol{O}(\boldsymbol{r}-\boldsymbol{r}') - \boldsymbol{O}(\boldsymbol{r}-\bar{\boldsymbol{r}}') + \delta\boldsymbol{G}^{im}(\boldsymbol{r},\bar{\boldsymbol{r}}'). \end{aligned} \tag{2.5}$$

Here, $\bar{\boldsymbol{r}}'$ is the position of the anti-stokeslet source, i. e., the stokeslet source at $\boldsymbol{r}'$ mirrored at the bounding plane, and $\delta\boldsymbol{G}^{im}(\boldsymbol{r},\bar{\boldsymbol{r}}')$ denotes the stokes-doublet and source-doublet contributions.

2.2.2 Rotne-Prager Approximation

To treat hydrodynamic interactions between the beads in the bead-spring chain, one cannot approximate them by point forces since the distance between neighboring beads is of the order of the bead radius. The Rotne-Prager approximation for the mobilities considers the finite size of the moving beads in leading order by adding the next higher term in the expansion of the mobilities into $1/r$. We first derive the Rotne-Prager approximation for an unbounded fluid and then comment on its form near a plane boundary.

The finite size of the beads leads to two corrections. First, the flow field initiated by an isolated spherical bead j at position $\boldsymbol{r}_j$ is

$$\boldsymbol{u}(\boldsymbol{r}) = (1 + \frac{1}{6}a^2\boldsymbol{\nabla}_j^2)\boldsymbol{O}(\boldsymbol{r}-\boldsymbol{r}_j)\boldsymbol{F}_j, \tag{2.6}$$

where $\boldsymbol{v}_j = \mu_0\boldsymbol{F}_j$ is the bead's velocity and $\mu_0 = 1/(6\pi\eta a)$ its Stokes mobility. The nabla operator acts on the position vector $\boldsymbol{r}_j$ of particle j. For large distances from the bead this reproduces the stokeslet field of a point force. In the presence of other beads, correction terms of higher order than $1/r^3$ occur and can be disregarded here. Second, according to Faxén's theorem, particle i placed into the flowfield $\boldsymbol{u}_j(\boldsymbol{r})$ of particle j assumes the velocity[52]

$$\boldsymbol{v}_i = \mu_0\boldsymbol{F}_i + (1 + \frac{1}{6}a^2\boldsymbol{\nabla}_i^2)\boldsymbol{u}_j(\boldsymbol{r}_i). \tag{2.7}$$

The last term on the right-hand side of eqn (2.7) adds a correction to the point-particle approximation. Note that Faxén's theorem is valid for any flow field $\boldsymbol{u}_j(\boldsymbol{r})$ satisfying the Stokes equations.

Inserting eqn (2.6) into eqn (2.7) and comparing to eqn (2.2), one can directly read off the Rotne-Prager mobilities as an expansion up to terms $1/r_{ij}^3$,

$$\boldsymbol{\mu}_{ii} = \mu_0\mathbf{1},\ \mu_0 = 1/(6\pi\eta a) \tag{2.8}$$

$$\begin{aligned} \boldsymbol{\mu}_{ij} &= (1 + \frac{1}{6}a^2\nabla_i^2)(1 + \frac{1}{6}a^2\nabla_j^2)\boldsymbol{O}(\boldsymbol{r}_i - \boldsymbol{r}_j) \\ &= \mu_0\left[\frac{3}{4}\frac{a}{r_{ij}}(\mathbf{1} + \hat{\boldsymbol{r}}_{ij}\otimes\hat{\boldsymbol{r}}_{ij}) + \frac{1}{2}\left(\frac{a}{r_{ij}}\right)^3(\mathbf{1} - 3\hat{\boldsymbol{r}}_{ij}\otimes\hat{\boldsymbol{r}}_{ij})\right], i \neq j, \end{aligned} \tag{2.9}$$

where $\boldsymbol{r}_{ij} = \boldsymbol{r}_i - \boldsymbol{r}_j$ and $\hat{\boldsymbol{r}}_{ij} = \boldsymbol{r}_{ij}/r_{ij}$. Higher-order corrections to the Rotne-Prager mobilities arise since the flow field initiated by particle i acts back on particle j. In addition, many-body interactions due to the presence of additional particles occur. The method of induced forces provides a systematic expansion of the mobilities in $1/r_{ij}$.[56] For particles in close contact, one has to use lubrication theory to determine the mobilities.[57,58] A program exists that incorporates all these effects and calculates mobilities for a given cluster of spherical particles.[59] However, we checked that the Rotne-Prager approximation agrees well with the more exact values of the mobilities down to distances of $3a$.

Near to a planar surface with no-slip boundary condition, the Rotne-Prager mobilities have to be modified. Replacing the Oseen tensor in eqn (2.6) by the Blake tensor $\boldsymbol{G}^{Blake}$, one approximates the flow field initiated by a spherical bead close to a bounding plane. In doing so, a non-uniform contribution to the force distribution on the bead's surface is neglected. Again, Faxén's theorem stated in eqn (2.7) gives the cross mobilities $\boldsymbol{\mu}_{ij}$ as in eqn (2.7) but now with $\boldsymbol{O}(\boldsymbol{r}_i - \boldsymbol{r}_j)$ replaced by $\boldsymbol{G}^{Blake}(\boldsymbol{r}_i, \boldsymbol{r}_j)$. Furthermore, we have to add the flow field initiated by the image system $\boldsymbol{G}^{im}$ of particle i to $\boldsymbol{u}_j(\boldsymbol{r}_i)$ in the Faxe′n's theorem. Then, through the interaction of particle i with its image system, the self-mobilities $\boldsymbol{\mu}_{ii}$ become:

$$\boldsymbol{\mu}_{ii} = \mu_0 \mathbf{1} + \left(1 + \frac{1}{6}a^2\nabla_i^2\right)\left(1 + \frac{1}{6}a^2\nabla_{\bar{i}}^2\right)\boldsymbol{G}^{im}(\boldsymbol{r}_i, \bar{\boldsymbol{r}}_i), \tag{2.10}$$

where $\nabla_{\bar{i}}$ means gradient with respect to the image coordinate $\bar{\boldsymbol{r}}_i$. Concrete formulas for the mobilities are given, for example, in Ref. 36.

2.2.3 Stretching and Bending Forces

We derive the forces acting on the single beads of the bead-spring model from its total free energy

$$H = H^S + H^B + H^A, \tag{2.11}$$

where H^S is the stretching energy obeying Hooke's law, H^B is the discretized bending free energy of an elastic rod, and H^A an energy connected to the actuation force. We will introduce it for the superparamagnetic filament in section 2.3.1. Twisting of the filament occurs only for three-dimensional strokes. In the following we will neglect twisting or assume that the filament does not resist twisting. We now present stretching and bending free energy in discretized form.

According to Figure 2.1, we introduce $\boldsymbol{t}_i = \boldsymbol{r}_i - \boldsymbol{r}_{i-1}$ as the vector connecting the centers of two adjacent beads. A deviation $l_i = |\boldsymbol{t}_i|$ of the beads from their equilibrium spacing l_0 then gives the total stretching free energy

$$H^S = \frac{1}{2}k\sum_{i=0}^{N-1}(l_i - l_0)^2, \tag{2.12}$$

where k is the stretching constant and N the total number of beads.

We derive the bending free energy of the bead-spring chain from the continuous bending energy of an elastic rod or worm-like chain,[51,60]

$$H^B = \frac{1}{2} A \int_0^L ds \left(\frac{d\hat{\boldsymbol{t}}}{ds} \right)^2, \tag{2.13}$$

where L is the total length of the elastic rod, s the arclength along it, and $\hat{\boldsymbol{t}}$ the unit tangent at location s. Writing the bending stiffness as $A = k_B T l_p$, where $k_B T$ is thermal energy, one obtains the persistence length l_p that gives the length scale on which the rod becomes flexible. Replacing $d\hat{\boldsymbol{t}}/(ds)$ by $(\hat{\boldsymbol{t}}_{i+1} - \hat{\boldsymbol{t}}_i)/l_0$, one arrives at the discretized version of the bending free energy,

$$H^B = \frac{A}{l_0} \sum_{i=-1}^{N-2} (1 - \hat{\boldsymbol{t}}_{i+1} \cdot \hat{\boldsymbol{t}}_i), \tag{2.14}$$

where $\hat{\boldsymbol{t}}_i = \boldsymbol{t}_i / l_i$.

The force $\boldsymbol{F}_i$ acting on bead i follows from

$$\boldsymbol{F}_i = \boldsymbol{F}_j^S + \boldsymbol{F}_j^B + \boldsymbol{F}_j^A = -\boldsymbol{\nabla}_{\boldsymbol{r}_i}(H^S + H^B + H^A), \tag{2.15}$$

where $\boldsymbol{\nabla}_{\boldsymbol{r}_i}$ is the gradient operator with respect to $\boldsymbol{r}_i$. Ref. 22 gives explicit expressions for all forces.

2.2.4 The Sperm Number

By writing the equations of motion (2) in reduced units, one is able to extract the relevant parameters of the problem. Therefore we rescale length, time, velocity, force, and mobilities by

$$\tilde{\boldsymbol{r}} = \boldsymbol{r}/L, \quad \tilde{t} = \omega t, \quad \frac{d\tilde{\boldsymbol{r}}}{d\tilde{t}} = \frac{1}{\omega L}\frac{d\boldsymbol{r}}{dt}, \quad \tilde{\boldsymbol{F}}_i = \frac{\boldsymbol{F}_i}{A/L^2}, \quad \tilde{\boldsymbol{\mu}}_{ij} = 6\pi\eta a \frac{L}{l_0} \boldsymbol{\mu}_{ij}, \tag{2.16}$$

where we have introduced a characteristic frequency ω of the actuating force. The factor A/L^2 is the magnitude of the bending forces and $6\pi\eta a L/l_0$ is the total friction coefficient of the filament when hydrodynamic interactions are neglected. In these reduced units, the equation of motion (2.2) for bead i becomes

$$\frac{d\tilde{\boldsymbol{r}}_i}{d\tilde{t}} = S_p^{-4} \sum_j \tilde{\boldsymbol{\mu}}_{ij} \tilde{\boldsymbol{F}}_j \quad \text{with} \quad \tilde{\boldsymbol{F}}_j = -\boldsymbol{\nabla}_{\tilde{\boldsymbol{r}}_j}(\tilde{H}^B + k_s \tilde{H}^S/2 + B_s^2 \tilde{H}^A). \tag{2.17}$$

We shortly discuss the relevant parameters in eqn (2.17). The sperm number[61,62]

$$S_p = \left(\frac{6\pi\eta \frac{a}{l_0} \omega L^4}{A} \right)^{1/4} = \frac{L}{l_h} \tag{2.18}$$

compares bending to frictional forces in the bead-spring chain. It is well-known from the continuum model of an elastic rod, where $6\pi\eta a/l_0$ is replaced by $\gamma_\perp$, which is the perpendicular friction constant per unit length of a slender body.[61–64] The characteristic length l_h is the penetration length of the elastohydrodynamics of a slender body.[63,64] An oscillation with frequency ω started at one of the ends of a sufficiently long filament penetrates into the filament on the distance l_h If the filament is much shorter than l_h, i.e., if $L \ll l_h$, the filament oscillates like a rigid rod over its whole length. Since hydrodynamic interactions are not included in our definition of l_h, our length only gives the order of magnitude of the penetration length.

The second parameter in eqn (2.17) is the reduced spring constant

$$k_s = \frac{kL\,/\,N}{A\,/\,L^2} = \frac{N^2 l_0^3}{A} k. \tag{2.19}$$

It compares stretching to bending forces. We typically choose it much larger than one so that changes in the length of the bead-spring chain are well below 10 %. The third parameter B_s^2 compares actuating to bending forces. For the superparamagnetic filament we will introduced it in section 2.3.1.

2.3 A Superparamagnetic Filament for Fluid Transport

Recently a superparamagnetic filament was constructed with the goal of using it as an artifcial cilium or flagellum. Micron-sized superparamagnetic colloidal particles are linked to each other by double-stranded DNA and form an elastic filament which resists bending and stretching[31–33] (for similar systems see references 65 and 66). Dreyfus *et al.* attached the superparamagnetic filament to a red-blood cell and created the first artificial micro-swimmer.[18] In a sperm cell bending waves propagate along the flagellum and push the cell forward. The waves are generated by the collective motion of internal molecular motors. Here, an oscillating external magnetic field actuates the superparamagnetic filament and induces a non-reciprocal stroke pattern that pushes the artificial swimmer forward. Bibette and Dreyfus will discuss the system in more detail in Chapter 10.

In the following, we extend the bead-spring model by the magnetic actuation. We have used the model to carefully study the artificial swimmer.[22] Different approaches for modeling the superparamagnetic filament and the swimmer exist based on resistive-force theory.[18,67–72] When attached to a planar surface, the superparamagnetic elastic filament is an ideal system for investigating strategies for generating fluid transport at low Reynolds number. We shortly review here our studies on two-dimensional[23] and three-dimensional beating patterns.[24] In particular, we demonstrate how transport and recovery strokes are realized with the superparamagnetic filament and show how metachronal waves in collectively beating cilia enhance the amount of pumped fluid. We start with explaining how the filament is actuated.

2.3.1 Actuation of the Filament

A magnetic field induces strong magnetic dipole moments in the beads. Since dipoles align parallel to each other along the dipole direction, the whole filament orients parallel to the field. When the field direction rotates slowly, the straight filament just follows the magnetic field. However, the filament cannot follow fast rotations since it is hindered by hydrodynamic friction with the fluid. Thus, the filament bends when dipolar forces try to line up the magnetic moments. This is the idea of actuating the superparamagnetic filament.

All superparamagnetic beads subject to a uniform external magnetic field B develop a dipole moment with identical orientation and strength,

$$p = \frac{4\pi a^3}{3\mu_0}\chi B, \tag{2.20}$$

where a is bead radius, χ the magnetic susceptibility, and $\mu_0 = 4\pi \times 10^{-7}\mathrm{N}/\mathrm{A}^2$ the permeability of free space. In eqn (2.20) we neglect dipolar contributions to the local magnetic field from all other dipole moments. The dipoles of the beads give rise to the total dipole-dipole interaction energy

$$H^A = \frac{4\pi a^6}{9\mu_0}(\chi B)^2 \sum_{i<j} \frac{1 - 3(\hat{\boldsymbol{p}} \cdot \hat{\boldsymbol{r}}_{ij})}{r_{ij}^3}, \tag{2.21}$$

where $r_{ij} = |\ \boldsymbol{r}_j - \boldsymbol{r}_i\ |$ and $\hat{\boldsymbol{r}}_{ij} = (\boldsymbol{r}_j - \boldsymbol{r}_i)/r_{ij}$. From this energy, we determine the dipolar or actuating force on bead i as already mentioned in eqn (2.15), $\boldsymbol{F}_i^A = -\boldsymbol{\nabla}_{r_i} H^A$.

To compare dipolar to bending forces, we introduced the relative strength B_s^2 in eqn (2.17) by defining the reduced magnetic field strength,

$$B_s = \frac{2}{3}\left(\frac{\pi}{\mu_0 A}\right)^{1/2} \frac{\chi a^3}{l_0} NB. \tag{2.22}$$

It is proportional to the magnetoelastic number introduced in Refs. 18, 67, 68. An alternative dimensionless number for characterizing the influence of the magnetic field is the Mason number, which one finds in the literature on magnetorheological suspensions,[73,69]

$$M_a = S_p^4/B_s^2. \tag{2.23}$$

The Mason number is the ratio of frictional to magnetic forces and determines the behavior of the superparamagnetic filament when magnetic forces dominate over bending forces. Finally, we note that all simulation results presented in the following were obtained with realistic parameter values close to experiments.[22,23]

2.3.2 Pumping Performance

Now, we introduce a unitless number called the pumping performance to identify stroke patterns which are optimal for pumping fluid. We choose the bounding surface at $z=0$ and integrate the time-averaged fluid flow (initiated by the beating filament) over a whole plane parallel to the bounding surface and situated above the beating filament. The integrated fluid flow, for example, along the y axis is determined by the laterally averaged Blake tensor,[74]

$$\bar{G}(z,z') = \int dxdy G_{yy}^{Blake}(x,y,z,z') = \frac{\min(z,z')}{\eta}. \tag{2.24}$$

Each force with component F_{yi} in y direction acting on bead i at distance z_i from the bounding wall creates a stokeslet flow. Summing over all stokeslets gives the integrated flow $\mathcal{F}$ generated by all the beads in the filament:

$$\mathcal{F} = \frac{1}{\eta}\sum_i z_i F_{yi}. \tag{2.25}$$

We already considered the case $z > z_i$ for all beads so that $\bar{G}(z,z_i) = z_i/\eta$. Contributions to $\mathcal{F}$ from the forces F_{xi} and F_{zi} vanish by symmetry. Interestingly, $\mathcal{F}$ does not depend on the position z of the integrated flow. Note that $\mathcal{F}/\pi$ agrees with the volume flow rate initiated by the filament through a plane perpendicular to the flow direction.[75,76] Since the filament is actuated periodically in time, we calculate the time-averaged fluid flow

$$\bar{\mathcal{F}} = \frac{1}{T}\int_t^{t+T} \mathcal{F} dt', \tag{2.26}$$

where T is the period of one actuation cycle. Finally, we introduce the unit less pumping performance,

$$\xi = \bar{\mathcal{F}} \,/\, \bar{\mathcal{F}}^{\text{ref}}, \tag{2.27}$$

where $\bar{\mathcal{F}}^{\text{ref}}$ is a reference value typical for a filament of length L. One possibility is the volume flow rate $\bar{\mathcal{F}}^{\text{ref}} = L^3/T$ that follows by dimensional analysis. Instead, in Ref. 23 we have introduced an idealized stroke pattern for which we expect optimum fluid transport with a filament of length L. During the transport stroke, one orients the filament perpendicular to the bounding surface and drags it parallel to the surface along a distance L. Then, the filament is rotated by 90°. In the recovery stroke one drags the filament along its long axis to its original position keeping it always close to the surface. Again we find $\bar{\mathcal{F}}^{\text{ref}} \approx L^3/T$.

2.3.3 Two-dimensional Stroke

We initiate a two-dimensional or planar stroke by a magnetic field of strength B whose direction oscillates in the yz plane about the normal of the bounding surface as illustrated in Figure 2.1,

$$B(t) = (0, B\sin\varphi(t), B\cos\varphi(t)), \tag{2.28}$$

where $\varphi(t)$ is the angle between the field vector and the z axis. In our modeling, the angular amplitude φ_{max} was always 60°.

In order to accomplish net fluid transport along the y direction, a non-reciprocal beating pattern of the filament is required meaning that under time-reversal the pattern is not identical.[5] In addition, the motion of the filament in positive and negative y direction has to be asymmetric. For the magnetically actuated filament we achieve both requirements with the time protocol of the angle $\varphi(t)$ illustrated in Figure 2.2. During the transport stroke with duration τ_l, the filament rotates slowly about its anchoring point following the magnetic field. Due to the small frictional forces the filament hardly bends. To complement the beating cycle by the fast recovery stroke, the field direction quickly rotates in the reverse direction during the smaller time intervall τ_s and the filament bends due to increased hydrodynamic friction. Note that for real cilia in nature the speeds are just reversed: the transport stroke is fast and the recovery stroke is slow. To quantify the asymmetry in the actuation cycle and in the resulting beating pattern of the filament, we define the asymmetry parameter using the respective durations of the transport and recovery strokes,

$$\varepsilon = \frac{\tau_l - \tau_s}{\tau_s + \tau_l}. \tag{2.29}$$

The asymmetry parameter is zero for $\tau_s = \tau_l$ and tends to one in the limit of $\tau_l \gg \tau_s$.

Figure 2.3 illustrates how the pumping performance ξ of a single filament depends on he sperm number S_p and the asymmetry parameter ε at a fixed magnetic field strength. The most striking feature is the pronounced peak for ε close to one and at $S_p \approx 3$. A similar peak occurs in the swimming velocity of the

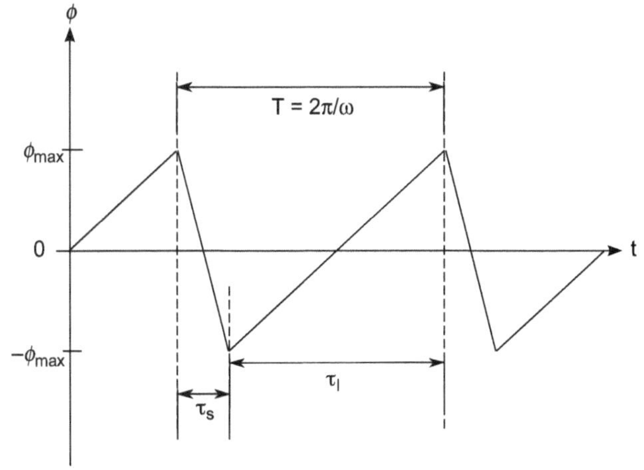

Figure 2.2 The angle φ enclosed by the magnetic field $\boldsymbol{B}(t)$ and the z axis is shown as a function of time. φ has different velocities when decreasing and increasing. (Figure reprinted from Gauger *et al.*,[23] with kind permission of Springer Science + Business Media.)

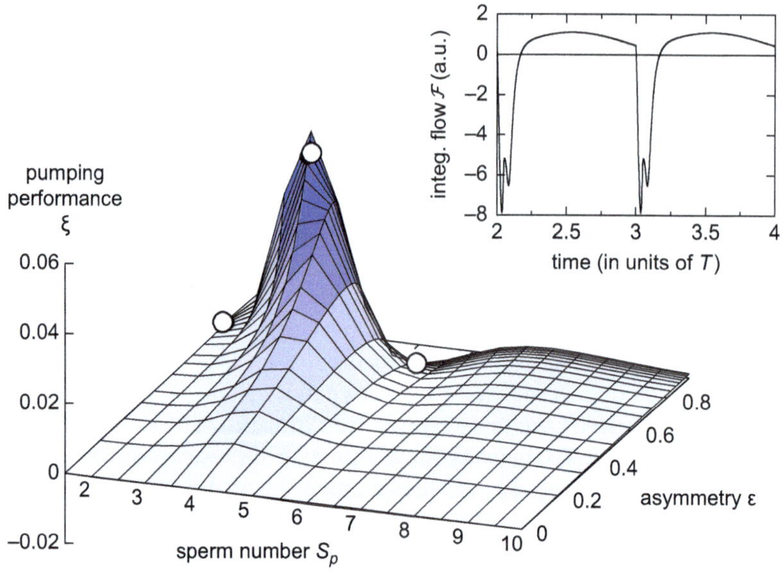

Figure 2.3 Pumping performance ξ for a single filament plotted versus sperm number S_p and asymmetry parameter ε. The reduced magnetic field strength is $B_s = 2.5$. The white dots mark parameters for which stroboscopic snapshots of the filament are shown in Figure 2.4a. Inset: Integrated flow $\mathcal{F}$ in arbitrary units as a function of time in units of T for the parameters $S_p = 3$, $B_s = 3$ and $\varepsilon = 0.9$. (Figure reprinted from Gauger *et al.*,[23] with kind permission of Springer Science + Business Media.)

artificial micro-swimmer.[18,22,68] The corresponding stroke pattern for $S_p \approx 3$ is illustrated in the middle picture of Figure 2.4a). During the slow transport stroke, the filament rotates clockwise being nearly straight whereas in the fast recovery stroke the filament bends due to large hydrodynamic friction forces and then relaxes back to the initial configuration. The inset of Figure 2.3 demonstrates that fluid transport is also noticeable in the recovery stroke ($\mathcal{F} < 0$). So, even for the most efficient stroke pattern, the pumping performance is the result of a small asymmetry in the amount of fluid transported to the right and left. In the example of the inset, which is close to the optimum stroke pattern, only 4.3% of the total amount of moved fluid is effectively transported in the positive y direction. As a result, the maximum pumping performance in Figure 2.3 is only 6% of the reference stroke described above. As expected, the pumping performance vanishes for symmetric beating ($\varepsilon = 0$) about the z axis and for $S_p \to 0$. Here the filament follows the actuating magnetic field instantaneously. It hardly bends and performs the reciprocal stroke pattern illustrated in the left picture of Figure 2.4a. Around $S_p \approx 5.5$, the pumping direction even reverses ($\xi < 0$) albeit with a rather weak performance. Finally, the pumping performance tends to zero for increasing S_p or frequency since the filament can no longer follow the actuating field as the right picture of

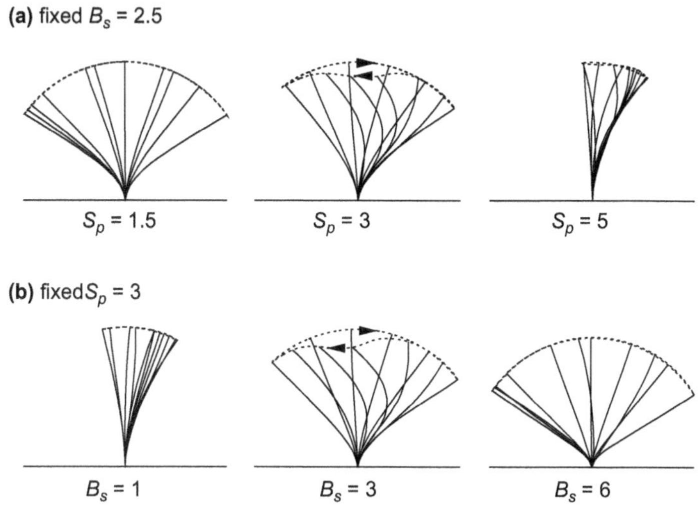

Figure 2.4 Stroboscopic snapshots of the filament at different times during the beating cycle for $\varepsilon = 0.9$. The trajectory of the top bead during one beating cycle is also indicated. During the slow transport stroke the filament rotates clockwise and the fast recovery stroke occurs to the left as the arrows indicate. The filament bends significantly only at intermediate sperm numbers S_p and magnetic field strengths B_s. (Figure reprinted from Gauger *et al.*,[23] with kind permission of Springer Science + Business Media.)

Figure 2.4a shows. Hence, optimal pumping performance is only achieved for intermediate values of S_p.

Figure 2.5a demonstrates a pronounced dependence of the pumping performance ξ on the strength B_s of the actuating magnetic field for different sperm numbers S_p. When the magnetic field increases from zero, the pumping perfomance first remains close to zero. Beyond a threshold value, it starts to grow, goes through a maximum, and even becomes negative. The snapshots in Figure 2.4b again help to clarify this behavior. Small field strengths B_s induce actuating forces that are too small to overcome hydrodynamic friction and the filament moves with a small angular amplitude. On the other hand, at large strengths B_s the filament is always straight and thus performs a reciprocal motion. So the optimal stroke exists in an intermediate regime for the strength B_s. The optimal performance shifts with increasing S_p to larger values of B_s since at higher frequencies a larger field is needed to move the filament through the fluid. Increased B_s also means the filament is stiffer and the optimum stroke occurs at higher frequencies $\omega \propto S_p^4$. Then, frictional forces are larger which also enhances the pumping performance as Figure 2.5a demonstrates. When magnetic forces acting on the filament exceed bending forces, the dynamics of the filament should be determined by the ratio of hydrodynamic friction to magnetic forces, which we introduced in eqn (2.23) as Mason number M_a. To demonstrate this, we rescaled curves $\xi(S_p)$ for different $B_s \geq 2$ by their

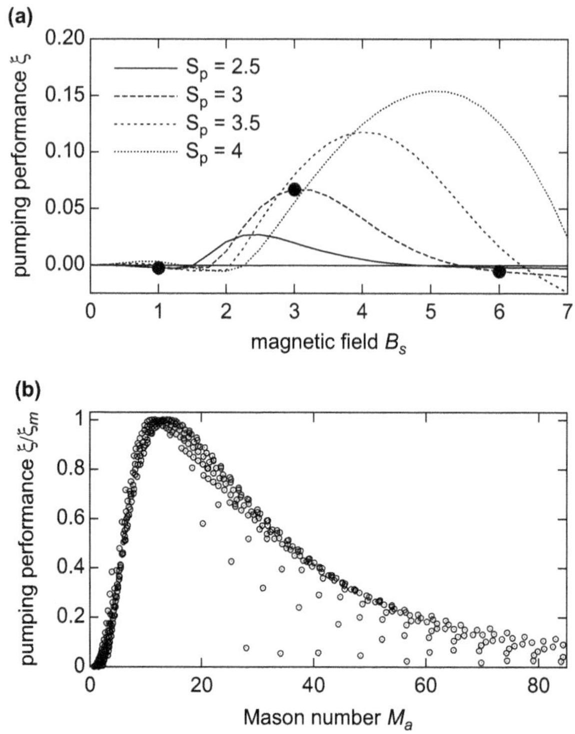

Figure 2.5 (a) Pumping performance ξ versus B_s for different values of S_p and $\varepsilon = 0.9$. The black dots mark parameters for which stroboscopic snapshots of the filament are shown in Figure 2.4b. (b) Pumping performance ξ plotted as a function of the Mason number $M_a = S_p^4 / B_s^2$, where ξ is given in units of the maximum values ξ_m when B_s is kept constant. (Figure reprinted from Gauger *et al.*,[23] with kind permission of Springer Science + Business Media.)

respective maximum values ξ_m. When plotted as a function of the Mason number M_a, the data points indeed fall on a master curve as Figure 2.5b illustrates. Deviations occur for data points with B_s close to 2.

We close with a final remark. We also investigated the influence of defects in the lower part of the filament, where the bending stiffness A is strongly reduced, and found that defects increase the pumping performance significantly.[24]

2.3.4 Metachronal Waves

We already mentioned in the introduction that nature often uses arrays of beating cilia rather then a single isolated cilium for generating fluid transport or to propel microorganisms such as a paramecium.[1] Hydrodynamic interactions between single cilia synchronize their beating with a phase lag of neighboring cilia which leads to traveling waves (see, for example, 78–86). Of course, in magnetically actuated cilia reviewed in this section the phase is determined by the oscillating external field. It cannot adjust itself in response to the motion of neighboring cilia. Nevertheless, we can use the artificial-cilium system to investigate in a systematic way how the pumping performance depends on the phase lag between neighboring cilia by actuating each cilium with a separate magnetic field. Two cilia beating in phase either placed in-line or parallel to each other decrease the pumping performance since their effective friction with the surrounding fluid is reduced and therefore they pump less fluid.[23] In the following we demonstrate that metachronal waves can indeed enhance the pumping performance.

We place several cilia in-line and actuate each of them by a separate magnetic field using a prescribed phase shift $\Delta\varphi$ between neighboring cilia. In the single rows in Figure 2.6a, a cilium lags behind its neighboring cilium to the left by the small phase shift $\Delta\varphi$. The slow transport stroke with a straight cilium goes to the right as the snapshots at different times show whereas the fast recovery goes

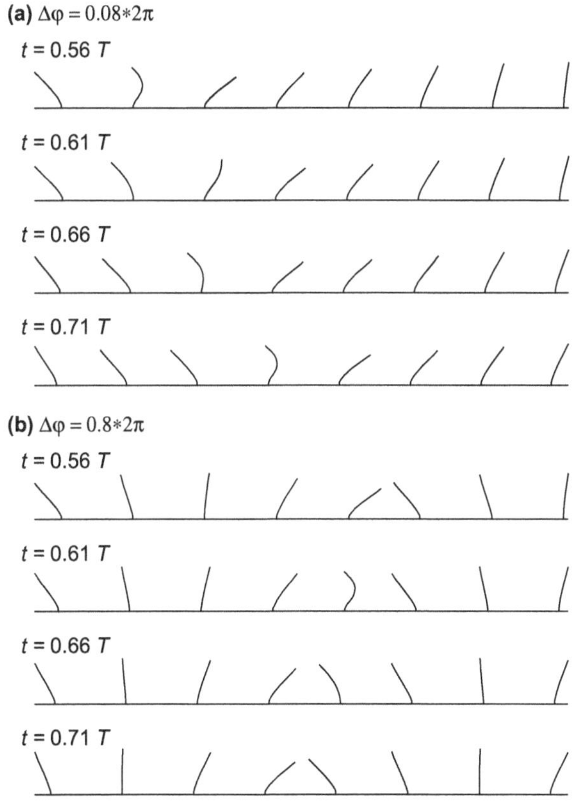

Figure 2.6 Snapshots of eight cilia beating in-line with different phase shifts. (a) $\Delta\varphi = 0.08 \times 2\pi$ (minimal pumping performance); and (b) $\Delta\varphi = 0.8 \times 2\pi$ (optimum pumping performance). Times t are given in units of the time period T. (Figure reprinted from Gauger *et al.*,[23] with kind permission of Springer Science + Business Media.)

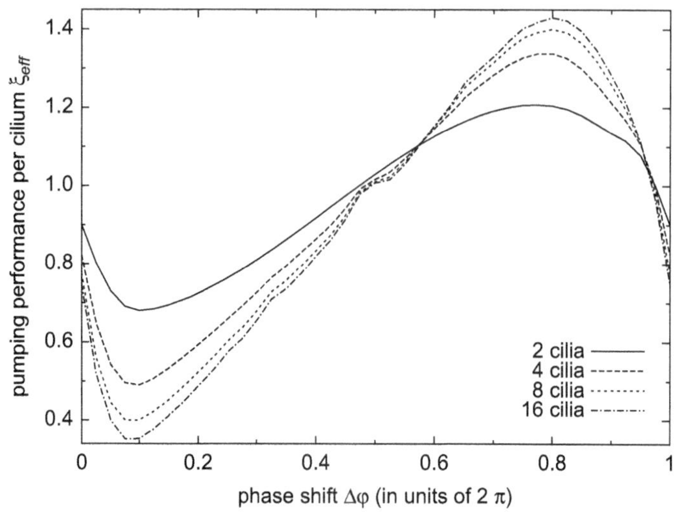

Figure 2.7 Average effective pumping performance ξ_{eff} for multi-cilia systems beating in-line plotted versus the phase shift between neighbouring cilia. ξ_{eff} is given in units of the performance ξ of a single filament actuated with the same parameters $S_p = 3$, $B_s = 3$ and $\varepsilon = 0.9$. (Figure reprinted from Gauger *et al.*,[23] with kind permission of Springer Science + Business Media.)

to the left. Furthermore, when time proceeds, the strongly bent cilium in the recovery stroke moves to the right. So metachronal wave propagation and transport stroke occur in the same direction which is also termed symplectic metachronism in literature.[1] Figure 2.6b shows the situation for phase shifts close to 2π, where a cilium moves ahead of its neighbor to the left. Now, the metachronal wave propagates opposite to the transport stroke and is therefore termed antiplectic.

According to Figure 2.7, the average effective pumping performance per cilium, ξ_{eff}, falls below the reference value of a single cilium for small $\Delta\varphi$ while it assumes a maximum value at around $\Delta\varphi = 0.8 \times 2\pi$ which means the antiplectic mode. The maximal pumping performance ξ_{eff} increases with the number of cilia and slightly moves to larger $\Delta\varphi$. Even at the relatively large distance of $1.5L$ between the cilia, ξ_{eff} increases by more than 40% relative to a single cilium. Bringing the cilia closer together, this value strongly increases. Hence, we conclude that metachronal coordination of ciliary beating at the right phase shift significantly enhances the ability to transport fluid. We can infer an explanation for this behavior from the snapshots in Figure 2.6. In the optimum stroke [see Figure 2.6b], the fifth cilium from the left performs the recovery stroke against the neighboring fourth cilium which hinders the fluid flow initiated by recovery stroke and therefore increases fluid transport to the right. On the other hand, in the metachronal wave with lowest pumping performance [see Figure 2.6a], the cilium performing the recovery stroke is further away from the neighboring cilia and therefore its fluid transport is less hindered.

2.3.5 Three-dimensional Stroke

In nature also three-dimensional stroke patterns exist, for example, for cilia that cover the surfaces of the protozoans paramecium or opalina. In Ref. 24 we showed how the actuating magnetic field can initiate three-dimensional stroke patterns in the superparamagnetic filament. Subsequently, three-dimensional strokes were realized in Ref. 21, 34, 35, 36. In the "cone stroke" the magnetic field vector follows the surface of a cone with opening angle φ which is tilted by an angle θ against the surface normal [see Figure 2.8a]. The idea is that the cone tilted towards the x axis generates a clear asymmetry between fluid transport in positive and negative y direction. Cilia producing the nodal flow for establishing the left-right asymmetry in mammals use exactly this stroke pattern.[76,77] Figure 2.8b illustrates the alternative "hybrid stroke". The transport stroke is realized by the field vector rotating in the yz plane which is then followed by a rotation around the z axis again on the surface of a cone with opening angle ϕ. Both parts of the cone stroke are executed in equal time.

Driving the filament slowly, it will remain straight. In this case we can calculate analytically an approximation for the pumping performance by assuming that the strengths of the forces in eqn 2.25 acting on the single beads are proportional to the beads' velocities. For the cone stroke, for example, these velocities are determined by geometry. One assumes that the fluid flow is produced by forces whose strengths increase linearly along the filament starting from the anchoring point. Then eqn 2.25 gives the following scaling law for the pumping performance: $\xi \propto \sin^2(\phi)\sin(\theta)$. It was also derived in Ref. [76]. In the parameter space of the cone stroke, $\theta + \phi \leq \pi/2$, where the equal sign means that the filament just touches the surface at its lowest point, the pumping performance assumes its maximum value at $\phi = \arccos(1/\sqrt(3)) \approx 54°$ and $\theta = \pi/2 - \phi$. We have checked the approximation for ξ with computer simulations and found excellent agreement.[24]

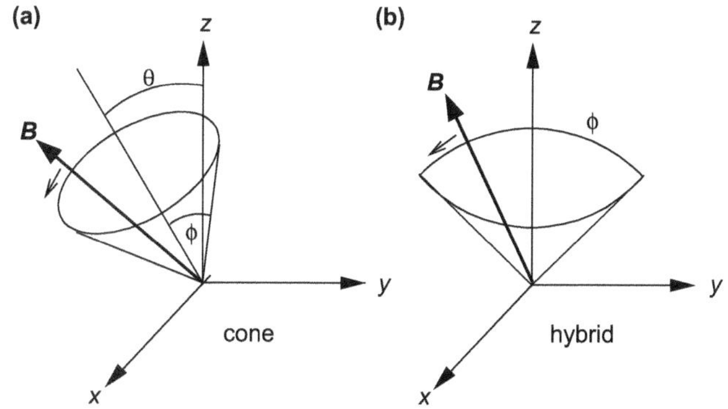

Figure 2.8 Kinematics of the actuating magnetic field for three-dimensional stroke patterns. (a) Cone stroke, and (b) hybrid stroke.

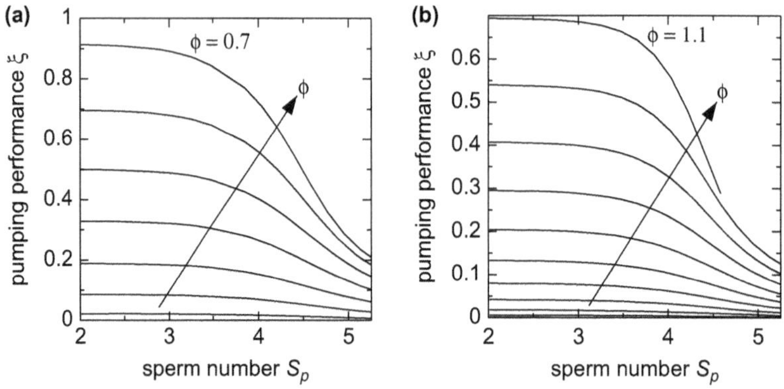

Figure 2.9 Pumping performance ξ plotted versus S_p. (a) Tilted cone ($\theta = \pi/4$): The opening angle ϕ increases from the lower curve at $\phi = 0:1$ (in radians) in steps of $\Delta\phi = 0:1$ to a maximum $\phi = 0:7$ for the upper curve. (b) Hybrid stroke: The opening angle ϕ increases from the lower curve at $\phi = 0.1$ (in radians) in steps of $\Delta\phi = 0.1$ to a maximum $\phi = 1.1$ for the upper curve.

Figure 2.9 shows the pumping performance of cone (a) and hybrid (b) stroke for several opening angles ϕ as a function of S_p. It is immediately obvious that three-dimensional strokes are much more efficient in pumping fluid than planar strokes. For the largest ϕ values the pumping performance is more than a factor 10 larger compared to the planar case in Figure 2.3. Increasing S_p or the beat frequency, the pumping performance decreases since the filament starts to bend due to increased hydrodynamic friction. For example, we find that in the cone stroke the bent filament rotates on a cone which is narrower than the magnetic-field cone and therefore fluid transport is reduced.[24]

2.4 Summary

The chapter introduced a numerical model for artificial cilia based on a bead-spring chain with bending rigidity whose beads interact hydrodynamically. I explained in detail how one implements the model in its viscous fluid environment using self- and cross-mobilities. In particular, I described how one treats the surface with its no-slip boundary condition to which the cilium is attached. The characteristic sperm number compares bending to frictional forces and helps to classify the dynamic behavior of the actuated filament.

The second part on the superparamagnetic filament addressed a very attractive realization of an artificial cilium. In our numerical work, we explored its potential for pumping fluid. Our investigations demonstrate that three-dimensional strokes have a considerably higher pumping performance than their two-dimensional counterparts. A second very interesting result shows that metachronal waves increase the pumping performance for an appropriate phase lag between neighboring cilia. Thus, the superparamagnetic elastic filament is attractive in two directions. First, it helps to elucidate biological features of beating flagella and cilia. Second, it offers exciting applications for transporting and mixing fluids in the field of microfluidics.

References

1. C. Brennen and H. Winet, *Ann. Rev. Fluid Mech.*, 1977, **9**, 339.
2. R. W. Linck, Cilia and flagella, in *Encyclopedia of Life Sciences*, Wiley, Chichester, 2001, www.els.net.
3. D. Bray, *Cell Movements: From Molecules to Motility*, Garland Publishing, New York, 2nd edn, 2001.
4. E. Lauga and T. R. Powers, *Rep. Prog. Phys.*, 2009, **72**, 096601.
5. E. M. Purcell, *Am. J. Phys.*, 1977, **45**, 3.
6. S. L. Tamm, *J. Cell Biol.*, 1972, **55**, 250.
7. H. Machemer, *J. Exp. Biol.*, 1972, **57**, 239.
8. K.-I. Okamoto and Y. Nakaoka, *J. Exp. Biol.*, 1994, **192**, 61.
9. M. J. Sanderson and M. A. Sleigh, *J. Cell Sci.*, 1981, **47**, 331.
10. D. Eshel and Z. Priel, *J. Physiol.*, 1987, **388**, 1.
11. M. A. Sleigh, *Cell Motility*, 1982, **2**, 19.
12. P. Satir and M. A. Sleigh, *Annu. Rev. Physiol.*, 1990, **52**, 137.
13. L. Gheber and Z. Priel, *Cell Motility and the Cytoskeleton*, 1994, **28**, 333.
14. I. Ibañez-Tallon, A. Pagenstecher, M. Fliegauf, H. Olbrich, A. Kispert, U.-P. Ketelsen, A. North, N. Heintz and H. Omran, *Hum. Mol. Genet.*, 2004, **13**, 2133.
15. (a) C. D. Stern, *Nature*, 2002, **418**, 29; (b) J. J. Essner, K. J. Vogan, M. K. Wagner, C. J. Tabin, H. J. Jost Yost and M. Brueckner, *Nature*, 2002, **418**, 37; (c) S. Nonaka, H. Shiratori, Y. Saijoh and H. Hamada, *Nature*, 2002, **418**, 96.
16. T. M. Squires and S. R. Quake, *Rev. Mod. Phys.*, 2005, **77**, 977.
17. J. den Toonder, F. Bos, D. Broer, L. Filippini, M. Gillies, J. de Groede, T. Mol, M. Reijme, W. Talen, H. Wilderbeek, V. Khatavkar and P. Anderson, *Lab Chip*, 2008, **8**, 533.
18. R. Dreyfus, J. Baudry, M. L. Roper, M. Fermigier, H. A. Stone and J. Bibette, *Nature*, 2005, **437**, 862.
19. B. A. Evans, A. R. Shields, R. Lloyd Carroll, S. Washburn, M. R. Falvo and R. Superfine, *Nano Lett.*, 2007, **7**, 1428.
20. F. Fahrni, M. W. J. Prins and L. J. van IJzendoorn, *Lab Chip*, 2009, **9**, 3413.
21. A. Babataheri, M. Roper, M. Fermingier and O. du Roure, *J. Fluid Mech.*, 2011, **678**, 5.
22. E. Gauger and H. Stark, *Phys. Rev. E*, 2006, **74**, 021907.
23. E. M. Gauger, M. T. Downton and H. Stark, *Eur. Phys. J. E*, 2009, **28**, 231.
24. M. T. Downton and H. Stark, *Europhys. Lett.*, 2009, **85**, 44002.
25. M. C. Lagomarsino, I. Pagonabarraga and C. P. Lowe, *Phys. Rev. Lett.*, 2005, **94**, 148104.
26. X. Schlagberger and R. R. Netz, *Europhys. Lett.*, 2005, **70**, 129.

27. M. Manghi, X. Schlagberger, Y.-W. Kim and R. R. Netz, *Soft Matter*, 2006, **2**, 653.
28. M. Manghi, X. Schlagberger and R. Netz, *Phys. Rev. Lett.*, 2006, **96**, 068101.
29. Y. Kim and R. Netz, *Phys. Rev. Lett.*, 2006, **96**, 158101.
30. I. Llopis, I. Pagonabarraga, M. C. Lagomarsino and C. P. Lowe, *Phys. Rev. E.*, 2007, **76**, 061901.
31. C. Goubault, P. Jop, M. Fermigier, J. Baudry, E. Bertrand and J. Bibette, *Phys. Rev. Lett.*, 2003, **91**, 260802.
32. L. Cohen-Tannoudji, E. Bertrand, L. Bressy, C. Goubault, J. Baudry, J. Klein, J.-F. Joanny and J. Bibette, *Phys. Rev. Lett.*, 2005, **94**, 038301.
33. A. Koenig, P. H1ebraud, C. Gosse, R. Dreyfus, J. Baudry, E. Bertrand and J. Bibette, *Phys. Rev. Lett.*, 2005, **95**, 128301.
34. N. Coq, S. Ngo, O. du Roure, M. Fermigier and D. Bartolo, *Phys. Rev. E*, 2010, **82**, 041503.
35. N. Coq, A. Bricard, F.-D. Delapierre, L. Malaquin, O. du Roure, M. Fermigier and D. Bartolo, *Phys. Rev. Lett.*, 2011, **107**, 014501.
36. M. Vilfan, A. Potočik, B. Kavčič, N. Osterman, I. Poberaj, A. Vilfan and D. Babič, *Proc. Natl Acad. Sci. U.S.A.*, 2010, **107**, 1844.
37. G. Kokot, M. Vilfan, N. Osterman, A. Vilfan, B. Kavčič, I. Poberaj and D. Babič, *Biomicrofluidics*, 2011, **5**, 034103.
38. M. Vilfan, G. Kokot, A. Vilfan, N. Osterman, B. Kavčič, I. Poberaj and D. Babič, *Beilstein J. Nanotechnol.*, 2012, **3**, 163.
39. C. E. Sing, L. Schmid, M. F. Schneider, T. Franke and A. Alexander-Katz, *Proc. Natl Acad. Sci. U.S.A.*, 2010, **107**, 535.
40. F. Fahrni, M. W. J. Prins and L. J. van IJzendoorn, *J. Magn. Magn. Mater.*, 2009, **321**, 1843.
41. S. N. Khaderi, C. B. Craus, J. Hussong, N. Schorr, J. Belardi, J. Westerweel, O. Prucker, J. Rühe, J. M. J. den Toonder and P. R. Onck, *Lab Chip*, 2011, **11**, 2002.
42. J. Belardi, N. Schorr, O. Prucker and J. Rühe, *J. Adv. Funct. Mater.*, 2011, **21**, 3314.
43. S. N. Khaderi, M. G. H. M. Baltussen, P. D. Anderson, D. Ioan, J. M. J. den Toonder and P. R. Onck, *Phys. Rev. E*, 2009, **79**, 046304.
44. M. Baltussen, P. Anderson, F. Bos and J. den Toonder, *Lab Chip*, 2009, **9**, 2326.
45. S. N. Khaderi, M. G. H. M. Baltussen, P. D. Anderson, J. M. J. den Toonder and P. R. Onck, *Phys. Rev. E*, 2010, **82**, 027302.
46. S. N. Khaderi, J. M. J. den Toonder and P. R. Onck, *J. Fluid Mech.*, 2011, **688**, 44.
47. S. N. Khaderi, J. M. J. den Toonder and P. R. Onck, *Biomicrofluidics*, 2012, **6**, 014106.
48. C. L. van Oosten, C. W. M. Bastiaansen and D. J. Broer, *Nat Mater.*, 2009, **8**, 677.
49. A. Bhattacharya, G. A. Buxton, O. B. Usta and A. C. Balazs, *Langmuir*, 2012, **28**, 3217.

50. R. Ghosh, G. A. Buxton, O. B. Usta, A. C. Balazs and A. Alexeev, *Langmuir*, 2010, **26**, 2963.
51. O. Kratky and G. Porod, *Recl. Trav. Chim. Pays-Bas*, 1949, **68**, 1106.
52. J. K. G. Dhont, *An Introduction to Dynamics of Colloids*, Elsevier, Amsterdam, 1996.
53. H. A. Lorentz, *Zittingsverlag Akad. v. Wet.*, 1896, **5**, 168.
54. J. R. Blake, *Proc. Camb. Phil. Soc.*, 1971, **70**, 303.
55. J. R. Blake and A. T. Chwang, *J. Eng. Math.*, 1974, **8**, 23.
56. B. Cichocki, B. U. Felderhof and K. Hinsen, *J. Chem. Phys.*, 1994, **100**, 3780.
57. D. J. Jeffrey and Y. Gnishi, *J. Fluid Mech.*, 1984, **139**, 261.
58. S. Kim and S. J. Kanila, *Microhydrodynamics: Principles and Selected Applications*, Butterworth-Heinemann, Boston, 1991.
59. K. Hinsen, *Comput. Phys. Commun.*, 1995, **88**, 327.
60. L. D. Landau and E. M. Lifschitz, *Lehrbuch der Theoretischen Physik Band VII: Elastizitätstheorie*, Akademie Verlag, Berlin, 1991.
61. C. P. Lowe, *Phil. Trans. R. Soc. Lond. B*, 2003, **358**, 1543.
62. M. C. Lagomarsino, F. Capuani and C. P. Lowe, *J. Theor. Biol.*, 2003, **224**, 215.
63. C. H. Wiggins and R. E. Goldstein, *Phys. Rev. Lett.*, 1998, **80**, 3879.
64. C. H. Wiggins, D. Riveline, A. Ott and R. E. Goldstein, *Biophys. J.*, 1998, **74**, 1043.
65. S. L. Biswal and A. P. Gast, *Phys. Rev. E*, 2003, **68**, 021402.
66. S. L. Biswal and A. P. Gast, *Phys. Rev. E*, 2004, **69**, 041406.
67. M. L. Roper, R. Dreyfus, J. Baudry, M. Fermigier, J. Bibette and H. A. Stone, *J. Fluid Mech.*, 2006, **554**, 167.
68. M. L. Roper, R. Dreyfus, J. Baudry, M. Fermigier, J. Bibette and H. A. Stone, *Proc. Royal Soc. A*, 2008, **464**, 877.
69. A. Cebers, *J. Phys.: Condens. Matter*, 2003, **15**, S1335.
70. A. Cebers, *Curr. Opin. Coll. Interface Sci.*, 2005, **10**, 167.
71. A. Cebers, *Magnetohydrodynamics*, 2005, **41**, 63.
72. M. Belovs and A. Cebers, *Phys. Rev. E*, 2006, **71**, 051503.
73. S. Melle, O. G. Calderón, M. A. Rubio and G. G. Fuller, *Phys. Rev. E*, 2003, **68**, 041503.
74. J. Blake, *J. Fluid Mech.*, 1972, **55**, 1.
75. N. Liron, *J. Fluid Mech.*, 1978, **86**, 705.
76. D. J. Smith, J. R. Blake and E. A. Gaffney, *J. R. Soc. Interface*, 2008, **5**, 567.
77. S. Gueron, K. Levit-Gurevich, N. Liron and J. J. Blum, *Proc. Natl. Acad. Sci. U.S.A.*, 1997, **94**, 6001.
78. S. Gueron and K. Levit-Gurevich, *Proc. Natl. Acad. Sci. U.S.A.*, 1999, **96**, 12240.
79. M. C. Lagomarsino, B. Bassetti and P. Jona, *Eur. Phys. J. E*, 2002, **26**, 81.
80. M. C. Lagomarsino, P. Jona and B. Bassetti, *Phys. Rev. E*, 2003, **68**, 021908.
81. B. Guirao and J.-F. Joanny, *Biophys. J.*, 2007, **92**, 1900.

82. T. Niedermayer, B. Eckhardt and P. Lenz, *CHAOS*, 2008, **18**, 037128.
83. N. Osterman and A. Vilfan, *Proc. Natl. Acad. Sci. U.S.A.*, 2011, **108**, 5727.
84. C. Wollin and H. Stark, *Eur. Phys. J. E*, 2011, **34**, 1.
85. F. Kendelbacher and H. Stark, in *Engineering of Chemical Complexity*, ed. G. Ertl and A. Mikhailov, World Scientific, Singapore, 2012.
86. N. Hirokawa, Y. Tanaka, Y. Okada and S. Takeda, *Cell*, 2006, **125**, 33.

CHAPTER 3

Computational Design of Magnetic Artificial Cilia

S. N. KHADERI,[a] J. M. J. DEN TOONDER[b] AND P. R. ONCK*[a]

[a] University of Groningen, The Netherlands; [b] Eindhoven University of Technology, The Netherlands
*Email: p.r.onck@rug.nl

3.1 Introduction

To design propulsion mechanisms to transport fluids in micro-channels that form an integral component of a device, such as a lab-on-a-chip, is a challenging task. This is because at these length scales viscous forces usually dominate over the inertial forces[1] and the conventional macroscopic fluid transport mechanisms offer only little help. Interestingly, nature has developed propulsion mechanisms that can operate in both these regimes - inertia and viscosity dominated at large and small length scales, respectively. At large length scales, examples include blood transport using the heart, feeding of jelly fish and locomotion of scallops. At small length scales, fluid transport is achieved using organelles such as cilia and flagella. The key feature of cilia and flagella is that their motion is periodic but non-reciprocal in nature. The flagella achieve a non-reciprocal motion by propagating waves of lateral displacement along their body. In cilia, the non-reciprocal motion is achieved by their special asymmetric motion, consisting of an effective and a recovery stroke. In this article we use the principles used by cilia to design mechanical actuators that can create fluid transport in micro-channels.

RSC Nanoscience & Nanotechnology No. 30
Artificial Cilia
Edited by Jaap M J den Toonder and Patrick R Onck

Published by the Royal Society of Chemistry, www.rsc.org

During the last 5 years many examples of magnetically actuated artificial cilia have appeared in the literature. From an experimental point of view, these studies aimed to advance the fabrication techniques and to develop methodologies to characterise the fluid transport.[2–14] From a theoretical point of view, computational methods have been developed to understand the related physical phenomena and to optimise the fluid propelled.[12,13,15–24] Currently many different forms of artificial cilia are present, a considerable subset of which are discussed in the current volume. We focus our attention on plate-like magnetically-driven artificial cilia that consist of polymer films filled with magnetic nanoparticles, which can be fabricated using lithographic techniques.[7,9,17]

The output or the performance of a fluid propulsion device is characterised by the flow and pressure it can generate. The performance depends on several factors, such as the viscosity and density of the fluid, the applied magnetic field, the shape and size of the cilia, their elastic and magnetic properties and the channel geometry. With the help of dimensional analysis, we can group these factors into a set of dimensionless parameters through which the response of the ciliary system can be systematically studied using numerical models. Once we obtain a complete understanding of the dependence of flow and pressure generated on the dimensionless parameters, we can explore these combinations of parameters for which the fluid transport and pressure generated are maximal.

This chapter is organised as follows. The equations governing the dynamics of the ciliary motion and fluid flow, the solution procedure and the set of dimensionless parameters are given in section 3.2. We then show two configurations of cilia in section 3.3 that exhibit an asymmetric motion and create fluid transport in microchannels. This fluid transport is systematically studied in terms of dimensionless parameters and the region of maximum fluid transport is identified in the parametric space in section 3.4. In section 3.5, we study the effect of channel height and cilia spacing on the pressure generated and fluid transported. In section 3.6, it is illustrated how the flow can be enhanced by the out-of-phase beating of the cilia. Finally, in section 3.7 we elucidate how to design artificial cilia that can generate large flows and high pressure gradients.

3.2 Equations of Motion

3.2.1 Governing Equations

In this section we briefly describe the numerical model used to study the fluid transport induced by magnetically-actuated artificial cilia. We consider plate-like artificial cilia that exhibit a planar beat motion in typical microfluidic channels with a height H that is much smaller than its out-of-plane width (see Figure 3.1). Therefore, any variation in the out-of-plane direction can be neglected and under these assumptions it is sufficient to model the artificial cilia and the resulting flow in a two-dimensional setting.

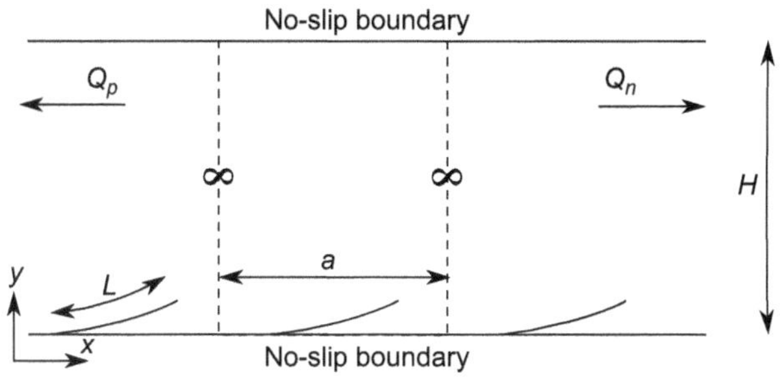

Figure 3.1 Schematic side view of the micro-fluidic channel used in the simulations. The unit-cell used is shown using dashed lines.
(Reproduced from Khaderi *et al.*[25])

3.2.1.1 Solid Dynamics Model

The artificial cilia have a typical thickness $h = 2\,\mu\text{m}$, a width $b = 20\,\mu\text{m}$ and a length L of 100 μm. These structures are modeled using Euler-Bernoulli beam elements taking into consideration geometric non-linearity in an updated Lagrangian framework. The principle of virtual work is used to arrive at the equations of equilibrium: at any time $t+\Delta t$ the virtual work of the external forces ($\delta W_{\text{ext}}^{t+\Delta t}$) equals the internal virtual work done by the elastic forces ($\delta W_{\text{int}}^{t+\Delta t}$). The internal virtual work is given by

$$\delta W_{\text{int}}^{t+\Delta t} = \int_V (\sigma \delta \epsilon + \rho(\ddot{u}\delta u + \ddot{v}\delta v))dV, \tag{3.1}$$

where u and v are the axial and transverse displacements of a point on the beam and ρ is the density of the beam. Furthermore, σ is the axial stress and ϵ is the corresponding strain, given by

$$\epsilon = \frac{\partial u}{\partial x} + \frac{1}{2}\left(\frac{\partial v}{\partial x}\right)^2 - y\frac{\partial^2 v}{\partial x^2}.$$

The external virtual work is

$$\delta W_{\text{ext}}^{t+\Delta t} = \int \left(f_x \delta u + f_y \delta v + N_z \frac{\partial \delta v}{\partial x}\right) A dx + \int \left(t_x \delta u + t_y \delta v\right) b dx, \tag{3.2}$$

where f_x and f_y are the magnetic body forces in the axial and transverse directions, N_z is the magnetic body couple in the out-of-plane direction, t_x and t_y are the surface tractions, A is the cross sectional area and b is the out-of-plane thickness of the cilia.

The linearised and discretised principal of virtual work can be written as,[26]

$$\delta \boldsymbol{p}^T (\boldsymbol{K}\Delta \boldsymbol{p} + \boldsymbol{M}_S \ddot{\boldsymbol{p}}^{\text{t}+\Delta\text{t}} - \boldsymbol{F}_{\text{ext}}^{\text{t}+\Delta\text{t}} + \boldsymbol{F}_{\text{int}}^{\text{t}}) = 0, \tag{3.3}$$

where $\boldsymbol{K}$ is the stiffness matrix that combines both material and geometric contributions, $\boldsymbol{M}_S$ is the mass matrix, $\boldsymbol{F}_{\text{ext}}^{\text{t}+\Delta\text{t}}$ is the external force vector, $\boldsymbol{F}_{\text{int}}^{\text{t}}$ is the internal force vector, $\Delta\boldsymbol{p}$ is the nodal displacement increment vector and $\ddot{\boldsymbol{p}}$ is the nodal acceleration vector. The nodal acceleration vector is discretised in time using Newmark's algorithm[27] (using parameters $\gamma = 1.0$ and $\beta = 0.5$) so that Eqn. 3.3 can be written in terms of the velocity of the beam. The complete discretised equations of motion for the solid mechanics model can be found elsewhere.[25]

3.2.1.2 Magnetostatics

The Maxwell's equations for the magnetostatic problem with no external currents are

$$\boldsymbol{\nabla} \cdot \boldsymbol{B} = 0, \quad \boldsymbol{\nabla} \times \boldsymbol{H} = 0, \tag{3.4}$$

with the constitutive relation $\boldsymbol{B} = \mu_0(\boldsymbol{M} + \boldsymbol{H})$, where $\boldsymbol{B}$ is the magnetic flux density, $\boldsymbol{H}$ is the magnetic field, $\boldsymbol{M}$ is the magnetization, and μ_0 is the permeability of vacuum. Eqn 3.4 is solved for $\boldsymbol{M}$ and $\boldsymbol{B}$ using the boundary element method.[25] The magnetic couple per unit volume is given by $\boldsymbol{N} = \boldsymbol{M} \times \boldsymbol{B}_0$. As the simulations are two-dimensional, the only non-zero component of the magnetic body couple is N_z which is the only contribution to the external virtual work in Eqn. 3.2. Since (as shown later) we apply only a uniform magnetic field over each cilium, the magnetic body forces due to field gradients are absent.

3.2.1.3 Fluid Dynamics and Solid-fluid Coupling

The fluid is assumed to be Newtonian and incompressible. The physical behaviour of the fluid is governed by the Navier-Stokes equation and incompressibility condition:

$$\begin{aligned} -\boldsymbol{\nabla} p + 2\mu \boldsymbol{\nabla} \cdot \boldsymbol{D} &= \rho^f \frac{d\boldsymbol{v}}{dt}, \\ \boldsymbol{\nabla} \cdot \boldsymbol{v} &= 0, \end{aligned} \tag{3.5}$$

where ρ^f is the fluid density, $d(\boldsymbol{v})\,/\,dt$ is the total derivative, p is the pressure in the fluid, $\boldsymbol{D}$ is the rate of deformation tensor, $\boldsymbol{v}$ is the velocity of the fluid and μ is the viscosity of the fluid. The set of equations in Eqn. 3.5 is solved using quadrilateral Eulerian finite elements that quadratically and linearly interpolate the velocity and pressure of the fluid, respectively. The solid and fluid domains are coupled by imposing the constraint that the velocity at the nodes of the solid beam are equal to the velocity of the surrounding fluid (commonly referred to as the point collocation method). This coupling is established with the help of Lagrange multipliers using the fictitious domain method.[25,28]

3.2.1.4 Solution Procedure

The solution procedure is as follows. The Maxwell's equations are solved at every time instant to obtain the magnetic field. From the magnetic field, the magnetic body couple acting on the cilia is calculated and is provided as an external load to the coupled solid-fluid model, which simultaneously solves for the cilia velocity, and the velocity and pressure of the fluid. The velocity of the cilia is integrated using Newmark's algorithm to obtain its new position, and the procedure is repeated.

3.2.2 Dimensionless Parameters

The parameters that govern the behaviour of the ciliary system can be identified by nondimensionalizing the principle of virtual work for the fluid and solid domain. The following set of dimensionless parameters are found to be of importance: the fluid number (F_n) - the ratio of the viscous forces to the elastic forces, the magnetic number M_n - the ratio of magnetic forces to the elastic forces and the Reynolds number Re - the ratio of fluid inertial forces to the viscous forces. The form of these parameters is:

$$\begin{aligned} F_n &= 12\frac{\mu}{Et_{\text{beat}}}\frac{L^3}{h^3} \\ M_n &= 12\frac{N_z}{E}\frac{L^2}{h^2} \\ \text{Re} &= \frac{\rho^f L^2}{\mu t_{\text{beat}}}. \end{aligned} \tag{3.6}$$

In the above expressions t_{beat} is the cycle time of the ciliary beat. In the above, E and ρ are the elastic modulus and density of the cilia, ρ^f and μ are the density and viscosity of the fluid, L and h are the length and thickness of the cilia, N_z is the magnetic couple and t_{beat} is the time taken by the cilia to complete one beat cycle. For permanently magnetic cilia, the magnetic couple $N_z = M_r B_0$ and for super-paramagnetic cilia, $N_z = B_0^2/\mu_0$, where μ_0 is the magnetic permeability of free space and M_r is the remanent magnetization.

3.3 Asymmetric Configurations

We use the numerical model outlined in the previous section to study the motion of a periodic arrangement of cilia in a microfluidic channel of height $H = 5L$, with the cilia spaced $5L$ apart, i.e. $a = H = 5L$ in Figure 3.1. From this channel a square unit-cell consisting of one cilium is identified for the simulations, where L is the length of the cilia. No-slip boundary conditions are applied at the top and bottom boundaries of the channel and periodic boundary conditions at the left and right ends of the unit-cell (see Figure 3.1). The fluid has a viscosity $\mu = 1$mPas. The cilia have a thickness $h = 2$ μm, an elastic modulus $E = 1$ MPa and a density $\rho = 1600$ kg/m^3, unless mentioned otherwise. In this section, we assume the inertial forces in the fluid to be negligible, i.e. Re $= 0$.

3.3.1 Curled Permanently Magnetic Cilia

This configuration is based on the buckling-induced recovery stroke of permanently magnetic cilia. The initial geometry of the cilia is a quarter of a circle with radius 100 μm fixed at the bottom of the channel, see instant 1 in Figure 3.2a. The direction of the remnant magnetization is along the cilia with the magnetization vector pointing from the fixed end to the free end. The remnant magnetization of the cilia is taken to be $M_r = 15$ kA/m. A uniform external magnetic field of magnitude $B_0 = 13.3$ mT is applied at 225° to the x-axis from $t = 0$ ms to $t = 1$ ms and then linearly reduced to zero in the next 0.2 ms. When the external field is applied, clockwise moments (N_z) are acting on the portion near the fixed end of the cilia while near the free end counter-clockwise moments develop (see instant 1 in Figure 3.2b). Under the

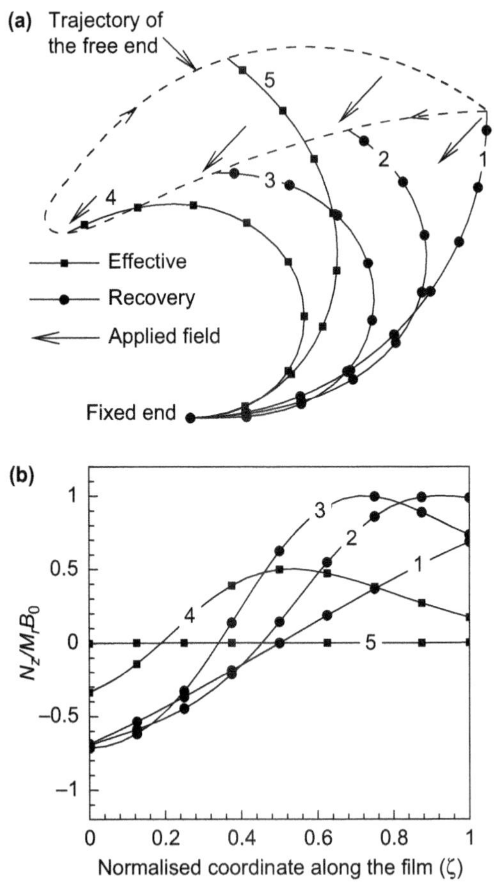

Figure 3.2 Buckling of a curled permanently magnetic (PM) film as a result of magnetic actuation, during the propulsion of fluid (Reproduced from Khaderi *et al.*[15]) (a) Snapshots of the film at 0 ms, 0.3 ms, 0.6 ms, 1.1 ms and 3 ms. (b) Normalized torque distribution along the film corresponding to the snapshots shown in (a).

influence of such a system of moments, the cilia undergo a buckling-kind of instability and snap-through to configurations 3 and 4 during which the zero-couple position travels to the fixed end. Then, the applied field is reduced to zero and the cilia return to the initial position through instance 5 in Figure 3.2a. Note that the propulsive action in the effective stroke takes place during the elastic relaxation of the cilia, whereas the cilia stay close to the bottom boundary during the recovery stroke because of the buckling-induced snap-through. The main criterion for the buckling-induced recovery stroke to take place is the initial opposing directions of the remnant magnetization and the applied magnetic field. This criterion will not be met in the case of para-magnetic cilia where the magnetization depends on the direction and magnitude of the applied magnetic field.

3.3.2 Super-paramagnetic Cilia

Tapered super-paramagnetic cilia that are subjected to a rotating magnetic field can exhibit a pronounced asymmetric motion. To elucidate this we consider super-paramagnetic cilia whose thickness decreases linearly from 2 μm at the fixed end to 1 μm at the free end. The cilia posses an anisotropic magnetic susceptibility of 4.6 and 0.8 in the tangential and normal directions, respectively.[29] They are subjected to a magnetic field of magnitude $B_0 = 31.5$ mT that is rotated from 0° to 180° in $t = 10$ ms and then held constant. The motion of one cilium is shown in Figure 3.3a. From the instants 1, 2 and 3 we see that during the effective stroke, the cilia nicely follow the applied magnetic field. When the cilia have reached position 4, the interaction between the magnetization and the applied magnetic field causes clockwise couples to act on the fixed end, while counter-clockwise couples act on the free end. From Figure 3.3b it can be seen that during the recovery stroke (shown using circles) the position of zero torque propagates from the fixed end to the free end (from instance 4 to 5). The tapering causes the magnetic couple per unit length to be higher at the fixed end (see instant 5 in Figure 3.3b), allowing the cilia to recover to the initial position (instant 1). The generated asymmetric motion is very similar to that of natural cilia.[30] It is to be noted that the cilia recover in the presence of an applied magnetic field and that the cilia complete one beat cycle for a 180° rotation of the magnetic field.

3.4 Fluid Transport

We now analyse the fluid flow caused by the ciliary motion of the super-paramagnetic cilia (Figure 3.3) when it is subjected to a magnetic field rotating in the counter-clockwise direction with a time period t_{ref}. Note that this magnetic field is not significantly different from that mentioned in section 3.3.2. The cilia complete one beat cycle in $t_{\text{beat}} = t_{\text{ref}}/2$. To analyse the fluid transport we choose a channel of height $H = 2L$ in which cilia are placed at an interval $a = 2L$, where L is the length of the cilia. Throughout this section the magnetic number is set to 10.89 indicating that the elastic forces are low compared to the

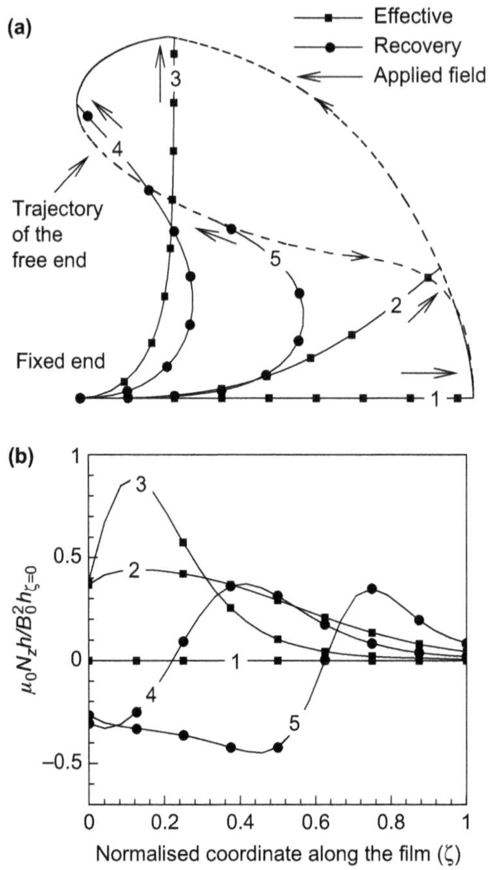

Figure 3.3 Motion of a super-paramagnetic (SPM) film in a rotating magnetic field, during the propulsion of fluid. (Reproduced from Khaderi *et al.*[15]) (a) Snapshots of the film at 0 ms, 2.5 ms, 5.0 ms, 7.5 ms and 8.5 ms. (b) Normalized torque distribution along the film corresponding to the snapshots shown in (a). Here μ_0 is the permeability of vacuum, h is the thickness at position ζ and $h_\zeta = 0$ is the thickness at the fixed end.

magnetic forces. In the following we will study the effect of Re and F_n for fixed M_n. The magnetic susceptibilities and the geometry of the cilia are the same as in section 3.3.2.

The fluid propelled is characterised by two parameters: the net volume of the fluid transported during a ciliary beat cycle and the effectiveness. The velocity field in the fluid, at any x position, integrated along the channel height gives the instantaneous flux through the channel. This flux when integrated in the direction of the effective and recovery strokes gives the positive (Q_p) and negative (Q_n) flows, respectively (see Figure 3.1). The generated net area flow per cycle is $Q_p - Q_n$ in the direction of the effective stroke. The effectiveness, defined as $(Q_p - Q_n)/(Q_p + Q_n)$, indicates which part of the totally displaced fluid is effectively converted into a net flow. An effectiveness of unity represents unidirectional flow.

3.4.1 Basic Mechanism

To elucidate the mechanism of fluid transport we first analyse the case of low Reynolds numbers, i.e. Re = 0.001 (at $F_n = 0.015$). The results are shown in Figure 3.4. At low Reynolds numbers the viscous forces are very large compared to the fluid inertia forces, leading to an instantaneous momentum diffusion from the cilia to the fluid. During the effective and recovery stroke the cilia move to the left and to the right, respectively. This creates a positive flux (see the instances 1 and 2 in Figure 3.4g and a negative flux (see the instances 3, 4 and 5 in Figure 3.4g). The flow created by the cilia increases during the effective stroke and decreases during the recovery. Due to the asymmetric motion of the cilia, the positive flow is larger than the negative flow; the cilia create a net flow of magnitude 0.2 (in units of $\pi L^2 / 2$) to the left by the end of the cycle ($t = t_{\text{beat}}$). So the fluid is effectively displaced to the left in one beat cycle of the cilia. This can also be seen by comparing the position of the fluid particles at the beginning and end of the cycle. In addition the motion of the particles reveals that the fluid transport is fluctuating in nature, which leads to a low value of effectiveness. It is to be noted that due to the low viscous forces (compared to the elastic forces) the cilia complete the beat cycle before t_{beat}. The cilia remain inactive in this 'dead' position until the start of the next cycle. Also, the velocities of the cilia and fluid are much larger during the recovery stroke compared to the effective stroke. However, as the Reynolds number is low the large kinetic energy input to the fluid is instantly dissipated and there is no further fluid transport during the dead position. However, this is not the case when the Reynolds number is finite, as will be studied in the next section.

3.4.2 Effect of Fluid Inertia

Having understood the fluid transport in the absence of fluid inertia, we now explore how the fluid transport and the effectiveness change when the Reynolds number increases. The area flow is plotted as a function of the Reynolds number (Re) in Figure 3.5. It can be seen that at high Reynolds numbers the fluid transport is significantly larger than that in the Stokes regime. However, at moderate Reynolds numbers the fluid transport is negative, which means the fluid is being propelled in the direction of the recovery stroke. This suggests a method, which will be detailed later, to control the direction of the fluid flow. To understand the cause of the direction reversal with a change of the Reynolds number we plot the positive and negative flow created by the cilia. At low and high Reynolds numbers, the positive flow is larger than the negative flow, while for moderate Reynolds numbers the opposite is the case. We now investigate what mechanisms are involved that drive the flow direction reversal at

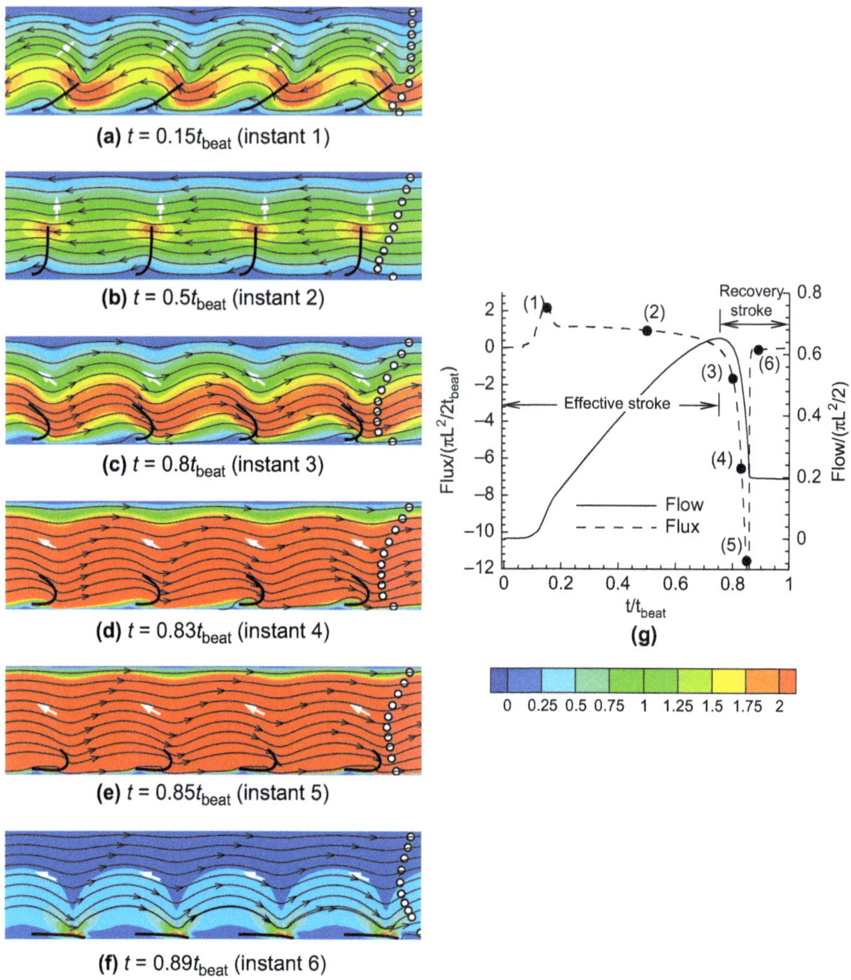

Figure 3.4 (a)–(f) Contours of absolute velocity (normalized with L/t_{beat}) at different time instants (Reproduced from Khaderi *et al.*[25]). The direction of the velocity is given by the streamlines and the white arrows represent the magnetic field at the respective time instances. The circles represent fluid particles. The parameters used are $F_n = 0.015$, $Re = 0.001$ and $M_n = 10.89$. Four unit-cells are shown for clarity. (g) Instantaneous flux and accumulated flow as a function of time. The time instances corresponding to (a)–(f) are duly marked.

moderate Reynolds number (Re ≈1) and what causes a higher flow at high Re compared to the Stokes regime.

Let us analyse what happens at moderate Reynolds numbers, by specifying Re = 1. At Re = 1, the negative flow is larger than that of the Stokes regime, while the positive flow is smaller. To analyse the reason behind this we plot the velocity contours, the flux and the flow in Figure 3.6. Two important

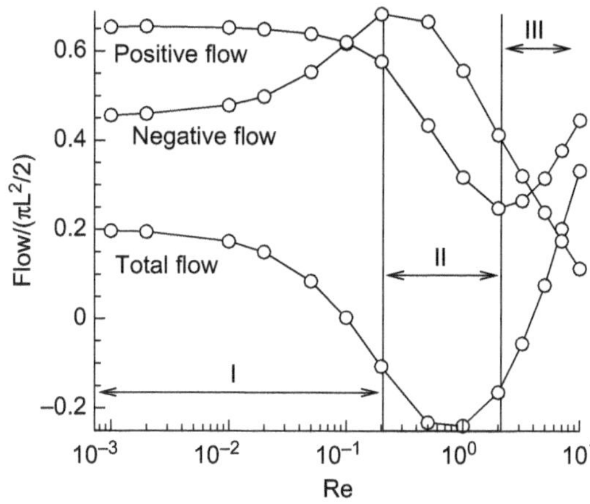

Figure 3.5 Flow as a function of Reynolds number *Re* for $F_n = 0.015$ and $M_n = 10.89$, with different regions of significance marked. Region I: The large kinetic energy input during the recovery stroke leads to a negative flow during the dead position. Region II: The negative flow decreases due to the delayed momentum diffusion, and the positive flow decreases because the effective stroke has to first cancel the negative flow. Region III: The positive flow increases because of the continuous velocity in the direction of the effective stroke.
(Reproduced from Khaderi *et al.*[25])

observations can be made from Figure 3.6. First, the maximum flux during the recovery stroke is much smaller than that in the Stokes regime. Second, the fluid transport takes place even when the cilia are in the dead position, as seen from instants 3 and 4 in Figure 3.6. The first observation is a direct consequence of the finite Reynolds number, whereas the second is the result of the combined effect of finite Reynolds number and low fluid number. When Re is finite, the momentum diffusion from the cilia to the fluid is not instantaneous. So the velocity is concentrated only in the lower half of the channel (compare Figures 3.6b and 3.4a) leading to lower values of the instantaneous flux compared to the Stokes regime. The low value of the fluid number, as seen before, leads to a dead position of the cilia. Now the kinetic energy input to the fluid during the fast recovery stroke is not dissipated instantly, but creates a large negative flow during the dead position and during the initial part of the effective stroke (Figures 3.6d and 3.6e). Consequently, during the first half of the subsequent effective stroke, the cilia have to spend energy in canceling the negative flow, instead of creating a positive flow. Hence, the positive flow is lower than that of the Stokes regime.

Next, we analyse the flow behaviour at a Reynolds number Re = 10. As the Re is increased from Re = 1 in Figure 3.5 the negative flow continues to decrease, whereas the positive flow starts to increase. To understand these trends we again analyse the velocity contours, the flux and the flow as a function of time (see Figure 3.7f). Two key features are observed here that were

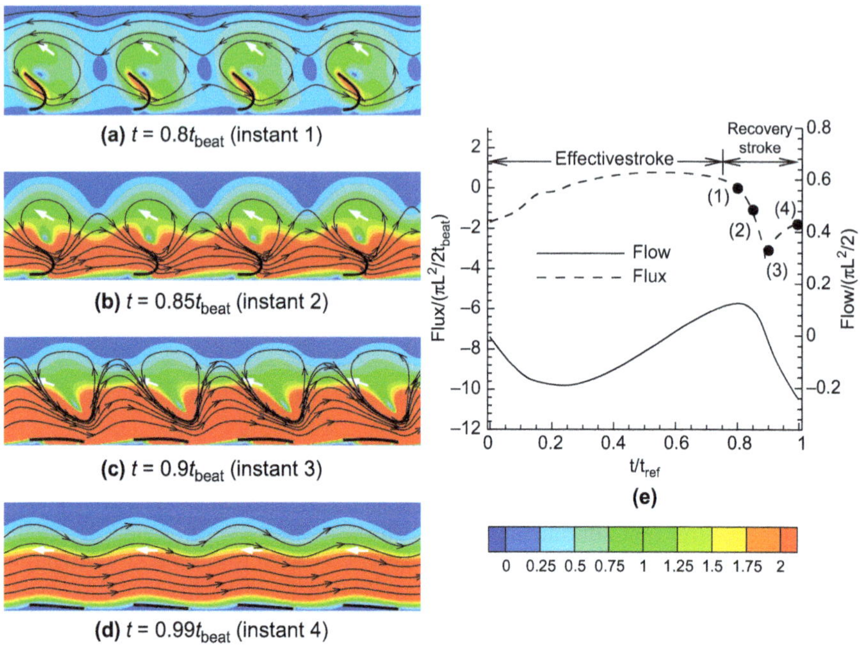

Figure 3.6 (a)–(e) Contours of absolute velocity (normalized with L/t_{beat}) at different time instants. The direction of the velocity is given by the streamlines and the white arrows represent the magnetic field at the respective time instances. The circles represent fluid particles. The parameters used are $F_n = 0.015$, $M_n = 10.89$ and $Re = 1$. Four unit-cells are shown for clarity. The legend for the contours is the same as that of Figure 3.4e. Instantaneous flux and accumulated flow as a function of time. The time instances corresponding to (a)–(d) are duly marked.
(Reproduced from Khaderi *et al.*[25])

absent in the cases studied before. First, we see flow in the direction of the effective stroke during recovery (see Figure 3.7d–3.7f). Second, the dead position of the cilia is absent (see Figure 3.7f). Because the viscous forces are so low compared to inertial forces, they do not dissipate the energy input to the fluid during the effective stroke and this leads to a flow in the direction of the effective stroke during recovery. At high Reynolds numbers, the fluid inertia forces also contribute to the drag on the cilia leading to the absence of the dead position, despite of the low fluid number. The combined effect of the absence of the dead position and enhanced localization of the flow during the recovery stroke (see Figures 3.7e and 3.7f) leads to the decrease of the negative flow compared to Re = 1. The total flow created by the cilia at the end of the cycle is 0.32 in units of $\pi L^2/2$ (see Figure 3.7g), which is higher than that of the Stokes regime (0.2). This can also be observed from the position of the fluid particles at the end of the beat cycle (compare Figure 3.7f to Figure 3.4f).

The observations of this section can be summarised as follows (see Figure 3.5). The amount of the negative flow is due to the competition between

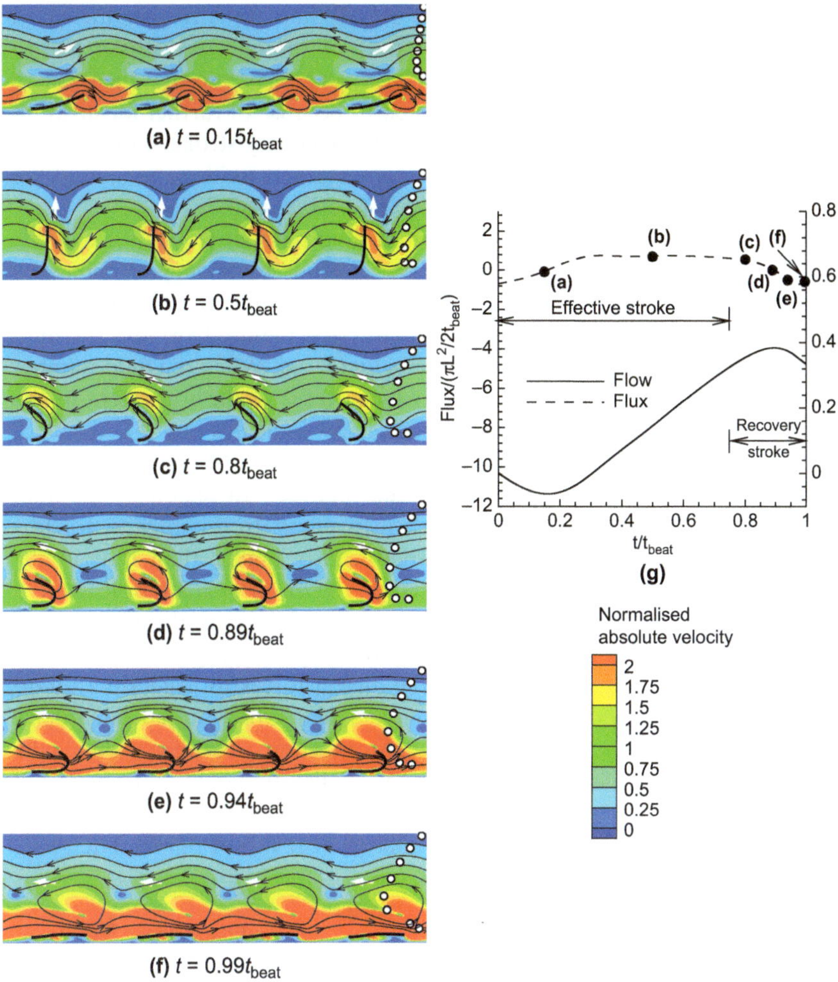

Figure 3.7 (a)–(f) Contours of absolute velocity (normalized with L/t_{beat}) at different time instants. The direction of the velocity is given by the streamlines and the white arrows represent the magnetic field at the respective time instances. The circles represent fluid particles. The parameters used are $F_n = 0.015$, $Re = 10$ and $M_n = 10.89$. Four unit-cells are shown for clarity. (g) Instantaneous flux and accumulated flow as a function of time. The time instances corresponding to (a)–(f) are duly marked.
(Reproduced from Khaderi *et al.*[25])

the inertia-induced flow in the recovery direction and its obstruction by the next effective stroke. At small Re, a large negative flux is observed only for a short duration of time (Figure 3.4g). When Re is increased, the inertia of the fluid localizes this flow, which takes place over a longer duration so that the total negative flow increases (see Figure.3.5 and Figure 3.6). The net effect is to reduce the total flow, which eventually becomes negative (i.e., flow is in the recovery direction, see region I of Figure 3.5). When Re is further increased (i.e., region II of Figure 3.5), the initial part of the effective stroke has to overcome the negative flow of the previous recovery stroke, so that the positive flow decreases. The negative flow also decreases because of further localization of the negative velocity near the cilia during the recovery stroke. In the region III, inertia is so dominant that the flow becomes nearly unidirectional and the total flux is positive during most of the cycle (see Figure 3.7).

The flow and effectiveness as a function of fluid number (F_n) and Reynolds number (Re) are shown in Figure 3.8. The contour plots suggest three key regions, namely A, B and C. In region A, the Stokes regime, the effectiveness of the fluid propulsion remains constant at ≈ 0.2 (see Figure 3.8b) and the fluid propelled decreases when the fluid number is increased. When the fluid number increases the viscous forces also increase, which decreases the area swept by the cilia tip and the flow. In region B, the negative flow indicates that it takes place in the direction of the recovery stroke. Now, the flow can be made positive by increasing the Reynolds number and the fluid number. In region C, the fluid inertia forces are large, leading to a high effectiveness. In this regime, for low fluid numbers the fluid transported is significantly higher than that in the Stokes regime. When the fluid number is large, the lower area swept leads to a flow that is comparable to that in the Stokes regime. However, the effectiveness of the fluid transport is high even in this region, thus creating a unidirectional flow. This study shows that we can use the inertia forces to generate a unidirectional flow using periodically beating artificial cilia. In natural ciliary systems, however, hydrodynamic synchronization between the cilia causes metachronal waves, which leads to unidirectional fluid transport.[31] In artificial ciliary systems also, as shown in section 3.6, we can achieve a metachrony by magnetically forcing the cilia to beat out-of-phase. The study

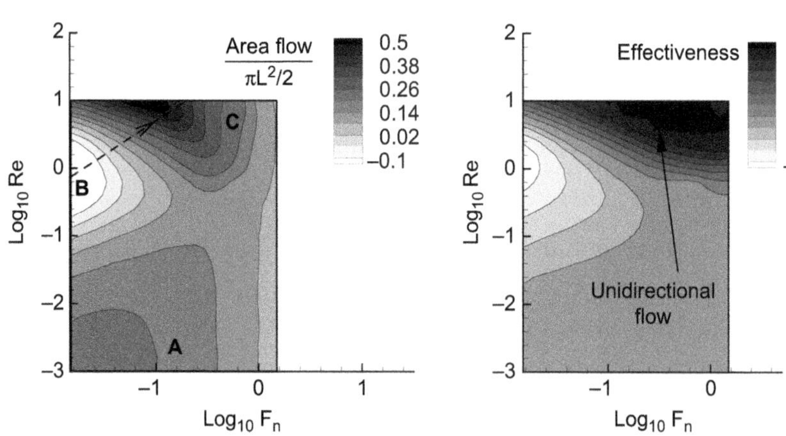

(a) Area flow

(b) Effectiveness

Figure 3.8 Area flow and effectiveness as a function of fluid number F_n and Reynolds number *Re* at a magnetic number of 10.89. (Reproduced from Khaderi *et al.*[25])

performed in the current section also suggests that we can change the direction of flow by changing the frequency of the ciliary beat (see the dashed arrow in Figure 3.8a, which shows the direction of increasing frequency or decreasing t_{beat}). For example, a ciliary system propels fluid in the direction of the recovery stroke in region B. Now, when the frequency of magnetic actuation is increased, the fluid number and the Reynolds number will increase. This results in a large unidirectional flow in the direction of the effective stroke (see Figure 3.8).

3.5 Effect of Channel Dimensions

Having studied the effect of the physical dimensionless parameters we turn our attention towards the geometric parameters, which are the height of the channel H and the cilia spacing a (see Figure 3.1). We study the effect of these parameters for the configurations that can be encountered in lab-on-a-chip systems; namely a closed-loop channel (see Figure 3.9a), and an open-loop channel (see Figure 3.9b). For applications such as polymerase chain reactions (PCR), where the fluid has to undergo a series of processes a closed-loop channel can be used.[32] In an open-loop channel we have well-defined inlet and outlet points for the fluid to enter and leave the channel. To perform two-dimensional simulations we assume that the radius of the loop of the closed-loop channel is much larger than the channel width and the cilia spacing, so that the analysis can be performed by using a periodic unit-cell which contains one cilium, as used in the previous section. However, due to the presence of a well-defined inlet and outlet in the case of the open-loop channel there is no periodicity, and the analysis has to be performed with the channel containing multiple cilia.

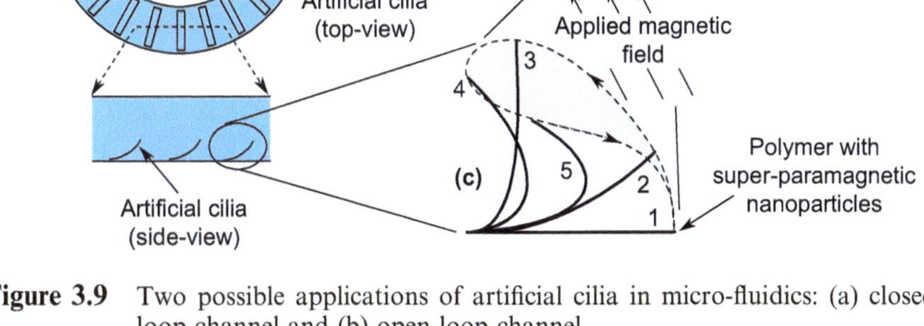

Figure 3.9 Two possible applications of artificial cilia in micro-fluidics: (a) closed-loop channel and (b) open-loop channel.
(Reproduced from Khaderi *et al.*[17])

3.5.1 Closed-loop Channel

The unit-cell for the closed-loop channel has a height H and width equal to the cilia spacing a, containing one cilium of length L. The top and bottom surfaces of the unit-cell are no-slip boundaries, while the left and right boundaries are periodic in velocity. This results in a pressure distribution which is also periodic, so that no pressure gradient can be generated by the closed-loop channel.

Figure 3.10 shows the area flow per cycle normalized by $(\pi L^2 / 2)$, as a function of the cilia spacing a / L for various channel heights H / L. Also shown is the corresponding volume flow assuming an out-of-plane channel width of 1 mm. The fluid transported by the cilia increases when the channel height is increased as well as when the cilia spacing is decreased. This increase is not because of the increase of the asymmetric area swept by the cilia. The swept area, as can be seen in Figure 3.10, decreases when the height of the channel is increased and when the cilia spacing is decreased. It is to be noted that the flow scales linearly with the height of the channel. This is due to the fact that the cilia create a shear flow in the channel which scales linearly with the channel height. When the cilia spacing is decreased, the fluid drag force (to be overcome per cilium) decreases, leading to a higher fluid propulsion.

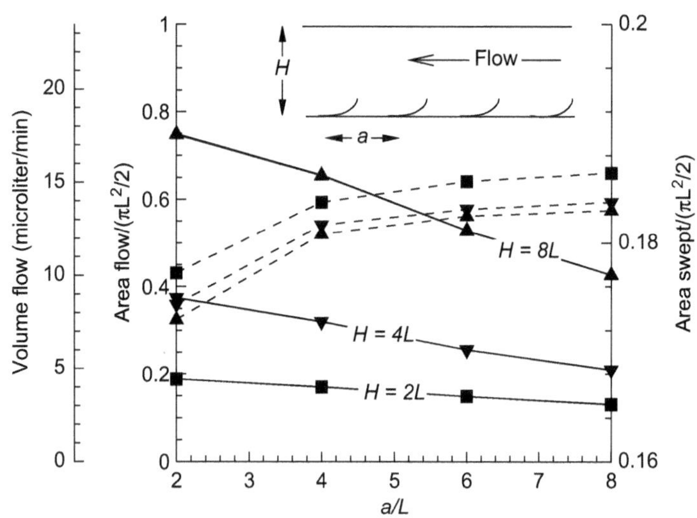

Figure 3.10 Volume flow (left axis), area flow (left axis) and area swept (right axis) as a function of the cilia spacing (a/L) for different channel heights (H/L) for the closed-loop channel, L is the cilia length. The solid lines correspond to the left axes, and the dashed lines correspond to the right axis. The volume flow is calculated by taking the out-of-plane width of the channel to be 1 mm.
(Reproduced from Khaderi *et al.*[17])

3.5.2 Open-loop Channel

It was shown in[17] that the flow created by the open-loop channel converges to that of the closed-loop channel as the number of cilia in the open-loop channel is increased. Hence, in the following we only focus on the pressure generated by the open-loop channel.

The experimental method to measure the pressure generated is illustrated in Figure 3.11 (top).[33,34] The two ends of the microfluidic channel containing the cilia are connected to vertical tubes. When the cilia are not operating the levels of the fluid in the tubes is the same. However, when the cilia are actuated using a magnetic field, the fluid level in the tube which is in the direction of the effective (recovery) stroke will increase to h_1 (decrease to h_2). The difference between these levels $h_1 - h_2$ when multiplied by the specific gravity of the fluid will give the pressure generated by the cilia. In the numerical analysis, we neglect the transient phase during which the fluid levels are changing, but focus on the steady state. In the steady state, the fluid will be in motion only near the artificial cilia. At a distance L_D away from the cilia the velocity will be zero (see Figure 3.11, top). Within this region a circulating flow is established that generates a mean pressure of $\rho g h_1$ and $\rho g h_2$ at the left and right of this region, respectively. This gives a pressure difference $\Delta p = \rho g(h_1 - h_2)$, with ρ the density of the fluid and g the acceleration due to gravity.

Motivated by this, the computational domain is chosen to consist of a closed channel containing multiple cilia, with the distance between the cilia and the left and right channel boundaries being L_D (see Figure 3.11, bottom). By performing simulations for different heights and L_D values, we found that the

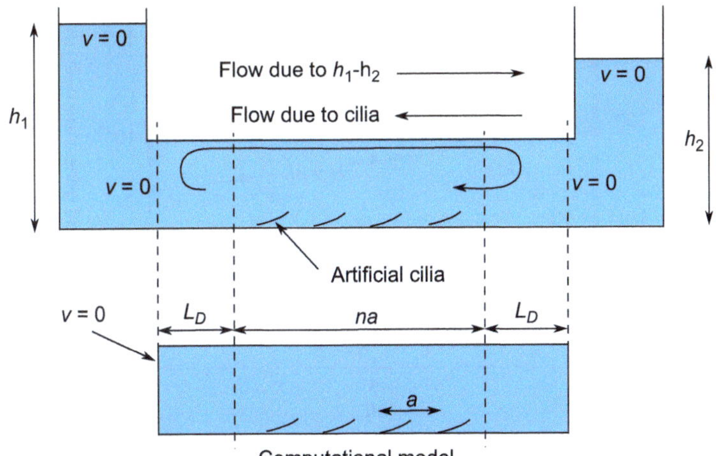

Figure 3.11 Schematic illustration representing the approach taken to calculate the pressure for an open loop channel.
(Reproduced from Khaderi *et al.*[17])

minimum L_D required to ensure a zero fluid velocity near the left and right boundaries, is equal to the height of the channel (H).

We now calculate the pressure gradient by integrating the pressure difference Δp (between the left and right boundaries) over a cycle and divide by the channel length na. Figure 3.12 shows the normalised pressure gradient $L \int_{\text{cycle}} \Delta p dt / n\mu a$ as a function of normalised cilia spacing (a/L) for different channel heights. It can be seen that as the cilia spacing and channel height decrease, the pressure gradient increases. The slope of the lines in Figure 3.12 is -1, indicating that the pressure jump Δp generated by a cilium is independent of the cilia spacing a. As a result, the total pressure difference generated by an array of n cilia is simply $n\Delta p$. On the right axis of Figure 3.12 the pressure head is shown, assuming a total channel length of 2 cm and fluid density of 1000 kg/m^3. We can see that the maximum pressure head created by the cilia can be around 3 mm. This pressure jump per cilium decreases with the height of the channel. To investigate the height dependence we plot the pressure gradient as a function of height for $a = 2L$ in the inset of Figure 3.12. The dashed line is a fit to the data in Log-Log scale. The slope of the fitted line is 2.4. This relatively strong dependence on the height of the channel may be related to two contributions. Firstly, the fluid can more freely flow backwards as the height H is increased, so the cilia have to exert less force. Secondly, the pressure on the boundaries is due to the force exerted by the cilia divided by the boundary area. Since the boundary area scales with H, the pressure is further decreased.

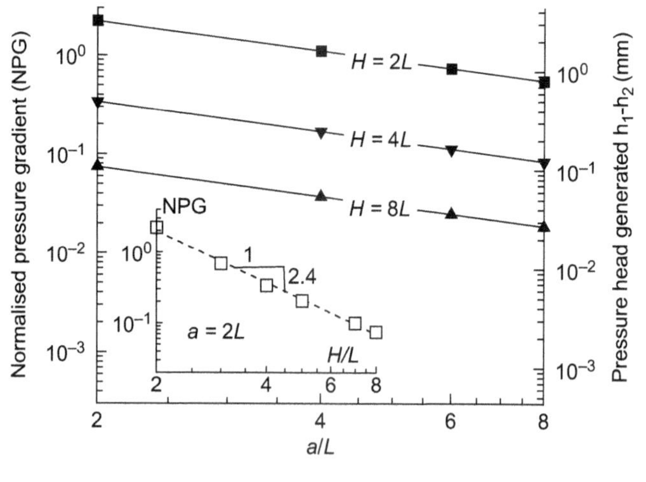

Figure 3.12 Variation of the pressure gradient generated with cilia spacing a/L for different channel heights (left axis). Also shown (right axis) is the corresponding pressure head generated assuming a total channel length of 2 cm and fluid density of 1000 kg m^{-3}.
(Reproduced from Khaderi *et al.*[17])

3.6 Out-of-phase Motion of Cilia

So far we have only studied the fluid transport created by artificial cilia that beat uniformly (simultaneously and in-phase). However, natural cilia beat out-of-phase in a wave-like manner called metachronal motion.[19,20,35] A similar metachronal motion of artificial cilia can be achieved by forcing adjacent cilia to beat out-of-phase by tuning the applied magnetic field. The metachronal wave created by the cilia motion can be either in the direction of the effective stroke or recovery stroke, called symplectic and antiplectic metachrony, respectively. The direction of the wave can be controlled by tuning the phase of the magnetic field applied to each cilia.

To generate an out-of-phase motion of the cilia we apply the following magnetic field to the ith cilium

$$B_{xi} = B_0 \cos(\omega t - \phi_i), \quad B_{yi} = B_0 \sin(\omega t - \phi_i),$$

having a phase $\phi_i = 2\pi(i-1)/n$ and an angular velocity $\omega = 2\pi/t_{\text{ref}}$, where t_{ref} is the cycle time. The chosen form of the phase ϕ_i makes the phase of the magnetic field at every n^{th} cilium identical. That is, the magnetic field is periodic after n repeats of cilia. This enables us to use a periodic unit-cell containing n cilia for the simulations. The phase difference $\Delta\phi$ between two adjacent cilia is $2\pi/n$. This has consequences for the motion of the cilia, both temporally and spatially. Firstly, the frequency of the magnetic couple, which causes the motion of the cilia, is doubled. This results in two cilia beats in one cycle. Secondly, the phase of the magnetic couple is twice that of the applied magnetic field. Due to this, when n is an even number, the magnetic couple is periodic after $n/2$ repeats of cilia. As a consequence, the metachronal wave travels opposite to the direction of the effective stroke (antiplectic metachrony, AM) when $0 < \Delta\phi < \pi/2$, and the metachronal wave travels in the direction of the effective stroke (symplectic metachrony, SM) when $\pi/2 < \Delta\phi < \pi$. When $\Delta\phi = 0$ or π the cilia beat uniformly (in-phase).

The fluid flow is plotted for different values of phase difference $\Delta\phi = 2\pi/n$ and cilia spacing a/L for $H = 2L$ in Figure 3.13 (bottom). When all the cilia are moving synchronously ($\Delta\phi = 0$ or π), the area flow will be approximately 0.22 for $a = 5L$. Now, as the cilia density is increased by decreasing a from $a = 5L$ to $a = 1.67L$, the viscous resistance to be overcome per cilium decreases, which causes the area flow to increase to 0.25 (similar to Figure 3.10). The fluid propelled shows a drastic increase once the cilia start beating out-of-phase. Then, the flux transported during the effective stroke remains nearly the same, whereas almost no flux is transported during the recovery stroke. As a result, the effectiveness sharply increases from around 0.3 (i.e., 30% of the totally displaced fluid is converted into flow) to 1 (fully unidirectional flow), see Figure 3.13 (top). Note that the increase in flow with a decrease in cilia spacing from $a/L = 5$ to 2.5 is much larger for out-of-phase motion compared to in-phase motion. Moreover, the fluid flow does not change

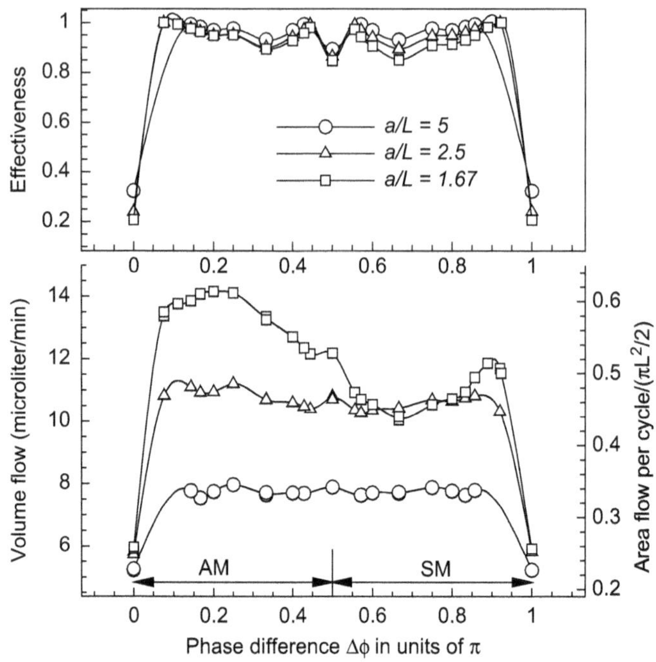

Figure 3.13 Flow (bottom) and effectiveness (top) as a function of phase difference $\Delta\phi$ for different inter-cilia spacings a/L in a channel of height $H=2L$. AM and SM refer to antiplectic metachrony (the wave direction is opposite to the direction of the effective stroke) and symplectic metachrony (the wave direction and the effective stroke direction are the same), respectively.
(Reproduced from Khaderi *et al.*[17])

considerably for different phase differences for these inter-cilia spacings. On the other hand, when the cilia spacing is small ($a = 1.67L$), we see a larger increase in the fluid flow when there is an antiplectic metachrony(AM) compared to symplectic metachrony(SM). It turns out that this is a consequence of hydrodynamic interactions between adjacent cilia.

These interactions for small inter-cilia spacing can be understood by analysing the velocity field for $a / L = 1.67$ corresponding to symplectic (Figure 3.14b) and antiplectic metachrony (Figure 3.14a) with $\Delta\phi = 5\pi / 6$ and $\Delta\phi = \pi / 6$, respectively, where cilium 5 is performing the effective stroke. It can be seen that the cilia 4 and 6 are farther from cilium 5 in the case of antiplectic metachrony, whereas the opposite is the case for symplectic metachrony. In antiplectic metachrony for small spacings, the close proximity of the adjacent cilia during the effective stroke leads to a vortex formation and limits the positive flow that can be generated by the effective stroke. Therefore the net fluid transport by the antiplectic metachrony is lower than the symplectic metachrony.

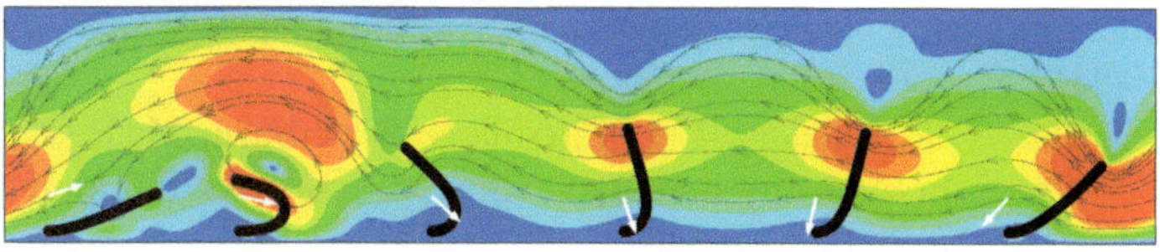

(a) Antiplectic metachrony: wave travels to the right

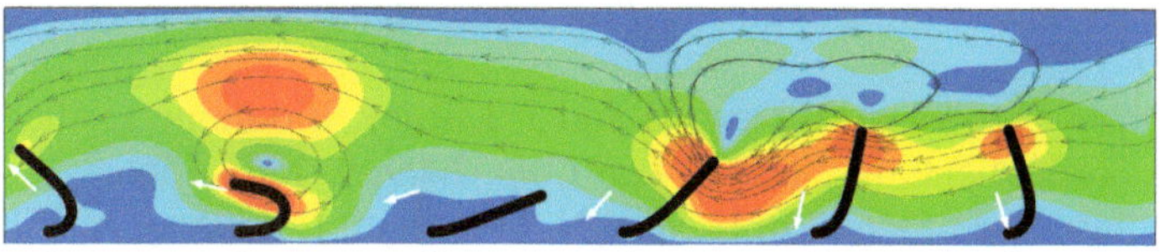

(b) Symplectic metachrony: wave travels to the left

Figure 3.14 Snapshots for antiplectic ($\Delta\phi = \pi/6$) and symplectic metachrony ($\Delta\phi = 5\pi/6$) for a wave speed of $6/t_{\text{beat}}$ cilia per second and cilia spacing $a/L = 1.67$. The contours represent the absolute velocity normalized with L/t_{beat} (blue and red colours represent a normalized velocity of 0 and 2, respectively). The direction of the velocity is represented by streamlines. The applied magnetic field is shown by the white arrows. (Reproduced from Khaderi *et al.*[20])

3.7 Design Guidelines

We would like to conclude by providing guidelines for the design of artificial ciliary systems based on the studies performed in this chapter. Let us start by assuming that the material properties of the cilia, the channel height and the fluid viscosity are known. The design parameters are assumed to be the geometry of the cilia (length, thickness and width), magnetic field, cilia spacing and frequency. As we want the cilia to operate under moderate pressure heads, we propose the length of the cilia to be half that of the channel height (section 3.5.2). From section 3.4.2, we find that for a given magnetic number M_n of around 10, the cilia create large and unidirectional flows for $\text{Re} > 1$ and $F_n < 1$. Using the first condition, we can find the frequency of operation. Using the second condition (on F_n), we can arrive at the ratio of the cilia length to thickness (the aspect ratio). Once we know the aspect ratio, we can find the magnetic field to be applied using the magnetic number. Any further enhancement in the fluid transport can be achieved by the design of the magnetic field that leads to an out-of-phase beat motion of the cilia.

References

1. E. M. Purcell, Life at low Reynolds number, *Am. J. Phys.*, 1977, **45**, 3–11.
2. J. den Toonder, F. Bos, D. Broer, L. Filippini, M. Gillies, J. de Goede, T. Mol, M. Reijme, W. Talen, H. Wilderbeek, V. Khatavkar and

P. Anderson, Artificial cilia for active micro-fluidic mixing, *Lab Chip*, 2008, **8**, 533–541.
3. B. A. Evans, A. R. Shields, R. L. Carroll, S. Washburn, M. R. Falvo and R. Superfine, Magnetically actuated nanorod arrays as biomimetic cilia, *Nano Lett.*, 2007, **7**, 1428–1434.
4. A. R. Shields, B. L. Fiser, B. A. Evans, M. R. Falvo, S. Washburn and R. Superfine, Biomimetic cilia arrays generate simultaneous pumping and mixing regimes, *Proc. Natl Acad. Sci. U.S.A.*, 2010, **36**, 15670–15675.
5. K. Oh, J. -H. Chung, S. Devasia and J. J. Riley, Bio-mimetic silicone cilia for microfluidic manipulation, *Lab Chip*, 2009, **9**, 1561–1566.
6. M. Vilfan, A. Potocnik, B. Kavcic, N. Osterman, I. Poberaj, A. Vilfan and D. Babic, Self-assembled artificial cilia, *Proc. Natl Acad. Sci. U.S.A.*, 2010, **107**, 1844–1847.
7. F. Fahrni, M. W. J. Prins and L. J. van IJzendoorn, Micro-fluidic actuation using magnetic artificial cilia, *Lab Chip*, 2009, **9**, 3413–3421.
8. C. L. van Oosten, C. W. M. Bastiaansen and D. J. Broer, Printed artificial cilia from liquid-crystal network actuators modularly driven by light, *Nat. Mater.*, 2009, **8**, 677–682.
9. J. Belardi, N. Schorr, O. Prucker and J. Ruhe, Artificial cilia: Generation of magnetic actuators in microfluidic systems, *Adv. Func. Mater.*, 2011, **21**, 3314–3320.
10. J. Hussong, N. Schorr, J. Belardi, O. Prucker, J. Ruhe and J. Westerweel, Experimental investigation of the flow induced by artificial cilia, *Lab Chip*, 2011, **11**, 2017–2022.
11. J. V. I. Timonen, C. Johans, K. Kontturi, A. Walther, O. Ikkala and R. H. A. Ras, A facile template-free approach to magnetodriven, multifunctional artificial cilia, *ACS Applied Materials and Interfaces*, 2010, **2**, 2226–2230.
12. E. M. Gauger, M. T. Downton and H. Stark, Fluid transport at low Reynolds number with magnetically actuated artificial cilia, *Eur. Phys. J. E*, 2009, **28**, 231–242.
13. M. T. Downton and H. Stark, Beating kinematics of magnetically actuated cilia, *Europhys. Lett.*, 2009, **85**, 44002.
14. Y. W. Kimand and R. R. Netz, Pumping fluids with periodically beating grafted elastic filaments, *Phys. Rev. Lett.*, 2006, **96**, 158101.
15. S. N. Khaderi, M. G. H. M. Baltussen, P. D. Anderson, D. Ioan, J. M. J. den Toonder and P. R. Onck, Nature-inspired microfluidic propulsion using magnetic actuation, *Phys. Rev. E*, 2009, **79**, 046304.
16. S. N. Khaderi, J. M. J. den Toonder and P. R. Onck, Fluid flow due to collective non-reciprocal motion of symmetrically-beating artificial cilia, *Biomicrofluidics*, 2012, **6**, 014106.
17. S. N. Khaderi, C. B. Craus, J. Hussong, N. Schorr, J. Belardi, J. Westerweel, O. Prucker, J. Ruhe, J. M. J. den Toonder and P. R. Onck, Magnetically-actuated artificial cilia for microfluidic propulsion, *Lab Chip*, 2011, **11**, 2002–2010.

18. S. N. Khaderi, M. G. H. M. Baltussen, P. D. Anderson, J. M. J. den Toonder and P. R. Onck, The breaking of symmetry in microfluidic propulsion driven by artificial cilia, *Phys. Rev. E*, 2010, **82**, 027302.
19. S. N. Khaderi and P. R. Onck, Fluid-structure interaction of three-dimensional magnetic artificial cilia, *J. Fluid Mech.*, 2012, **708**, 303–328.
20. S. N. Khaderi, J. M. J. den Toonder and P. R. Onck, Microfluidic propulsion by the metachronal beating of magnetic artificial cilia: a numerical analysis, *J. Fluid Mech.*, 2011, **688**, 44–65.
21. V. V. Khatavkar, P. D. Anderson, J. M. J. den Toonder and H. E. H. Meijer, Active micromixer based on artificial cilia, *Phys. Fluids*, 2007, **19**, 083605.
22. J. Hussong, W. P. Breugem and J. Westerweel, A continuum model for flow induced by metachronal coordination between beating cilia, *J. Fluid Mech.*, 2011, **684**, 137–162.
23. A. Alexeev, J. M. Yeomans and A. C. Balazs, Designing synthetic, pumping cilia that switch the flow direction in microchannels, *Langmuir*, 2008, **24**, 12102–12106.
24. R. Ghosh, G. A. Buxton, O. B. Usta, A. C. Balazs and A. Alexeev, Designing oscillating cilia that capture or release microscopic particles, *Langmuir*, 2010, **26**, 2963–2968.
25. S. N. Khaderi, J. M. J. den Toonder and P. R. Onck, Magnetically actuated artificial cilia: The effect of fluid inertia, *Langmuir*, 2012, **28**, 7921–7937.
26. R. K. Annabattula, W. T. S. Huck and P. Onck, Micron-scale channel formation by the release and bond-back of pre-stressed thin films: A finite element analysis, *J. Mech. Phys. Solids*, 2010, **58**, 447–465.
27. R. D. Cook, D. Malkus, M. Plesha, D. S. Malkus and M. E. Plesha, *Concepts and Applications of Finite Element Analysis*, John Wiley and Sons, Chichester, 2001.
28. R. van Loon, P. D. Anderson and F. N. van de Vosse, A fluid–structure interaction method with solid–rigid contact for heart valve dynamics, *J. Comput. Phys*, 2006, **217**, 806–823.
29. L. van Rijsewijk, Electrostatic and magnetic microactuation of polymer structures for fluid transport, Master's thesis, Eindhoven University of Technology, 2006.
30. M. Murase, *Dynamics of Cellular Motility*, John Wiley and Sons, Chichester, 1992.
31. P. Satir and M. A. Sleigh, The physiology of cilia and mucociliary interactions, *Annu. Rev. Physiol.*, 1990, **52**, 137–155.
32. J. West, B. Karamata, B. Lillis, J. P. Gleeson, J. Alderman, J. K. Collins, W. Lane, A. Mathewson and H. Berney, Application of magnetohydrodynamic actuation to continuous flow chemistry, *Lab Chip*, 2002, **2**, 224–230.
33. J. Jang and S. S. Lee, Theoretical and experimental study of magnetohydrodynamic micropump, *Sensors and Actuators A: Physical*, 2000, **80**, 84–89.

34. L. Chen, J. Ma, F. Tan and Y. Guan, Generating high-pressure sub-microliter flow rate in packed microchannel by electroosmotic force: potential application in microfluidic systems, *Sensors and Actuators B: Chemical*, 2003, **88**, 260–265.
35. S. Gueron, K. Levit-Gurevich, N. Liron and J. J. Blum, Cilia internal mechanism and metachronal coordination as the result of hydrodynamical coupling, *Proc. Natl Acad. Sci. U.S.A.*, 1997, **94**, 6001–6006.

CHAPTER 4

Modeling the Interaction of Active Cilia with Species in Solution: From Chemical Reagents to Microscopic Particles

PRATYUSH DAYAL,[a] OLGA KUKSENOK,[b] AMITABH BHATTACHARYA,[b] GAVIN A. BUXTON,[c] O. BERK USTA[d] AND ANNA C. BALAZS*[b]

[a] Chemical Engineering Department, Indian Institute of Technology, Gandhinagar, India; [b] University of Pittsburgh, Department of Chemical and Petroleum Engineering, 1249 Benedum Hall, Pittsburgh, PA, 15261, USA; [c] Robert Morris University, 6001 University Boulevard, Moon Township, PA 15108-1189, USA; [d] The Center for Engineering in Medicine at Massachusetts General Hospital/Harvard Medical School and Shriners Hospital for Children, 51 Blossom St., Boston, MA, 02114, USA
*Email: balazs@pitt.edu

4.1 Introduction

Within the theoretical and computational modeling communities, there is a rapidly growing number of studies aimed at analyzing the behavior of both biological and synthetic cilia. For example, as detailed in chapters of this book, researchers have investigated phenomena that range from the synchronized motion of biological cilia (metachronal waves) to the ability of synthetic cilia to drive fluid flow in micro-fluidic devices. There are, however, two important

RSC Nanoscience & Nanotechnology No. 30
Artificial Cilia
Edited by Jaap M J den Toonder and Patrick R Onck

Published by the Royal Society of Chemistry, www.rsc.org

areas that have received less attention from theoreticians: (1) understanding the chemosensory capabilities of cilia-like structures, and (2) probing the ways in which cilial structures could be harnessed to propel microscopic particles. Both these areas relate to the ability of cilia to interact with species in solution in 'smart' or meaningful ways. The first of these topics is quite intriguing since organisms from bacteria to mammals use cilia to sense local variations in the environment, amplify this local perturbation to a widespread 'message' and consequently, promote a large-scale, collective action. For example, motile cilia in the respiratory tract sense bitter compounds entering the airways and subsequently beat faster to eliminate the offensive substance.[1] Non-motile primary cilia also perform vital chemosensory functions.[2,3] Researchers have designed a range of artificial cilia in attempts to reproduce some of the sensory capabilities of these biological entities.[4–19] Driven by various applied fields and forces, these synthetic filaments have revealed remarkable dynamic behavior. What remains unclear, however, is how a flexible, cilium-like structure senses and transmits information about local chemical changes to a neighboring cilium and how this information is then propagated across multiple cilia. Addressing this issue can open new routes to controlling the self-organization and functionality of synthetic cilia, as well as provide insight into physico-chemical processes that play a role in the biological systems.

The second topic noted above also poses interesting challenges. It is worth recalling that a number of organisms harness cilia to propel microscopic particles, and thereby perform functions vital to their survival. For instance, marine suspension feeders use cilia to propel food into their bodies.[20] While the effectiveness of synthetic cilia in regulating fluid flow has been examined,[5,7] the possibility of using active, synthetic cilia to direct the movement of microscopic particles, such as biological cells and microcapsules, within microchannels has not been extensively explored.[21–25] Developing approaches for conveying cells or micro-carriers to specified locations within micro-fluidic devices is vital for performing accurate micro-scale analysis or chemical synthesis.[26,27] In contrast to propelling fluids,[7,19,28,29] it might be that the coordinated motion of the cilia is not sufficient to move micro-scale particulates; it is plausible that an adhesive interaction between the cilia and particulates is necessary for controlling the particle movement. Notably, in mammals, beating cilia located at the entrance of the oviduct cannot transport egg cells unless there is a critical level of adhesion between the cilia tips and cells.[30,31] The combined effects of such cilial adhesion and generated flows might be responsible for transport in other biological environments.[31]

Herein, we describe our use of computational modeling to investigate both of the issues described above. In the first study,[32] we focus on filaments formed from polymer gels undergoing the Belousov–Zhabotinsky (BZ) reaction[33] to uncover effective physical mechanisms for recognizing and transmitting chemical signals in biomimetic cilia. The BZ gels are unique because the polymer network can expand and contract periodically without external stimuli.[34–37] This autonomous, self-oscillatory behavior is due to a ruthenium catalyst, which is covalently bonded to the polymers.[34–37] The BZ reaction generates a periodic oxidation and reduction of the anchored metal ion; the hydrating effect of the oxidized metal catalyst induces an expansion of the gel, which then contracts when the catalyst is in the reduced state. In other words, the BZ gels autonomously transduce and amplify a chemical signal into mechanical pulsation. Due to this direct chemo-mechanical coupling, local variations in the concentration of reagents could affect the mechanical oscillation of a single cilium, and hence influence the interactions of multiple cilia. To address this issue, we develop a new computational model and investigate chemical communication in systems of multiple BZ cilia.

In the second study,[38] we use computer simulations to model the interactions between microscopic particles and actuated cilia that encompass a 'sticky' tip. To the best of our knowledge, there have been no prior computational or theoretical studies aimed at probing the role of adhesion between motile cilia and particles in solution. As we show below, for a critical adhesion strength, active cilia can propel particles from one neighbor to the next, significantly increasing the particle's velocity. On the other hand, if the adhesive interaction between a particle and this sticky tip is sufficiently strong, the particle can become trapped within the cilia layer. These results can facilitate the design of synthetic cilia that integrate adhesive and hydrodynamic interactions to selectively repel or trap particulates. Surfaces that are effective at repelling particulates are valuable for anti-fouling applications, while surfaces that can trap and, thus, remove particulates from the solution are useful for efficient filtration systems. Furthermore, the findings can yield insight into physical factors that influence adhesive interactions between biological cilia and particulates.

4.2 Modeling the Interaction of Belousov–Zhabotinsky Cilia and Reagents in Solution

Due to the mechanism described above, millimeter-sized pieces of BZ gels can oscillate autonomously for hours and the system can be refueled by adding more reagents to the solution.[34] This autonomous functionality and interconversion of chemical and mechanical energy makes the system truly biomimetic. There is another distinctive feature of the BZ cilia: the gels are photosensitive.[37] By selectively illuminating specific cilia in an array, we can control the pattern of wave propagation within the system. In effect, the motion of the tops of the cilia resembles the movement of piano keys, with the light serving to control the 'tune'. Below, we first describe the model we developed to capture this complex behavior.

4.2.1 Methodology

The dynamics of the classic BZ reaction in solution is described by the Oregonator model,[39] which captures the reaction kinetics in terms of just two dimensionless variables: the concentration of the key reaction intermediate

u ($HBrO_2$, the activator), and the concentration of oxidized metal-ion catalyst v (Ru^{3+} in the case considered here). Building on this model, we derived a set of equations that describe the elastodynamics and reaction–diffusion processes occurring within the BZ gels.[40,41] We then developed a lattice-based approach, the gel lattice spring model (gLSM),[40,41] which allows us to numerically solve these governing equations in a computationally efficient manner. Furthermore, we modified the model to account for the photosensitivity of the BZ gels.[42,43] The gLSM has proven to be a powerful approach for predicting new behavior, which was subsequently verified experimentally,[44,45] as well as replicating prior experimental findings.[41,46]

To model the BZ cilia in solution, we combined this gLSM technique with a finite difference approach that accounts for the diffusive exchange of u between the gels and fluid, and captures reaction–diffusion processes occurring in the solution.[32] Hence, we could capture the subtle interplay between the pulsations of the cilia and the dispersion of u, where the motion of the cilia affects the distribution of u, which in turn affects the movement of the cilia.

With the modifications described above, we obtained the following governing equations for the photo-responsive BZ gels:[40–43]

$$\frac{\partial \phi}{\partial t} = -\nabla \cdot (\phi \mathbf{v}^{(p)}) \tag{4.1}$$

$$\frac{\partial v}{\partial t} = -\nabla \cdot (v \mathbf{v}^{(p)}) + \varepsilon\left[(1-\phi)^2 u - (1-\phi)v\right] \tag{4.2}$$

$$\begin{aligned}\frac{\partial u}{\partial t} &= -\nabla \cdot (u \mathrm{v}^{(s)}) - \nabla \cdot \mathrm{j}^{(u)} \\ &+ \left((1-\phi)^2 u - u^2 - (1-\phi)\left[fv + \Phi\right] \frac{u - q(1-\phi)^2}{u + q(1-\phi)^2}\right)\end{aligned} \tag{4.3}$$

Here, ϕ is the polymer volume fraction; u and v are the respective concentrations of the activator and the oxidized catalyst. The vectors $\mathbf{v}^{(p)}$ and $\mathbf{v}^{(s)}$ are the velocities of the polymer network and solvent, respectively. The term $\mathbf{j}^{(u)}$ is the diffusive flux of the solvent through the gel and is given by $\mathbf{j}^{(u)} = -(1-\phi)\nabla(1-\phi)^{-1}u$. The last terms in eqns (4.2) and (4.3) describe the BZ reaction kinetics within the gel, with the parameters q, f and ε having the same meaning as in the original Oregonator model.[39] The variable Φ accounts for the additional production of bromide ions due to illumination by light of a specific wavelength, and is proportional to the light intensity.[42,43,47] Eqns (4.1)–(4.3) are solved together with the appropriate force balance equation, which is derived assuming that the polymer network undergoes purely relaxational dynamics.[41] Details of the full three-dimensional formulation of the gLSM are given by Kuksenok *et al.*[41]

The evolution of u outside the gels and within the external fluid is given by:

$$\frac{\partial u}{\partial t} = \nabla^2 u - u^2 \tag{4.4}$$

The last term on the right-hand side represents the decay of activator due to the disproportionation reaction.[48] Due to the slow dynamics of the gels and low viscous forces, hydrodynamic effects in the system are negligible.[32] Hence, we neglect the effects of hydrodynamics and focus only on reaction-diffusion processes (both in the gel and outer fluid).

Eqns (4.1)–(4.3) are solved solely inside the gel, on a Lagrangian grid. The deformable hexahedral elements of this grid are defined by the gel nodes.[41] Eqn (4.4) is solved solely in the fluid, on a fixed, regular, Eulerian grid. At every time step, the flux of activator from the fluid to the gel is interpolated from the Eulerian grid, while the flux of activator from the gel to the fluid is interpolated from the Lagrangian grid. For the surfaces of the Lagrangian grid that are attached to the wall, we impose no-flux boundary conditions for the activator. These boundary conditions allow for the exchange of activator between the fluid and gel only across the mobile gel–fluid interfaces. We use no-flux, Dirichlet or periodic boundary conditions for the external boundary on the fluid grid. We specify these boundary conditions in the text below.

Where possible, the system parameters were taken from known experimental data.[32] To determine the characteristic length and time scales of the system, we assumed that the diffusion coefficient of the activator,[40] $D_u = 2 \times 10^{-9} \mathrm{m}^2\,\mathrm{s}^{-1}$, remains the same in the gel and surrounding fluid.[49] Consequently, we find that our dimensionless units of time and length correspond to the respective physical values of $T_0 = 0.31$ s and $L_0 = \sqrt{D_u T_0} = 25$ µm.

4.2.2 Results and Discussion

Before examining how the diffusion of the activator, u, in the fluid results in the communication among multiple oscillating cilia, it is useful to first consider the simpler case of an isolated BZ gel immersed in solution and determine how the concentration of u affects the chemo-mechanical oscillations of the gel. We focus on a sufficiently small sample where traveling waves are not generated within the material,[41] *i.e.* the gel swells and de-swells essentially uniformly in all directions. This gel is placed in the center of the simulation box, with the initial separation between the walls and the gel surfaces being set to 5.5 dimensionless units. The diffusion of u takes place through all the gel surfaces and we set identical activator concentrations at all the six fluid boundaries ($u = u_b$).

Figure 4.1a clearly shows that the frequency of the gel's oscillation, ω, increases monotonically with an increase in u_b. In other words, the chemo-mechanical oscillations in the gel occur at a higher frequency when the concentration of the activator in the surrounding fluid has a higher value. This figure also shows that the frequency of oscillations is affected by the presence of light and ω for an illuminated sample remains lower than that for a non-illuminated one for a wide range of u_b values. The light intensity in Figure 4.1a (represented by the dimensionless value Φ) is significantly greater than the critical value required to suppress the oscillations in an isolated sample,[42,43]

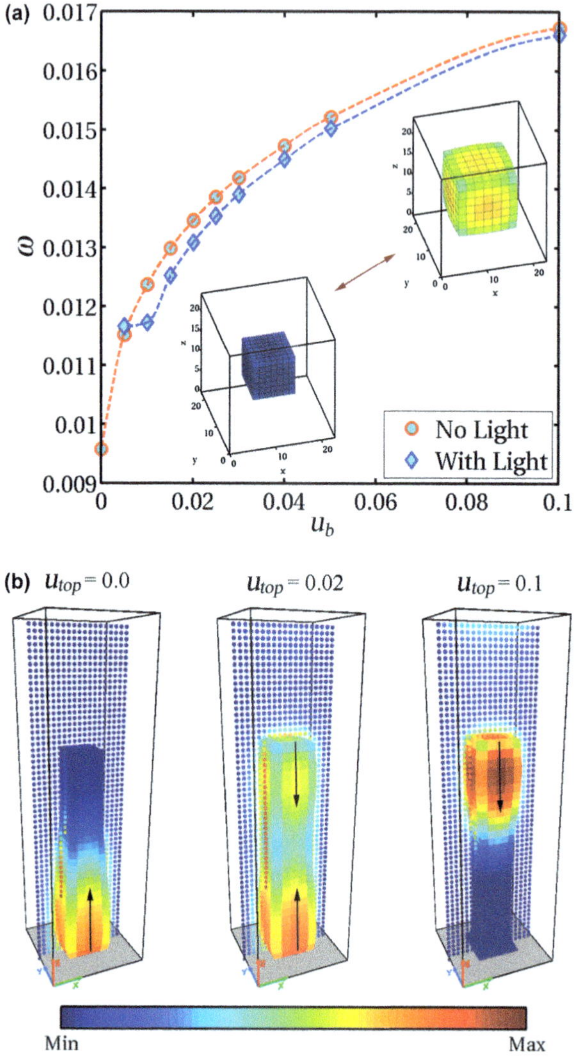

Figure 4.1 Effect of activator concentration on the oscillations of a single gel sample within a fluid. (a) The dependence of the frequency of the gel's oscillation, ω, on activator concentration at the fluid boundaries, u_b. Gel size is $8 \times 8 \times 8$ nodes. The activator concentration is $u = u_b$ at all the six fluid boundaries. We show two cases: with ($\Phi = 1.5 \times 10^{-3}$) and without light ($\Phi = 0$). (b) Late time snapshot of BZ cilium ($6 \times 6 \times 30$ nodes) immersed in solution; the activator concentration is specified at the top boundary of the box ($u = u_{\mathrm{top}}$) and is maintained at $u = 0$ at all the other fluid boundaries. Here and below, the oxidized catalyst concentration (v) on the gel's surface and the activator concentration (u) across the central plane of the fluid box are mapped using the color bar; the minimum values (in blue) are always set to zero, while we vary the maximum values (in red) as provided in each case. Here, we set $u_{\mathrm{max}} = v_{\mathrm{max}} = 0.3$ for $u_{\mathrm{top}} = 0.0$ and 0.02 and $v_{\mathrm{max}} = 0.18$ for $u_{\mathrm{top}} = 0.10$. Here and below, the black arrows represent the direction of the traveling wave.

which does not experience an influx of u from the fluid boundaries. Notably, it is due to this influx of u that the illuminated sample in Figure 4.1a continues to oscillate.

The observed dependence of ω on u in the surrounding fluid provides significant insight into the effect of the activator concentration on the propagation of traveling chemo-mechanical waves in a single BZ cilium (see Figure 4.1b). The results in Figure 4.1a also allow us to predict the dynamic behavior of traveling chemo-mechanical waves in the more complicated cases involving multiple, communicating cilia described further below.

Each BZ cilium in Figure 4.1b is immersed in solution and the activator concentration is specified at the top boundary of the box ($u = u_{\text{top}}$) and is maintained at $u = 0$ at all the other fluid boundaries. The images in Figure 4.1b represent three independent simulations. We observe that when the activator concentration is zero at the top fluid boundary ($u_{\text{top}} = 0$, left image), the traveling wave propagates from the fixed to the free end. This phenomenon is consistent with experiments on a single, long rectangular BZ gel attached to a wall,[35] and with earlier computer simulations on single, end-tethered BZ gels.[40,46,50] When the activator concentration is increased to moderate values ($u_{\text{top}} = 0.02$, central image), the traveling waves emerge simultaneously from both the ends of the sample. In each cycle, the waves travel towards the center of the gel, where they meet and annihilate each other. With a further increase in the activator concentration at the top boundary ($u_{\text{top}} = 0.1$, right image), the traveling waves switch direction and emerge from the free tip and travel towards the fixed end.

The phenomena observed in Figure 4.1b can be explained as follows. As the activator concentration at the top of the simulation box is increased, the frequency of the traveling waves emanating from the free end of the cilium increases (as can be rationalized by the data in Figure 4.1a). In a system containing multiple oscillators, the region with the highest frequency determines the ultimate direction of wave propagation.[51,52] In our BZ cilium, each element acts as an oscillator, and the effective intrinsic frequency[53] of this oscillator depends on the boundary values of u in the outer fluid and neighboring elements. For $u_{\text{top}} = 0.0$, the fixed end of the gel has the highest intrinsic frequency and hence, the traveling waves propagate bottom-up. In the case of $u_{\text{top}} = 0.1$, the free end of the gel has the highest intrinsic frequency, and consequently, the waves travel top-down. Figure 4.1b clearly shows that the local activator concentration in the fluid determines the directionality of wave propagation in the cilium.

4.2.2.1 Probing the Behavior of Two Interacting Cilia

Since our aim is to determine the factors that control the inter-cilial communication, we now focus on two BZ cilia immersed in fluid (Figure 4.2) and examine how the separation between these gels, d, affects the system's behavior. The size of each cilium is $(5 \times 5 \times 29)\lambda$, where λ denotes the gel's

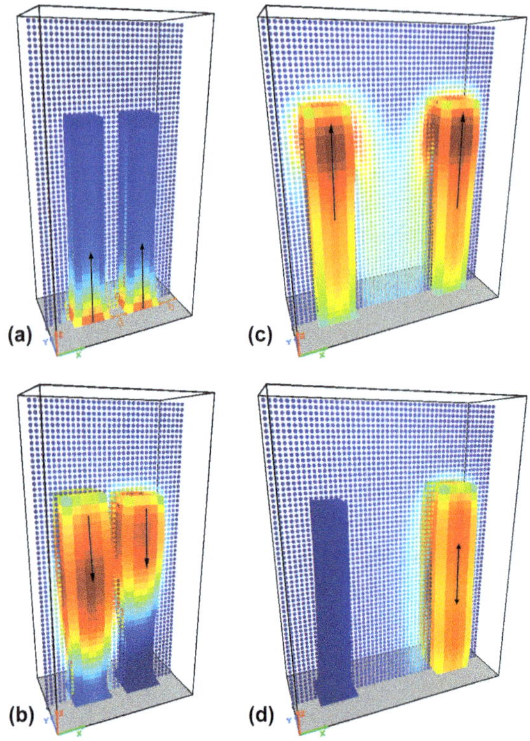

Figure 4.2 Effect of varying d, the inter-cilia spacing, on the dynamics of two cilia for $b = 5.5$. Panels (a) and (b) correspond to the respective early and late time snapshots for $d = 3.5$. Panels (c) and (d) correspond to the respective early and late time snapshots for $d = 20.5$. Here, we set $v_{max} = 0.12$ and $u_{max} = 0.33$.

initial degree of swelling. The diffusive exchange of u between the gel and fluid occurs through all the gel's surfaces in contact with the solution. The parameters b, the spacing between the outer cilia surfaces and fluid boundaries, d and the size of the gel define the length of the simulation box in the x-direction; the length of the box in the y and z directions is fixed at 20 and 63 dimensionless units, respectively. The concentration of u at $x = 0$ and $x = L_x$ is set to zero. Periodic boundary conditions are imposed in the y direction and no-flux boundary conditions are specified in the z direction. Based on the scaling in section 4.2.1, the initial size of the non-deformed gels corresponds to approximately 0.2 mm × 0.2 mm × 1.3 mm and the distance between the two cilia ranges from 0.09 mm to 0.51 mm (*i.e.* from 3.5 to 20.5 dimensionless units).

We focus on the two exemplar cases shown in Figure 4.2, where we fix $b = 5.5$. The color of the dots across the central plane of the simulation box indicates the activator concentration (see color bar in Figure 4.1b). At early times for both these cases, the chemo-mechanical waves generated by the BZ reaction within the gel travel bottom-up and the wave propagation in the neighboring cilia is synchronized. At late times, however, the behavior in these two examples becomes distinctly different. If the two cilia are placed relatively close together ($d = 3.5$), the waves travel from top to bottom at late times (Figure 4.2b). For the two cilia that are placed relatively far apart ($d = 20.5$), the traveling waves at late times are generated at the center of each cilium. Additionally, these waves oscillate out of phase with respect to each other (see Figure 4.2d).

Figure 4.3 shows the distribution of u around the cilia, revealing that this distribution is highly non-uniform and depends markedly on d. (The distribution is calculated at a time t when the catalyst is predominantly in the reduced state.) When the cilia are located relatively close to each other (Figure 4.2a), the highest concentration of u in the fluid is observed above the cilia (Figure 4.3a). The relatively low u values between the cilia at small d arises because the fraction of activator, which was produced during the oxidation phase and spread out from the cilia, diffuses back into the gels from the narrow inter-cilial spacing when the catalyst is in the reduced state. For cilia that lay further apart, however, the highest concentration of u is closer to the center of the simulation box (Figure 4.3b). With the relatively large space between the cilia, the fraction of u that diffuses back to the gels during the reduction phase constitutes only a small fraction of the total amount of u that is located between the cilia.

The plots in Figure 4.3 help explain the directionality of wave propagation in the two exemplar cases. For the case of the closely spaced cilia, the higher

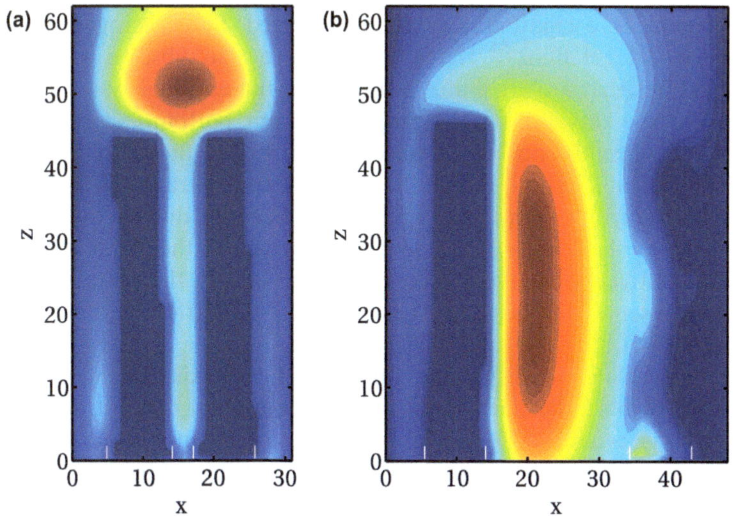

Figure 4.3 The distribution of u across the central plane of the simulation box when the catalyst in the cilium is predominantly in the reduced state; the value of b is fixed at 5.5. Panels (a) and (b) correspond to $d = 3.5$ and $d = 20.5$, respectively. The white lines at the bottom of each panel show the position of each cilium. Here, we set $u_{max} = 0.02$ in (a) and $u_{max} = 0.08$ in (b).

concentrations of u near the tops of the gels results in higher intrinsic oscillation frequencies close to the free ends and causes the observed top-to-bottom wave propagation, similar to the right most case of a single cilium in Figure 4.1b. (We observed a similar distribution of u as in Figure 4.3a when we kept the inter-cilia separation at $d = 3.5$ but decreased the distance between the cilia and outer fluid boundaries to $b = 3.5$; hence, we also observed the top-to-bottom wave propagation in this case.[32]) In the case where the cilia lie further apart ($d = 20.5$), the highest concentration of u and consequently, the highest intrinsic oscillation frequency corresponds to the center (in the z-direction) of each cilium. The latter observation helps rationalize the observed late-time wave propagation seen in Figure 4.2d, where the waves initiate from the center of the cilia.

The above examples demonstrate that the u-mediated communication between the two cilia affects the direction of wave propagation within each cilium.[32] The distribution of u also affects the relative bending of these filaments; this can be seen more clearly in the cases below.

4.2.2.2 Investigating Communication Among Multiple Belousov–Zhabotinsky Cilia

We now expand our system to encompass five cilia that are arranged along the x-direction with $d = 3.5$ and $b = 5.5$; we again set $u = 0$ at the boundaries. The early and late time snapshots in Figure 4.4a–c indicate that the chemo-mechanical oscillations in the cilia are synchronized. Similar synchronization of five chemical oscillators has been observed for heterogeneous, two-dimensional BZ gel films.[54] Here, however, we also observe a number of distinct features that are attributes of this three-dimensional system. First, we observe that the traveling waves propagate from bottom-to-top during the early stages, when the activator concentration in the fluid is low, and then switch direction at late times, when the activator concentration is highest at the top of cilia array, as shown in Figure 4.4d. Second, due to the symmetric arrangement of the cilia and boundary conditions in the x-direction, which result in a relatively high average concentration of u in the central portion of the box, we observe that the central cilia leads the oscillations within the system (see Figure 4.4b).

Third, we isolated a distinctive form of chemotaxis in the system; the anchored filaments collectively bend toward the highest concentration of u in the system. As can be seen in Figure 4.4c, the cilia numbered 1,2 and 4,5 (numbering from left to right) are effectively ‘bunched’ together as they tilt towards the high concentration of u in the central region. The bending of these cilia can be attributed to a non-uniform concentration of u in the lateral direction (see Figure 4.4d) and hence, a non-uniform degree of swelling along the x-direction within each gel. The gradients across the sample indicate that there is a component of wave propagation in the x-direction. (Note that the wave propagates predominately along the length of the cilia due to the high

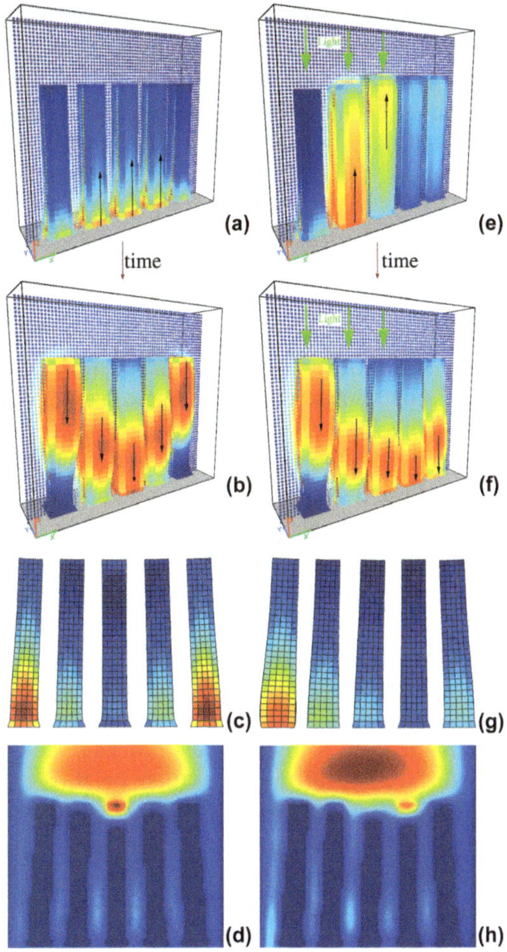

Figure 4.4 Dynamics of the five cilia system. The size of the simulation box is 69 × 20 × 63 (in x, y and z directions) with $d = 3.5$ and $b = 5.5$. (a) and (b–d) show respective early and late time snapshots in the absence of light ($\Phi = 0$). (e) and (f–h) show respective early and late time snapshots when cilia 1, 2 and 3 (numbered from left to right) are illuminated by light ($\Phi = 1.5 \times 10^{-3}$). Here, we set $u_{max} = 0.36$, $v_{max} = 0.26$ in (a), $u_{max} = 0.32$, $v_{max} = 0.20$ in (b), $u_{max} = 0.038$ in (d), $u_{max} = 0.32$, $v_{max} = 0.30$ in (e), $u_{max} = 0.30$, $v_{max} = 0.28$ in (f), and $u_{max} = 0.033$ in (h).

aspect ratio of these filaments.) In particular, the concentration of u is higher for the inner than the outer surface of the cilia due to $u = 0$ constraint at the boundaries of the simulation box (and consequently, these differences are most pronounced for cilia 1 and 5). Hence, the inner surface has a higher intrinsic frequency and the lateral direction of wave propagation is from the inner to the outer surface of each cilium. The net process causes the 'pumping' of some amount of solvent from the inner space between the cilia to the outer regions.

In previous studies on single BZ gel samples, we showed that the pumping of fluid in a given direction resulted in the net motion of the BZ gels in the opposite direction.[55] Here, however, the bottoms of the cilia are anchored to the substrate. Thus, the induced motion of the solvent (through the cilia) from the area between the cilia to the outer bath drives the cilia to bend in the opposite direction: namely, the cilial tips bend inwards and thus, towards each other.

Figure 4.4c depicts a novel form of chemo-mechanical transduction in this synthetic system, where dynamically evolving chemical gradients in the solution induce a deflection of the cilia and an effective attraction between these anchored gels. As noted above, these cilia display a distinct form of chemotaxis: the self-generated gradient in the activator for the chemical reaction ultimately draws the cilia towards the highest concentration of u.

The 'bunching' of the five cilia towards the center of the box is significantly more pronounced than the inward bending of the two cilia (see Figure 4.2d). The distinct clustering of the five cilia in Figure 4.4b and c can be attributed to the maximum in u being located over the top of the middle cilium (Figure 4.4d) rather than spread more uniformly over the tops of two cilia (much as in Figure 4.3a). The average values of u in the fluid are also significantly higher in Figure 4.4d than for the examples involving two cilia (see scale bars for respective maximum values). These factors contribute to relatively large lateral gradients in u, which drives the four outer cilia to cluster towards the central cilium.

The fourth distinctive feature of these three-dimensional systems is that the positions of the top surfaces of all the cilia (their heights, curvature, and the inter-cilia distance) change according to the observed synchronized oscillations. These dynamic changes can be visualized most dramatically by focusing on the vertical displacements of the tops of cilia and watching as the surfaces move up and down in a complex dynamical pattern, as will be seen in the example described below.

The BZ gels are photosensitive; illuminating a sample with light of a specific wavelength gives rise to an additional flux of bromide ions, which inhibits the oscillatory behavior.[37] By increasing the intensity of this light beyond a critical value, the oscillations can be completely suppressed (in the absence of the influx of u to the gels from the surrounding fluid).[37] We now exploit these photo-responsive properties to control the interactions in our five cilia array, including regulating the rhythmic height variations in the oscillating array.

For the case shown in Figure 4.4e, the geometric set-up and boundary conditions are the same as in Figure 4.4a, except that cilia 1, 2 and 3 (numbered from left to right) are illuminated by light. The effect of light is captured in our simulations by setting $\Phi = 1.5 \times 10^{-3}$ (see eqn (4.3)), which represents the additional flux of bromide ions.[47] This Φ value is greater than the critical value required to curb the production of u and suppress the oscillatory behavior [42,43] and hence, under these conditions, an isolated cilium would not exhibit chemo-mechanical oscillations. The oscillations in the five cilia array in Figure 4.4e would also be suppressed if the entire array were illuminated with uniform light corresponding to this value of Φ. In Figure 4.4e, however, the non-illuminated cilia (numbered 4 and 5) continue to produce the activator u, which diffuses through the fluid to cilia 1, 2 and 3. As the concentration waves travel from right to left (in the negative x-direction), the local activator concentration increases sharply and thereby switches 'on' each cilium (see Figure 4.4e).

Once the concentration of u builds ups in the system, the waves travel synchronously from top to bottom (see Figure 4.4f). Due to the light source, the leading wave front shifts towards the non-illuminated side. Here, all the cilia are effectively coming closer together, similar to the case in Figure 4.4c; however, the middle cilium is also somewhat tilted (see Figure 4.4g) because the distribution of u surrounding this central cilium is asymmetric, as seen in Figure 4.4h.

The main features of the collective dynamics of the illuminated cilia (Figure 4.4e–h) can be understood from the data in Figure 4.1a. On the one hand, the difference between ω for the non-illuminated and illuminated samples is rather small; hence, the overall change in the direction of wave propagation in both samples is driven by the increase in u near the free ends at late times. On the other hand, ω in the presence of light remains lower than that in the absence of light for a wide range of u_b. The latter observation explains both the shift in the position of the leading front towards the cilia in the dark (see Figure 4.4f), as well as the asymmetric 'bunching' of the cilia towards the non-illuminated region (Figure 4.4g).

The dynamic variations in the behavior of the cilia can also be seen from a top-down view. In particular, the height of each cilium increases, reaches the maximum value when the catalyst is in the oxidized state and then decreases when the catalyst is in the reduced state. We call this variation in height the 'piano effect' since the cilia appear to move like the keys of a piano. To illustrate this piano effect, in Figure 4.5 we plot the temporal behavior of the top cilial surfaces for the case in Figure 4.4f. This image in Figure 4.5 shows the pronounced transient pattern that occurs while the activator concentration in the fluid remains relatively low, *i.e.* until about $t \sim 10^3$, which corresponds to more than a dozen oscillation cycles. Each cilium reaches its corresponding maximum height in a sequential manner. The fact that the light source causes the leading wave front to shift towards the non-illuminated side is also reflected in the variations of heights of the cilia.

In summary, through these studies, we demonstrated that the BZ cilia exhibit a remarkable chemo-sensing capability and ability to autonomously translate this chemo-sensitivity into a distinct mechanical response. These attributes could be usefully exploited in a range of micro-fluidic applications, where, for example, an external light source could be utilized to tailor the movement of a soft 'conveyor belt'[56] and thus, regulate the transport of microscopic objects within the devices. Below, we describe how adhesive interactions between active cilia and microscopic particles could also be exploited to control the cilia-mediated motion of such microscopic objects.

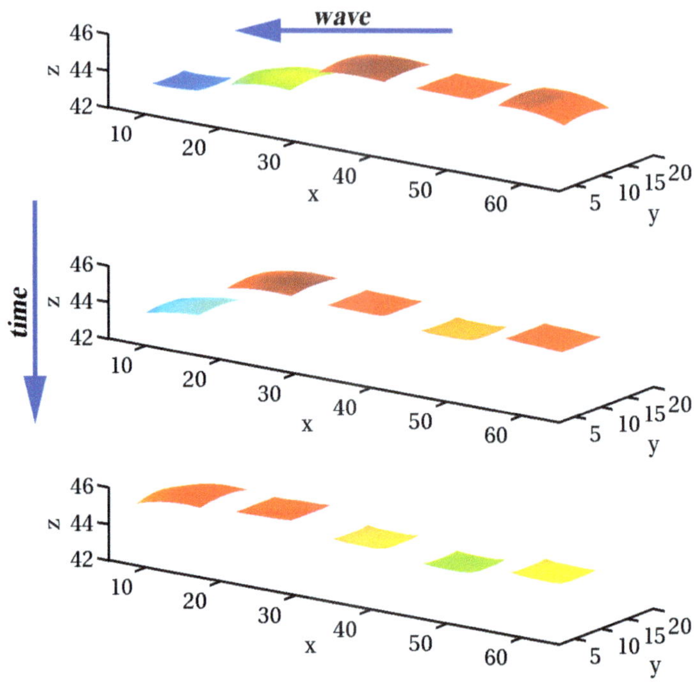

Figure 4.5 The 'piano' effect. Variation of the positions of the top cilia surfaces at early times when cilia 1, 2 and 3 (numbered from left to right) are illuminated by light ($\Phi = 1.5 \times 10^{-3}$). Each cilium reaches its corresponding maximum height in a sequential manner and appears to move like the keys of a piano.

4.3 Modeling the Interaction of Adhesive Cilia with Microscopic Particles

We now turn our attention to the second issue described in section 4.1: probing the interaction between active cilia and a microscopic particle. In the cilia-mediated conveyance of egg cells, it appears that adhesion between the cilia and cells is necessary for the effective transport of these particulates.[30,31] Cilia–particle adhesion also appears to be critical for marine suspension feeders to capture food particles from the surrounding fluid.[20] Furthermore, in some ciliary feeders, the capture of food particles depends on the sensory capabilities of the cilia, which temporarily and locally alter their stroke in the response to the presence of the food.[20]

Previously, we modeled the three-dimensional interactions between oscillating, synthetic cilia and a microscopic particle in a fluid-filled microchannel.[21] The synthetic cilia were simulated as elastic filaments that were tethered to a substrate and actuated by a sinusoidal force, which was applied to their free ends. The four cilia were arranged in a square pattern, and a neutrally buoyant particle was initially located between these filaments. These studies revealed that, depending on the frequency of the beating cilia, the particle was either driven downward toward the substrate or driven upward and expelled into the fluid above the cilial layer. Analogous to the behavior observed certain ciliary feeders,[20] a change in the stroke (*e.g.* frequency) enabled the system to exert considerable control over the motion of the particles.

In the prior work,[21] we did not, however, examine the effect of cilia–particle adhesion on the behavior of the system. Our more recent studies[38] were inspired by the importance of adhesive interactions in the biological examples noted above. To carry out the studies described below, we modified our computational model[21] so that we could consider larger cilial arrays. Using this new approach,[38] we examined the interactions between particles in solution and a regular array of beating elastic cilia, whose tips experience an adhesive interaction with the particle's surface. As we show below, the concerted beating of this cilial array causes the neutrally buoyant particle to be 'tossed' like a beach-ball, from one cilium to the next, and thereby propelled in a specific direction along the top of the ciliated surface. By varying the cilia–particle adhesion strength and the cilia stiffness, we also isolated the parameters for which the particle is 'trapped' or 'released' by the cilia. In addition to providing design criteria for regulating the trafficking of particulates in microchambers, such studies could provide insight into physical factors that influence the behavior of the biological cilia.

4.3.1 Methodology

Each cilium has a length L and is formed from a chain of N beads of radius a (Figure 4.6). The coordinates of each bead are specified by $\mathbf{r}^i$, where i varies from 1 to N, and the first bead in the chain ($i=0$) is tethered to the wall ($y=0$). The elastic potential energy of the chain is:[57,58] $U^{\text{chain}} = \sum_{i=1}^{N} (1/2)\left[k_s(|\mathbf{r}^i - \mathbf{r}^{i-1}| - l)^2 + k_b(\theta_i - \pi)^2\right]$. The variable θ_i is the

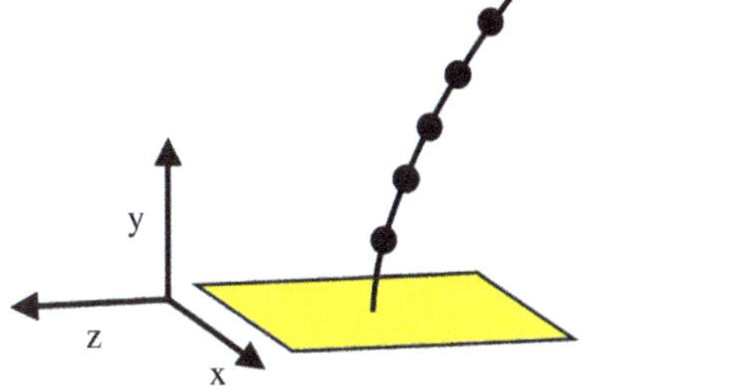

Figure 4.6 Schematic of cilium and actuation force (blue arrow). Only red beads in the cilium stick to the particle.

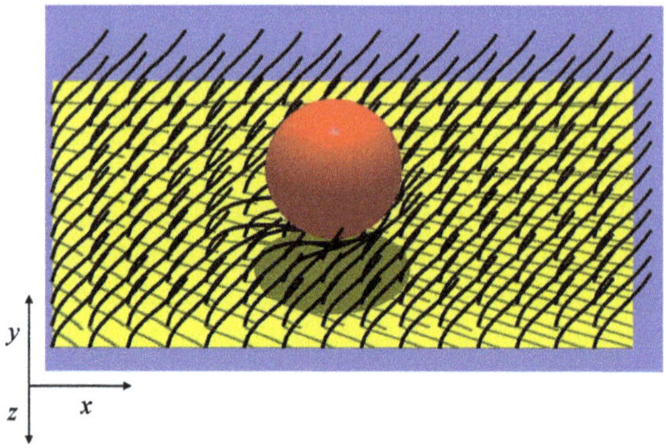

Figure 4.7 Three-dimensional snapshot from the simulation. Black filaments denote cilia and the red sphere is the particle.

angle between neighboring bonds and is given by $\theta_i = \cos^{-1}(\hat{\mathbf{b}}^{i+1,i} \cdot \hat{\mathbf{b}}^{i-1,i})$, where $\hat{\mathbf{b}}^{i,j} = [\mathbf{r}^i - \mathbf{r}^j]/|\mathbf{r}^i - \mathbf{r}^j|$, with $\hat{\mathbf{b}}^{0,1} = \mathbf{j}$ being the wall normal direction. The constants k_s, k_b and $l = L/N$ represent the stretching modulus, bending modulus and equilibrium distance between adjacent beads, respectively. Here, $k_b = EI/l$, where E is the Young's modulus, and $I = \pi a^4/4$. Figure 4.7 shows the 15×12 cilia array that forms the ciliated surface in our simulation; the inter-cilium separation is $\delta_x = \delta_z = L/3$.

The neutrally buoyant, spherical particle has a radius R and position $\mathbf{r}^s$. The three beads at the cilium tip are 'sticky' and hence, whenever the separation between the sphere and a sticky bead is less than a critical distance, (*i.e.* whenever $|\mathbf{r}^s - \mathbf{r}^i| - R < r^c$), we introduce a spring-like bond between the tip and nearest point on the particle surface (say, $\mathbf{r}^p$). In effect, the cilia 'sense' the presence of the particle through a short-range adhesive interaction. This adhesive interaction is modeled *via* an attractive Morse potential:

$$V^a(|\mathbf{r}^p - \mathbf{r}^i|) = D[1 - \exp\{\lambda(|\mathbf{r}^p - \mathbf{r}^i| - r_e)\}]^2 \tag{4.5}$$

which is applied only when $|\mathbf{r}^p - \mathbf{r}^i| > r_e$. The tip–particle bond is broken when $|\mathbf{r}^p - \mathbf{r}^i| \geq r^c$. We choose $r^c = r_e + \ln 2/\lambda$, the separation at which the attractive force is maximized, *i.e.* $\partial^2 V^a/\partial r^2 |_{r=r^c} = 0$. Each adhesive bead can form only one bond with the particle. The points of contact $\mathbf{r}^p$ on the particle are evolved in time using Quaternion equations.[59,60] A separate repulsive exclusion force acts on the bead and particle when $|\mathbf{r}^s - \mathbf{r}^i| - R < r_e$; similar exclusion forces are implemented for bead–bead and bead–wall interactions.

The fluid velocity, $\mathbf{u}(\mathbf{x})$, is evolved in the $L_x \times L_y \times L_z$ sized box using the Lattice Boltzmann Method (LBM).[61] For the i-th cilium bead, cilia–fluid coupling is implemented *via* a frictional force,[62,63] proportional to the slip velocity, $\dot{\mathbf{r}}^i - \mathbf{u}(\mathbf{r}^i)$. An equal and opposite force is interpolated onto the fluid grid in order to conserve momentum. No-slip boundary conditions are applied to the fluid at the particle surface, top wall, and bottom wall;[61] periodic boundary conditions are applied in the wall-parallel directions. The cilia and particle positions are evolved *via* a velocity Verlet algorithm.

An external, cyclic force, $\mathbf{F}^{ext}(t) = F_x^0 \cos(\omega t)\mathbf{i} + F_y^0 \sin(\omega t)\cos(\psi)\mathbf{j} + F_y^0 \sin(\omega t)\sin(\psi)\mathbf{k}$, is applied to the extreme tip of each cilium (*i.e.* at $i = N$); the force oscillates harmonically in a plane that is tilted by an angle ψ with respect to the x–y plane. The cilium tip is farther from the wall during the effective stroke, as compared to the recovery stroke (Figure 4.8a). This motion is responsible for net fluid flow in the direction of the effective stroke (*i.e.* the negative x-direction).

Given a fixed cilia geometry and low Reynolds number, the cilial motion depends solely on the sperm number, $S_p = L[4\pi\mu\omega/EI]^{1/4}$, which is the ratio of the viscous to the elastic forces on a cilium.[64] We first simulated actuated cilia without the particle, varying the value of E from E_0 to $5E_0$. Based on our values of E_0 and ω (see below), we obtain a sperm number of $S_p = 4.3$. It is important to note that the imposed actuation force is taken to be proportional to cilia stiffness, *i.e.* $F_x^0 \propto E$ and the ratio $F_y^0/F_x^0 = 1/4$ is fixed. The reason the tip forcing is proportional to cilia stiffness is as follows. To facilitate comparisons between cilia of different stiffness, we attempted to maintain a similarity between the trajectories of the different cilial tips. If a cilium is stiff (*i.e.* low

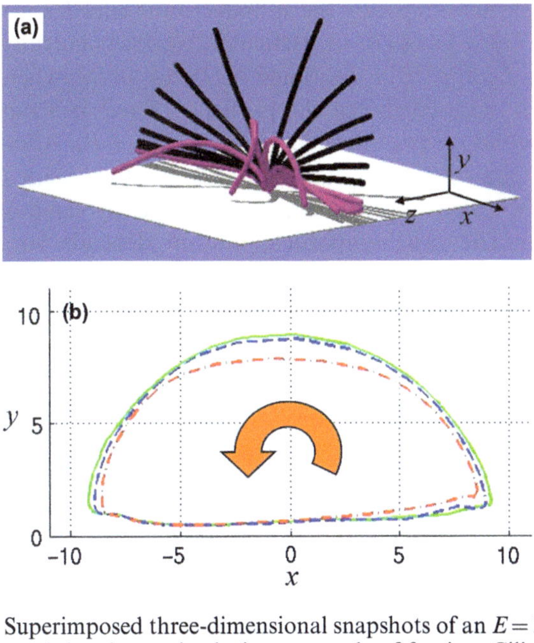

Figure 4.8 (a) Superimposed three-dimensional snapshots of an $E = 3E_0$ cilium, taken at equal time intervals, during one cycle of forcing. Cilium is black when tip moves in negative x direction (effective stroke), and magenta when tip moves in positive x direction (recovery stroke). (b) Path of cilium tip over one cycle, projected onto the x–y plane, for $E = E_0$ (red, dot-dashed), $E = 3E_0$ (blue, dashed), and $E = 5E_0$ (green, solid).

sperm number), the elastic forces dominate, so that the external force needs to be proportional to EI / L^2 to ensure that the cilium trajectory remains the same with increasing stiffness. Here, we keep L and I fixed and make the external force proportional to cilium stiffness, E, in order to obtain similar tip trajectories for the different cilia. Figure 4.8b clearly shows that the trajectory of the cilium tip does not vary over a large range of stiffness when we make this choice.

The respective dimensions of the computational box are $L_x \times L_y \times L_z = 45 \times 24 \times 36$ LB units, and the time period of actuation is $T = 30\,000$ LB time units. The length of each cilium is $L = 9$ LB units. Here, 1 LB unit of length denotes the grid spacing and 1 LB unit of time denotes the time step used in the Lattice Boltzmann scheme. The dimensionless values for the relevant parameters used in the simulations are $N = 9$, $a/L = 0.016$, $l/L = 1/9$, $Re = \omega L^2/\nu = 0.1$, $S_p(E_0) = L[4\pi\mu\omega/E_0 I]^{1/4} = 4.3$, $F_x^0 L^2/EI = 64$, $L_x/L = 5$, $L_z/L = 4$, $L_y/L = 2.7$, $R/L = 0.58$, $\lambda L = 189$, $(r^c - r_e)/L = 3.66 \times 10^{-3}$, and $r_e/L = 0.055$.

4.3.2 Results and Discussion

4.3.2.1 Particle Trajectories

We introduce a particle, and vary the stiffness of the cilia E and the strength of the adhesion, D (see eqn (4.5)). In particular, we systematically vary E in the range of E_0 to $5\,E_0$ and D in the range of D_0 to $10\,D_0$, where $D_0 = F_x^0 / 2\lambda$. Initially, the cilia is actuated, but the particle is fixed at a random position, within the interaction range of the cilia (*i.e.* $r_y^s(0) < L + R$). After $t > 10\pi/\omega$, the particle is freed, and the cilia–particle adhesion, with strength D, is activated. Four independent simulations are carried out at each value of (D,E) for 60 actuation cycles. The ensuing particle motion depends strongly on both adhesion strength and cilia stiffness. Our results show that any given particle trajectory converges to a 'released', 'propelled' or 'trapped' state. Figure 4.9a displays a trajectory where the particle escapes the cilia and, eventually, attains the 'released' state, at a height $r_y^s \approx R + L$ with respect to the lower wall. This particle is transported almost completely by fluid advection. Figure 4.9b shows a particle trajectory for a higher adhesion level, corresponding to the 'propelled' state. Here, cilia tips periodically attach themselves to the particle during the effective stroke, and the particle is thus pushed/pulled forward; the particle is released during the recovery stroke. Unlike the 'released' state, a 'propelled' state is characterized by a lower average particle height, and significant variation in particle height. Figure 4.9c shows a trajectory at an even higher adhesion level; here the particle remains 'trapped' within the cilia layer and thus, has zero time-averaged velocity.

Figure 4.10 shows the time-averaged particle velocity, $U(D,E) = \langle \dot{r}_x^s \rangle$, plotted with respect to D/D_0, for different cilia stiffness E, at fixed ω. The averaging was performed over the last 25 actuation cycles of each simulation.

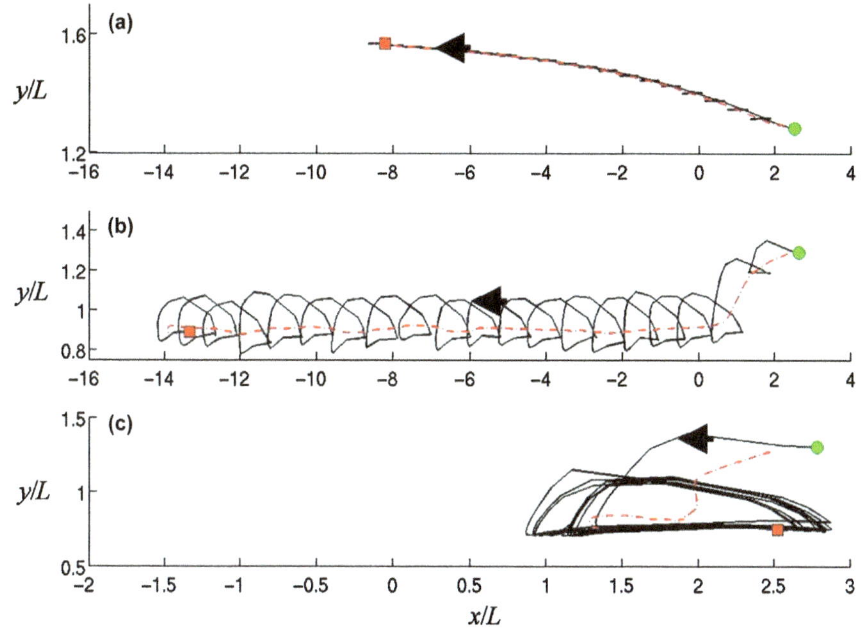

Figure 4.9 Particle trajectories (solid black curves), starting at green circles, ending at red squares, projected on the x–y plane, for $E = 3\,E_0$. Filtered particle trajectories, with height $h(t)$, shown by dashed red curves. Trajectory at (a) low adhesion, $D = 0.66\,D_0$, shows 'released' state, (b) medium adhesion, $D = 1.66\,D_0$, shows 'propelled' states, (c) high adhesion, $D = 5\,D_0$, shows particle 'trapping'.

$U(D,E)$ was normalized with respect to the average particle velocity at zero adhesion strength $U_0(E) = U(D,E)|_{D=0}$; the relative velocity $U^*(D,E) = U(D,E)/U_0(E)$ indicates the change in the average velocity due to effects of adhesion. For all curves, there is an optimal adhesion level where U^* is maximized. Notably, at $E = 6E_0$, the particle moves roughly 100% faster at this adhesion; significant increases also occur at other E values. The maxima in U^* occurs due to the 'propelled' states at these adhesion levels, where the cilia appear to 'toss' the particle from one neighbor to the next. In other words, the cilia periodically attach to the particle surface, transferring momentum in the process and thereby increasing the velocity of the particle.

During these 'propelled' states, the adhesive cilia also pulls the particle closer to the lower wall, where the net stream-wise fluid velocity is higher, leading to faster particle advection Notably, the average fluid velocity induced by the cilia has a Couette flow profile $U_x^{\text{Fluid}}(y) \propto (L_y - y)$, for $y \geq L$. Therefore, during the 'released' state, for which the average particle height is $\left\langle r_y^s \right\rangle \approx L + R$, the particle samples a lower fluid velocity than it would during the 'propelled' state, for which $\left\langle r_y^s \right\rangle \approx L$.

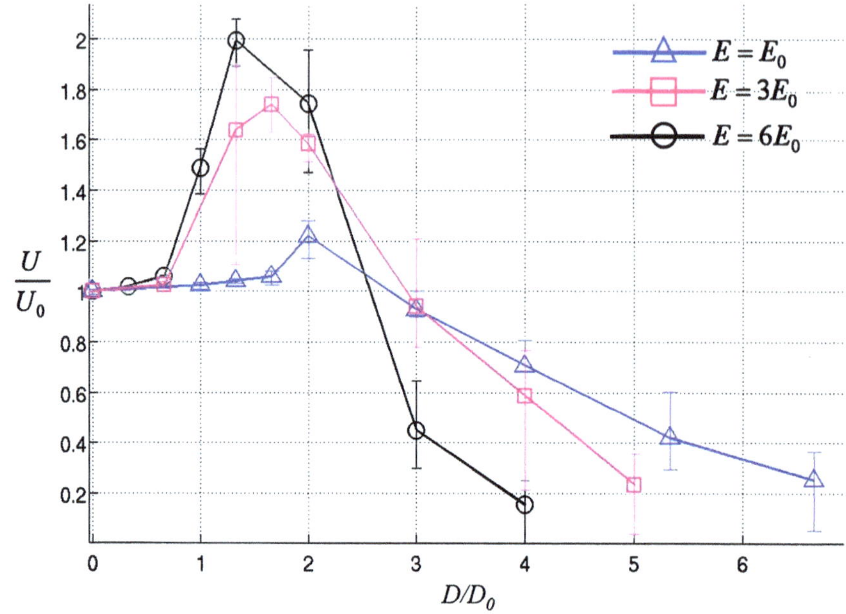

Figure 4.10 Average particle velocity over the trajectory, $U = \langle \dot{r}_x^s \rangle$, versus adhesion strength, for different cilia Young's modulus, E. Error bars at each (E,D) show maximum and minimum values of U for four independent simulations.

Increasing the adhesion above the optimal value eventually leads to the 'trapped' state. Here, the cilia remain attached to the sphere, and $U^* \sim 0$ for this particle.

A measure that differentiates a trajectory in the 'released' from that in the 'propelled' state is the variance $h' = \sqrt{\left\langle (r_y^s - h)^2 \right\rangle}$ of the particle height $r_y^s(t)$, with respect to its moving average $h(t) = (1/2T) \int_{t-T}^{t+T} r_y^s(t')\, dt'$, where $T = \pi/\omega$. The height $h(t)$ gives an estimate of the average height of the particle trajectory, and will not depend on time if $r_y^s(t)$ becomes periodic, with period $2T$. A trajectory in the 'released' state is characterized by a small variance, *i.e.* $h'/L \ll \varepsilon^h$, while a trajectory in 'propelled' state has a large variance, $h'/L > \varepsilon^h$. We choose $\varepsilon^h = 0.03$ as the threshold for sensing a 'propelled' state. A particle is 'trapped' if its average velocity is low, *i.e.* $U^* < \varepsilon^u$; here, we choose $\varepsilon^u = 0.2$.

4.3.2.2 *Phase Map*

Using the above definitions, we separate the (D,E) plane into 'released' (I), 'propelled' (II) and 'trapped' (III) states (Figure 4.11). The I/II boundary can be sensitive to the value of ε^h; therefore, in Figure 4.11, this boundary is also displayed *via* a contour plot of $h'(D, E)$. Interestingly, we find that the increase

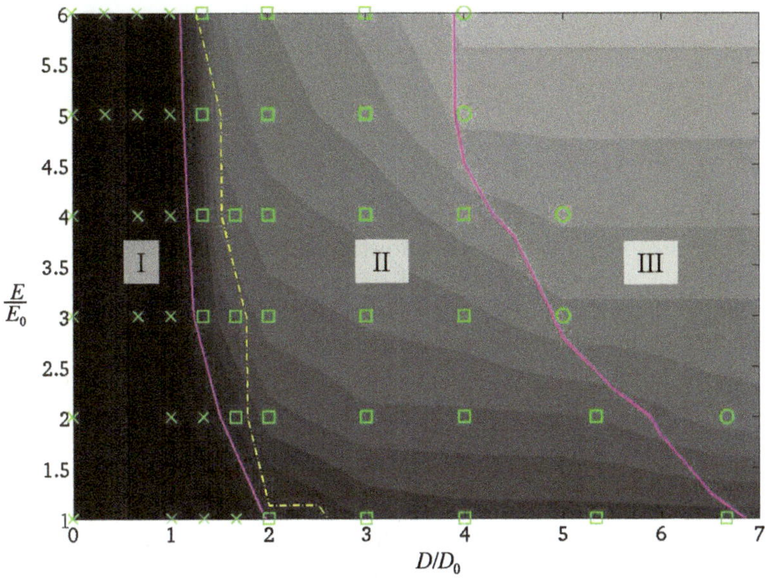

Figure 4.11 Phase map of particle motion as a function of (*D*,*E*). Simulations were run at each point marked by green symbol. Crosses mark phase I ('released' state), circles mark phase II ('propelled' state, squares mark phase III ('trapped' state). Solid magenta lines divide the phases. Yellow dot–dashed line marks I/II boundary predicted by low-order model. Contour plot of the height variance of the particle from LBM simulations, $h'(D,E)$, is shown in the background.

in h' with D is not gradual. Rather, for any given stiffness E, the I/II boundary marks a critical adhesion level $D_c(E)$, above which the particle trajectory makes a sharp transition in h', going from being in a 'released' state to a 'propelled' state, regardless of initial particle depth in the array.

From recent experimental studies on synthetic cilia, we can specify the length of the cilia to be $L = 20$ μm and the beating frequency to be $1/T = 40$ Hz.[6] If we assume the surrounding fluid to be water, then we can take the kinematic viscosity of the fluid to be $\nu = 10^{-6}$ m^2 s^{-1}and the density of the fluid to be $\rho = 10^3$ kg m^{-3}. It follows from the dimensionless parameters used in the simulation that the radius of the particle is $R = 12$ μm, the cross-sectional radius of each cilium is $a = 333$ nm, the inverse of length scale over which Morse potential decays is $\lambda = 9.6 \times 10^6$ m^{-1}, the moment of inertia of the cilium cross section is $I = \pi a^4/4 = 9.7 \times 10^{-27}$ m^4, and the scale of the Young's modulus is $E_0 = 0.132$ MPa. Based on these values, we can specify the I/II boundary in terms of surface energy $\gamma = D/\pi a^2$, and interfacial tensile strength $\sigma = D\lambda/2\pi a^2$ of the cilia tips, per unit area.[65] Our results indicate that, for $E = E_0 = 0.132$ MPa, the I/II boundary will occur at $\sigma = 293$ Pa, $\gamma = 6.1 \times 10^{-5}$ J m^{-2} (which follows from $D_c^* = D_c/D_0 = 2$ at $E = E_0$), while, for

$E = 6\,E_0 = 0.794$ MPa, the boundary will occur at $\sigma = 967$ Pa, $\gamma = 2 \times 10^{-4}$ J m^{-2} (which follows from $D_c^* = 1.1$ at $E = 6\,E_0$).

4.3.2.3 Lower-order Model

To obtain further insight into physical factors controlling the I/II boundary within the phase map, we derived a lower-order model for the cilia–particle interaction. We model the movement of the tips of both the free cilia and the attached cilia, which have formed bonds with the particle. The model consists of the following degrees-of-freedom: tip location of free cilia $\mathbf{r}$, average tip location of attached cilia $\bar{\mathbf{r}}$, radius of circular contact area d between the cilia array and particle, and particle height r_y^s. Here, the vectors $\mathbf{r}$ and $\bar{\mathbf{r}}$ denote the distance and orientation of the cilia tips with respect to their tethered ends.

The particle–cilia dynamics are solved only for the effective stroke, during which the tip of each cilium stays within the x–y plane. Therefore, we can approximate $\mathbf{r}(t) = X(t)\mathbf{i} + Y(t)\mathbf{j} = r(t)\hat{\mathbf{r}}$ using two degrees of freedom; a similar expression can be obtained for $\bar{\mathbf{r}}$. Each cilium in the array is effectively modeled as a rod with some elastic energy and viscous mobility. For any given cilium tip vector, $\mathbf{r}'$, we then approximate:

$$\dot{\mathrm{r}}' = \dot{r}'\,\hat{\mathrm{r}} + r'\dot{\theta}'\,\hat{\theta} = \mu^{-1}[-\nabla_{\mathrm{r}'}U^E(\mathrm{r}') + \mathrm{F}^{\mathrm{tip}}] \tag{4.6}$$

where $\mathbf{F}^{\mathrm{tip}}$ is a force at the tip, which could arise due to external actuation, as well as the cilia–particle Morse potential. Here, $U^E(\mathbf{r}) = [k_r^{\pm}(r - L)^2/L + k_\theta(\pi/2 - \theta)^2]/2$ is the elastic energy of this cilium, and μ^{-1} is its mobility matrix, defined as $\mu_{r\theta}^{-1} = \mu_{\theta r}^{-1} = 0$, $\mu_{rr}^{-1} = 2/(C_{\parallel}L)$, $\mu_{\theta\theta}^{-1} = 3/(C_{\perp}L)$. Using approximations from Resistive Force Theory,[66] we assume $C_{\parallel} = \alpha 2\pi\mu/\ln(L/a)$ along the filament, and $C_{\perp} = 2C_{\parallel}$ perpendicular to it. The evolution equations for $\mathbf{r}$, $\bar{\mathbf{r}}$, d, r_y^s are derived as follows.

The equation for free cilia tips, $\mathbf{r}$, is obtained by equating $\mathbf{F}^{\mathrm{tip}} = \mathbf{F}^{\mathrm{ext}}$. To evolve $\bar{\mathbf{r}}$, we must take into account the additional force from the particle, as well as the change in the number of bonds, $n = \rho_c \pi d^2$, in which $\rho_c = 1/\delta_x^2$ is the grafting density of the cilia. We obtain:

$$\dot{\bar{\mathrm{r}}} = 2(\mathrm{r} - \bar{\mathrm{r}})\dot{d}/d + \mu^{-1}[-\nabla_{\bar{\mathrm{r}}}U^E(\bar{\mathrm{r}}) + \mathrm{F}^{\mathrm{ext}}(t) + \mathrm{F}^{\mathrm{Morse}}(r_y^s, \bar{r}_y)] \tag{4.7}$$

where $\mathrm{F}^{\mathrm{Morse}}(r_y^s, \bar{r}_y) = -\,\partial V^a(|\,r_y^s - R - \bar{r}_y\,|)/\partial \bar{r}_y \mathrm{j}$ is the Morse potential force acting per cilia–particle bond. The particle undergoes over-damped motion in response to the Morse potential force from n bonds: $\dot{r}_y^s = -\,nF_y^{\mathrm{Morse}}(r_y^s, \bar{r}_y)/\beta 6\pi R\mu$.

If the particle is submerged in the array (*i.e.* $\delta = Y - (r_y^s - R) > 0$) and the free cilia tips are approaching the particle ($\dot{Y} - \dot{r}_y^s > 0$), then we model the contact radius d in terms of the depth of the sphere in the array, *i.e.* $d = \sqrt{R\delta}$. When the top surface of the array recedes from the sphere (*i.e.* $\dot{Y} - \dot{r}_y^s \leq 0$), we assume that the number of tips interacting with the particle stays constant (*i.e.* $\dot{d} = 0$).

After calibrating the relevant constants $k_r^\pm$, k_θ, α and β from LBM simulations, we numerically evolve the six degrees of freedom over several effective strokes. As in the LBM simulations, we find that $r_y^s(t)$ becomes periodic in time after a few cycles, and we can obtain its height variation $h'(E,D)$. Using the criterion $h'/L = \varepsilon^h$ to detect the propelled trajectories, we find that the I/II boundary is captured well by our lower-ordered model (see Figure 4.11). Taken together, the different approaches corroborate our findings that the particle motion can be controlled by tailoring the stiffness of the cilia or the stickiness of its tips.

4.4 Conclusions

Through these studies, we demonstrated that synthetic cilia can exhibit sensory capabilities that are analogous to the biological counterparts. In particular, by focusing on a single cilium formed from a BZ gel, we showed that this cilium is highly sensitive to the local concentration of the activator u, as is evident by the changes in the directionality of the wave propagation with variations in the local distribution of u. Thus, the traveling waves in the cilium can be used as an indicator for the local chemical environment. The distinct chemotactic behavior of the five cilia array, where the tethered gels bent towards higher concentrations of u, also highlighted the chemo-sensitivity of these synthetic cilia. By modeling the effects of an external light source, we gained further insight into the means by which multiple BZ cilia communicate through self-generated chemical gradients in the solution.

These attributes could be usefully exploited in a range of micro-fluidic applications, where, for example, an external light source could be used to direct the movement of a soft 'conveyor belt'[56] and thus, manage the transport of microscopic objects within the devices. The motion of the BZ cilia is relatively slow (on the order of 10 $\mu m\ s^{-1}$) and hence, this conveyor belt would be most effective in scenarios that involve diagnostic studies or analyses of the species as they are transported through the micro-channels.

We also formulated a computational approach to model the interaction between active, adhesive cilia and microscopic particles in solution. In the process, we developed a method that provides a general framework for predicting the interaction of particles with a range of active biological and synthetic cilia. Using this approach, we found that the particle velocity and trajectory depend strongly on the particle–cilia adhesion and cilia stiffness E (or, alternatively, the sperm number).

Notably, the simulations revealed that at an optimal adhesion level, the cilia can significantly increase the velocity U at which a particle is propelled along the top of the layer. This increase in U is due to a concerted pushing and pulling exerted by the respective cilia in front and behind the particle. This increase in the particle velocity was most pronounced for the stiffest cilia considered here.

We also determined a phase map as a function of E and D and thereby pinpointed the critical adhesion levels $Dc(E)$, above which a particle goes from being in a 'released' to a 'propelled' state, regardless of initial particle depth in the array. These results revealed that D_c/D_0 decreases with increasing E, at a rate much slower than $1/E$, implying (since D_0) that D_c increases with increasing E. In other words, floppier cilia can capture particles at lower adhesion strength. This result is consistent with the understanding that particles get stuck more easily on softer passive adhesive surfaces.[67] We also pinpointed the region in phase space where the particles become trapped within the active layer.

These results provide guidelines for tailoring the functionality of the ciliated surface by tailoring the stiffness of the cilia and the 'stickiness' of the cilia tips. Namely, for a range of E and D, the cilia are highly effective at preventing particles from penetrating the layer and thus, can provide valuable antifouling properties to a range of surfaces. Conversely, there is a range of E and D where the beating cilia can be used to extract particles from solution and, thus, could act as an active filtration system.

Acknowledgement

A.C.B. gratefully acknowledges NSF for partial support of P.D.; ONR for partial support of A.B. and DOE for partial support of O.K.

References

1. A. S. Shah, Y. Ben-Shahar, T. O. Moninger, J. N. Kline and M. J. Welsh, *Science*, 2009, **325**, 1131.
2. S. T. Christensen, L. B. Pedersen, L. Schneider and P. Satir, *Traffic*, 2007, **8**, 97.
3. R. A. Bloodgood, *J. Cell Sci.*, 2010, **123**, 505.
4. F. Liu, D. Ramachandran and M. W. Urban, *Adv. Funct. Mater.*, 2010, **20**, 3163.
5. A. R. Shields, B. L. Fiser, B. A. Evans, M. R. Falvo, S. Washburn and R. Superfine, *Proc. Natl. Acad. Sci. U.S.A.*, 2010, **107**, 15670.
6. B. A. Evans, A. R. Shields, R. L. Carroll, S. Washburn, M. R. Falvo and R. Superfine, *Nano. Lett.*, 2007, **7**, 1428.
7. J. den Toonder, F. Bos, D. Broer, L. Filippini, M. Gillies, J. de Goede, T. Mol, M. Reijme, W. Talen, H. Wilderbeek, V. Khatavkar and P. Anderson, *Lab Chip*, 2008, **8**, 533.
8. F. Fahrni, M. W. J. Prins and L. J. van IJzendoorn, *Lab Chip*, 2009, **9**, 3413.
9. C. L. van Oosten, C. W. M. Bastiaansen and D. J. Broer, *Nat. Mater.*, 2009, **8**, 677.
10. M. Vilfan, A. Potocnik, B. Kavcic, N. Osterman, I. Poberaj, A. Vilfan and D. Babic, *Proc. Natl. Acad. Sci. U.S.A.*, 2010, **107**, 1844.
11. T. Sanchez, D. Welch, D. Nicastro and Z. Dogic, *Science*, 2011, **333**, 456.

12. J. Aizenberg, L. D. Zarzar and P. Kim, *Adv. Mater.*, 2011, **23**, 1442.
13. A. Sidorenko, T. Krupenkin and J. Aizenberg, *J. Mater. Chem.*, 2008, **18**, 3841.
14. O. Tabata, H. Hirasawa, S. Aoki, R. Yoshida and E. Kokufuta, *Sens. Actuators, A*, 2002, **95**, 234.
15. P. Dayal, O. Kuksenok, A. Bhattacharya and A. C. Balazs, *J. Mater. Chem.*, 2012, **22**, 241.
16. E. M. Gauger, M. T. Downton and H. Stark, *Eur. Phys. J. E*, 2009, **28**, 231.
17. B. Pokroy, A. Epstein, M. C. M. Persson-Gulda and J. Aizenberg, *Adv. Mater.*, 2009, **21**, 463.
18. K. D. Harris, R. Cuypers, P. Scheibe, C. L. van Oosten, C. W. M. Bastiaansen, J. Lub and D. J. Broer, *J. Mater. Chem.*, 2005, **15**, 5043.
19. A. Alexeev, J. M. Yeomans and A. C. Balazs, *Langmuir*, 2008, **24**, 12102.
20. M. Sleigh, *Comp. Biochem. Physiol.*, 1989, **94A**, 359.
21. R. Ghosh, G. A. Buxton, O. B. Usta, A. C. Balazs and A. Alexeev, *Langmuir*, 2010, **26**, 2963.
22. H. Masoud and A. Alexeev, *Soft Matter*, 2011, **7**, 8702.
23. J. V. I. Timonen, C. Johans, K. Kontturi, A. Walther, O. Ikkala and R. H. A. Ras, *Appl. Mater. Interface.*, 2010, **2**, 2226.
24. K. Oh, J. Chung, S. Devasia and J. J. Riley, *Lab Chip*, 2009, **9**, 1561.
25. A. Gopinath and L. Mahadevan, *Proc. R. Soc. London Ser. A, Math. Phys. Eng. Sci.*, 2011, **467**, 1665.
26. S. Maeda, Y. Hara, R. Yoshida and S. Hashimoto, *Angew. Chem. Int.*, 2008, **47**, 6690.
27. J. Lahann, S. Mitragotri, T. N. Tran, H. Kaido, J. Sundaram, I. S. Choi, S. Hoffer, G. A. Somorjai and R. Langer, *Science*, 2003, **299**, 371.
28. Y. W. Kim and R. R. Netz, *Phys. Rev. Lett.*, 2006, **96**, 158101.
29. C. H. Wiggins and R. E. Goldstein, *Phys. Rev. Lett.*, 1998, **80**, 3879.
30. J. T. Norwood, C. E. Hein, S. A. Halbert and R. G. W. Anderson, *Proc. Natl. Acad. Sci. U.S.A.*, 1978, **75**, 4413.
31. P. Talbot and K. Riveles, *Reprod. Biol. Endocrin.*, 2005, **3**, 52.
32. P. Dayal, O. Kuksenok, A. Bhattacharya and A. C. Balazs, *J. Mater. Chem.*, 2012, **22**, 241.
33. A. N. Zaikin and A. M. Zhabotinsky, *Nature*, 1970, **225**, 535.
34. S. Sasaki, S. Koga, R. Yoshid and T. Yamaguchi, *Langmuir*, 2003, **19**, 5595.
35. R. Yoshida, E. Kokufuta and T. Yamaguchi, *Chaos*, 1999, **9**, 260.
36. R. Yoshida, *Bull. Chem. Soc. Japan*, 2008, **81**, 676.
37. S. Shinohara, T. Seki, T. Sakai, R. Yoshida and Y. Takeoka, *Angew. Chem. Int. Ed.*, 2008, **47**, 9039.
38. A. Bhattacharya, G. A. Buxton, O. B. Usta and A. C. Balazs, *Langmuir*, 2012, **28**, 3217.
39. J. J. Tyson and P. C. Fife, *J. Chem. Phys.*, 1980, **73**, 2224.
40. V. V. Yashin and A. C. Balazs, *J. Chem. Phys.*, 2007, **126**, 124707.
41. O. Kuksenok, V. V. Yashin and A. C. Balazs, *Phys. Rev. E*, 2008, **78**, 041406.
42. P. Dayal, O. Kuksenok and A. C. Balazs, *Langmuir*, 2009, **25**, 4298.

43. P. Dayal, O. Kuksenok and A. C. Balazs, *Soft Matter*, 2010, **6**, 768.
44. I. C. Chen, O. Kuksenok, V. V. Yashin, M. R. Moslin, A. C. Balazs and K. J. Van Vliet, *Soft Matter*, 2011, **7**, 3141.
45. I. C. Chen, O. Kuksenok, V. V. Yashin, A. C. Balazs and K. J. Van Vliet, *Adv. Funct. Mater.*, 2012, **22**, 2535.
46. O. Kuksenok, V. V. Yashin, M. Kinoshita, T. Sakai, R. Yoshida and A. C. Balazs, *J. Mater. Chem.*, 2011, **21**, 8360.
47. H. J. Krug, L. Pohlmann and L. Kuhnert, *J. Phys. Chem.*, 1990, **94**, 4862.
48. R. Aihara and K. Yoshikawa, *J. Phys. Chem. A*, 2001, **105**, 8445.
49. R. Yoshida, G. Otoshi, T. Yamaguchi and E. Kokufuta, *J. Phys. Chem. A*, 2001, **105**, 3667.
50. O. Kuksenok, V. V. Yashin and A. C. Balazs, *Phys. Rev. E*, 2009, **80**, 056208.
51. A. S. Mikhailov and A. Engel, *Phys. Lett. A*, 1986, **117**, 257.
52. O. U. Kheowan, E. Mihaliuk, B. Blasius, I. Sendina-Nadal and K. Showalter, *Phys. Rev. Lett.*, 2007, **98**, 074101.
53. By 'intrinsic frequency', we are referring to the frequency that an isolated element would oscillate with, given the same value of u in the outer fluid.
54. V. V. Yashin and A. C. Balazs, *Phys. Rev. E*, 2008, **77**, 046210.
55. O. Kuksenok, V. V. Yashin and A. C. Balazs, *Soft Matter*, 2007, **3**, 1138.
56. Y. Murase, S. Maeda, S. Hashimoto and R. Yoshida, *Langmuir*, 2009, **25**, 483.
57. J. G. Kirkwood, *J. Chem. Phys.*, 1939, **7**, 506.
58. P. N. Keating, *Phys. Rev.*, 1966, **145**, 637.
59. I. P. Omelyan, *Phys. Rev. E*, 1998, **58**, 1169.
60. N. S. Martys and R. D. Mountain, *Phys. Rev. E*, 1999, **59**, 3733.
61. A. J. C. Ladd, *J. Fluid Mech.*, 1994, **271**, 285.
62. P. Ahlrichs and B. Dunweg, *Int. J. Mod. Phys. C*, 1998, **9**, 1428.
63. O. B. Usta, A. J. C. Ladd and J. E. Butler, *J. Chem. Phys.*, 2005, **122**, 09490.
64. C. P. Lowe, *Phil. Trans. R. Soc. London B*, 2003, **358**, 1543.
65. The adhesion scale, D_0, is itself a linear function of the Young's modulus, E, so that $D_0(E) = F_x^{ext} / 2\lambda = 32EI/L^2\lambda$. Using the specified values of E_0, I, L and λ, we obtain $D_0(E_0) = 1.07\times10^{-17}$ J, $D_0(6E_0) = 6.42\times10^{-17}$ J. The latter values of D_0, along with the dimensional values, $\lambda = 9.6\times10^6$ m^{-1}, $\alpha = 333$ nm, yield estimates of $\gamma = D/\pi a^2$, $\sigma = D\lambda/2\pi a^2$ for $D = D_c(E_0) = 2D_0(E_0)$ and $D = D_c(6E_0) = 1.1D_0(6E_0)$.
66. C. Brennen and H. Winet, *Annu. Rev. Fluid Mech.*, 1977, **9**, 339.
67. A. Jagota and S. J. Bennison, *Integr. Comp. Biol.*, 2002, **42**, 1140.

Section 2
Experimental Approaches to Create Artificial Cilia

CHAPTER 5

Electrostatic Artificial Cilia

JAAP M. J. DEN TOONDER[a,b]

[a] Philips Research, High Tech Campus 34, 5656 AE Eindhoven, The Netherlands; [b] Eindhoven University of Technology, Postbox 513, 5600 MB Eindhoven, The Netherlands
Email: j.m.j.d.toonder@tue.nl

5.1 Introduction

Many microscopic organisms use cilia to propel themselves through a liquid. Cilia can be viewed as small hairs or flexible rods (with a typical length of 10 μm and a diameter of about 250 nm) covering the outer surface of the micro-organism, see Figure 5.1. An individual cilium makes an oscillating motion that is asymmetric, producing an effective stroke and a recovery stroke, with a frequency between 10 and 30 Hz.[1,2] The collective beating of the thousands of cilia is very effective in generating flow: the swimming speed of Paramecium, for example, can be approximately 1 mm s^{-1}.[3]

Inspired by nature, a number of groups have been developing artificial cilia for fluid manipulation—pumping and mixing—in micro-fluidic devices during the past decade. Several approaches have been taken. Examples are electrostatically actuated artificial cilia,[4,5] magnetically actuated artificial cilia,[6–14] optically responsive artificial cilia,[15] artificial cilia based on responsive hydrogels,[16,17] and artificial cilia driven by resonance.[18] In addition, a number of studies have been published on modelling the response and effect of (artificial) cilia, to understand, predict, and design their behaviour.[19–25] The nature and appearance of the artificial cilia is diverse. Sizes of artificial cilia range from micrometres to millimetres. Fabrication processes are very different for the various approaches. The variety of approaches shows that inspiration

RSC Nanoscience & Nanotechnology No. 30
Artificial Cilia
Edited by Jaap M J den Toonder and Patrick R Onck

Published by the Royal Society of Chemistry, www.rsc.org

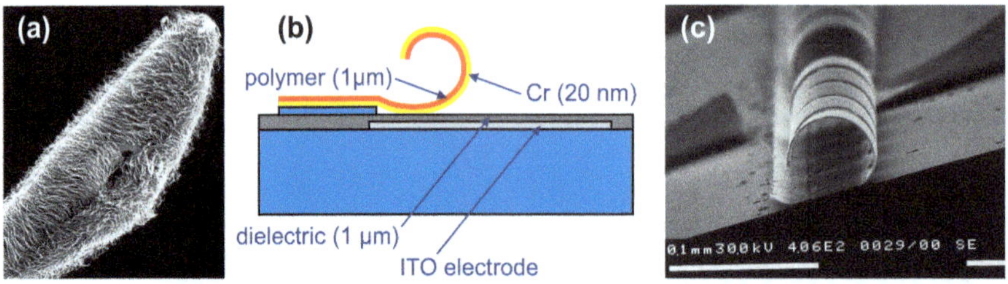

Figure 5.1 Real and artificial cilia. (a) Paramecium: a micro-organism that makes use of beating cilia on its surface for propulsion. Its cross section is about 40 mm. (b) Cross-sectional sketch of our artificial cilium. It is actuated by applying a voltage difference between the ITO electrode and the Cr film, upon which the cilium rolls out. After switching off the voltage, it rolls back by elastic recovery. (c) Electron micrograph showing our artificial cilia that have a length of 100 mm and a width of 20 mm. (Reproduced from den Toonder *et al.*[4])

by nature is not synonymous with literally copying; also the working mechanism of the artificial cilia may be completely different from how nature works, but still result in the anticipated effects. The fluid velocities induced by artificial cilia in experiments range from micrometres per second[6,7] to hundreds of micrometres per second.[4,8,9] In terms of potential flow rates, the latter range is certainly interesting for real micro-fluidic applications.

Inspired by nature, we have developed electrostatically actuated artificial cilia, consisting of polymer-based micro-actuators, that are integrated in a micro-fluidic channel.[4,5] In this chapter we show that our artificial cilia can generate substantial fluid flow, and that they can be used to mix fluids actively and effectively in micro-fluidic channels. Also, by combining experimental studies with a numerical model, we explain why our artificial cilia are so effective in generating flow.

5.2 Flow Generation and Micro-mixing by Electrostatic Artificial Cilia

5.2.1 Electrostatic Artificial Cilia

Our artificial cilia are shown in Figure 5.1. The typical structure is that of a curled micro-beam consisting of a double-layer of a thin polymer film, namely polyimide (PI) with a thickness of 1 µm, and a thin conductive chromium (Cr) layer with a thickness of 20 nm. Figure 5.1c shows a scanning electron microscopy (SEM) image of actual structures made, having a length of 100 µm and a width of 20 µm. As depicted in the schematic in Figure 5.1b, the glass substrate is covered with an indium tin oxide (ITO) electrode, and a dielectric silicon oxide–nitride stack with a total thickness of about 1 µm. The micro-actuators are made using micro-systems technology processing, as described in section 5.4. The radius of curvature of the micro-actuators is controlled by tuning the mechanical stress in the PI and Cr films by adjusting the deposition conditions and the relative thickness of both films. Note that the artificial cilia are significantly larger than biological cilia that have a typical length of 10–20 μm. The technology used allows the miniaturization of the actuators by an order of magnitude, but for the purpose of a proof-of-concept demonstration we show only results of the larger artificial cilia shown in Figure 5.1.

The actuation of the artificial cilia is done electrostatically: by applying a voltage difference between the ITO electrode and the Cr layer, an electrostatic attraction force is induced and the curled micro-actuator rolls out and extends over the surface. After switching off the voltage, the structure rolls back to its original curled shape by elastic recovery. We used an AC voltage with a frequency of 1 kHz, that was switched on and off with switching frequencies between 1 and 200 Hz. A minimum voltage is required to actuate the artificial cilia, determined by the materials and geometry applied. The rolling-out threshold voltage was around 70 V for our micro-actuators. Upon applying the AC voltage, the micro-actuators reacted almost instantaneously. The cilia did not react to the AC frequency but they rolled out and rolled back following the switching frequency. With the use of a high-speed camera, we determined the rolling-out and rolling-back times. Figure 5.2 shows snapshots taken from a high-speed recording of the artificial cilia movement in air, imaged from a side view. In this case, 80 V was applied to actuate the cilia. In air, the rolling-out time was very fast, namely 7 μs, and the rolling-back time was measured to be about 20 μs. In a silicone oil with a viscosity of 9.3 mPa·s, these numbers were less than 1 ms, and about 3 ms, respectively. The movement in silicone oil (viscosity, 9.3 mPa·s) of the tip of one artificial cilium versus time, for different applied voltages, is shown in Figure 5.3. The figure depicts the trace of the cilium tip, as projected onto the plane of view (see section 5.4 for an explanation). Indeed, the cilium unrolls in less than 1 ms, and the rolling-out time depends on the magnitude of the applied voltage V_m, namely 1 ms for $V_m = 80$ V and 0.2 ms for $V_m = 140$ V. The rolling-back time, on the other hand, is essentially independent of the applied voltage (beween 2.5 and 3.5 ms), which is understandable because rolling back is driven merely by elastic recovery.

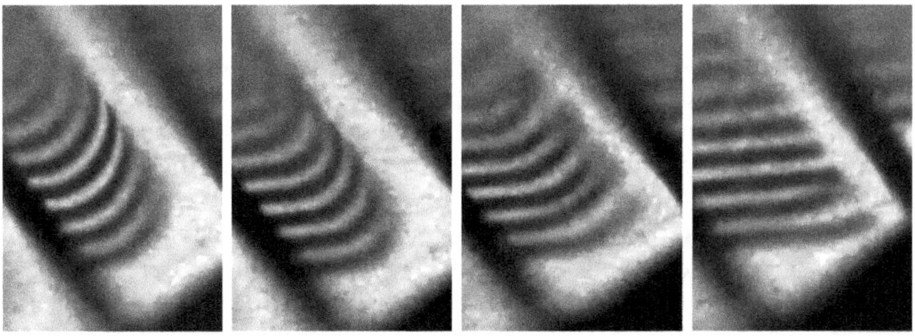

Figure 5.2 Stages observed during artificial cilia unrolling, imaged in air: 80 V kHz^{-1} was applied to the artificial cilia.

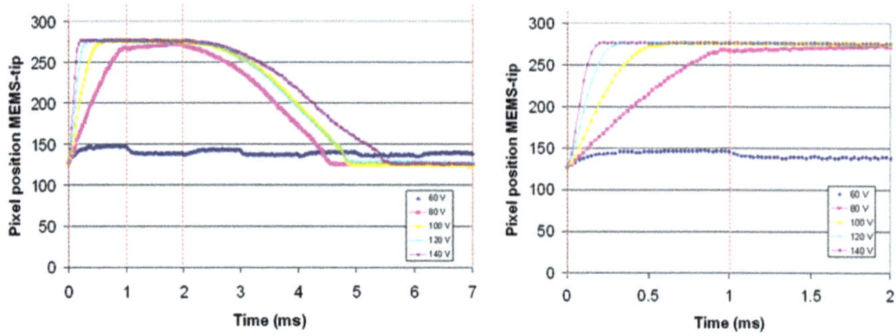

Figure 5.3 Left: the movement of the tip of one artificial cilium in silicone oil *versus* time, for various applied voltages. During the first 1 ms, a negative voltage is applied. From 1 to 2 ms, a positive voltage is applied. After 2 ms, no voltage is applied anymore. Right: detailed view of the first 2 ms of the left graph.

A detailed view of the first 2 ms of the artificial cilia movement is shown on the right in Figure 5.3.

An analysis of the whole process of unrolling and recovering of our artificial cilia thus clearly shows an asymmetric behaviour in time; however, hardly any asymmetry in shape could be observed between unrolling and rolling back.

We carried out durability tests by continuous actuation in air at a switching frequency of 1 Hz. Only after one week of actuation, the first micro-actuators started to fail, that is after about 600,000 switching cycles.

5.2.2 Particle Tracking Experiments

To investigate the effectiveness of fluid manipulation by the artificial cilia, we carried out particle tracking experiments in silicone oil, from which the induced flow velocities were estimated. The cilia were arranged on a substrate in square segments of 1 mm^2, as shown in a top view in Figure 5.4a. The segment contains five columns of 20 cilia, visible in the figure as black rectangles since they are in the curled state. The surface was covered with a 0.5 mm thick silicone oil film (viscosity 9.3 mPa·s), so that the cilia were completely immersed. To visualize the flow, we carried out various experiments using two kinds of tracer particles dispersed in the fluid, namely titanium dioxide (TiO_2) particles with a mean diameter of 0.5 μm and hollow glass spheres with an average diameter of 12 μm.

Figure 5.4 shows the estimated flow speeds as a function of switching frequency and applied voltage. Flow speeds up to 0.6 mm s^{-1} were generated. The flow direction is determined by the rolling-out direction of the micro-actuators. The induced velocity increases with both the switching frequency and the actuation voltage, and the agreement between the two types of particles is reasonably good.

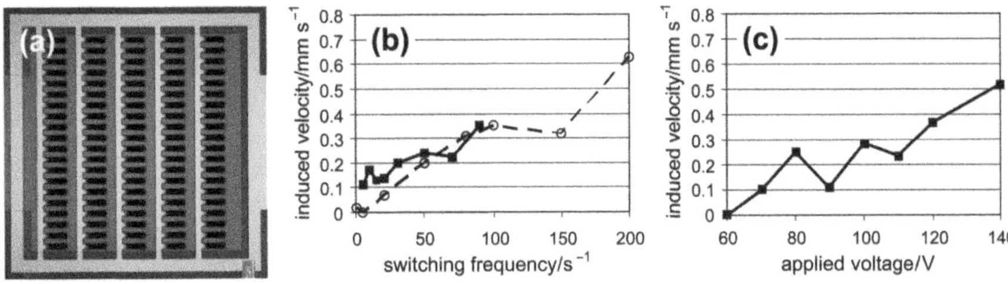

Figure 5.4 Results of particle tracking experiments. (a) Top view of the cilia layout: the cilia are arranged in five columns of 20 cilia, and they are covered with a 0.5 mm thick film of silicone oil (viscosity 9.3 mPa·s), containing either TiO_2 or hollow glass as tracer particles. The cilia were actuated with different switching frequencies and actuation voltages, and the movement of the tracer particles was recorded at 30 frames per second. Particle tracking was done manually from the obtained movies, and the induced flow velocities were estimated. (b) Induced velocity as a function of switching frequency, estimated from tracking of TiO_2 particles (solid lines) and hollow glass spheres (broken lines). The applied AC voltage is 75 V kHz^{-1}. (c) The effect of applied AC voltage (always 1 kHz) on the induced velocity, measured using hollow glass spheres. The switching frequency is here fixed at 50 Hz. The lines are drawn as a guide to the eye. (Reproduced from den Toonder *et al.*[4])

5.2.3 Mixing Experiments

Having established the effectiveness of flow generation by our artificial cilia in an open-faced channel, we tested the concept of mixing in a micro-fluidic channel. We designed and fabricated a Y-shaped mixing channel, shown in Figure 5.5. A polydimethylsiloxane (PDMS) cover, containing the Y-shaped channel structure, is mounted on top of the glass substrate with the artificial cilia. The bottom channel wall is thus covered with artificial cilia, which are arranged in 16 segments of 1×1 mm, containing various cilia lay-outs, typically in an A–B–A–B–... sequence. In this configuration, a mixing 'cycle' is made up by two consecutive segments (*e.g.* A–B or C–D in the figure), hence the channel contains eight cycles. The main flow is driven by two syringe pumps connected by tubes to the two inlets, whereas the movement of the artificial cilia induces a transverse flow.

The nature of the transverse flow depends on the layout of the cilia. In one design (A–B–A) shown in Figure 5.5, all cilia in one segment roll out in the same direction, perpendicular to the main flow, and in the subsequent segment the roll-out direction is reversed. In this design, the transverse flow will consist of vortices spanning the complete channel cross section. The fluid, as it travels through the channel, will therefore experience transverse vortical flow of which the direction of rotation changes each time a segment is passed. In the other design (C–D–C) shown in Figure 5.5, the cilia motion will create two counter-rotating transverse vortices with different sizes in a segment. In neighboring segments, the positions of the larger and the smaller vortex are interchanged. This leads to a transverse flow pattern which resembles that of the staggered

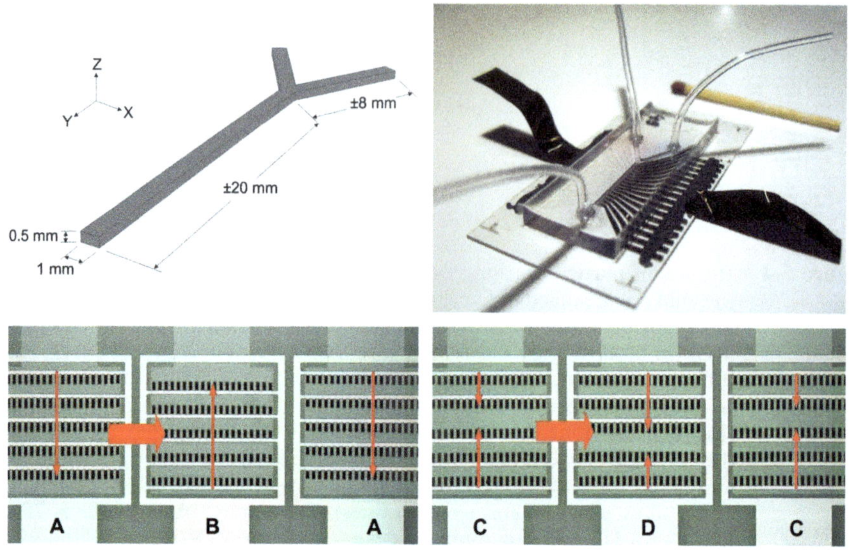

Figure 5.5 Our artificial cilia mixing Y-channel device. The artificial cilia were fabricated on a glass substrate. The cover containing the channel is made of PDMS. Artificial cilia are present on the bottom of the channel, and are arranged in square segments each having their own cilia configuration. The main flow is driven by a pump; the moving cilia create transverse flows. Two designs are shown, each as a top view of $1 mm^2$ segments with five rows of twenty cilia. The small black parts are cilia that are curled upwards from the surface. The A–B–A design will create transverse vortices covering the complete channel width, whereas the C–D–C design will induce two counter-rotating vortices of different size. (Reproduced from den Toonder *et al.*[4])

herringbone mixer.[26] In both designs, the basic idea is that mixing should occur by the repeated change in transverse flow as the fluid travels through the channel. Two consecutive segments make up one 'cycle'. A sufficient number of cycles should be passed for complete mixing to occur.[27] For the staggered herringbone mixer, for example, this number is between four and ten cycles, depending on the flow conditions.

We carried out flow visualization experiments using silicone oil with two colours. The fluid entering through the two inlets was coloured with a red and a blue dye, respectively. The main flow rate, driven by the external pumps, was varied between $10\,\mu L\,min^{-1}$ and $60\,\mu L\,min^{-1}$, corresponding to mean velocities between $0.33\,mm\,s^{-1}$ and $2\,mm\,s^{-1}$. The experiments were done with silicone oils of two viscosities, namely 9.3 mPa·s and 0.93 mPa·s. These conditions correspond to global Reynolds numbers between 0.03 and 2. The global Reynolds number is defined as $Re = \rho UL/\eta$, in which U is the mean velocity through the channel, L is a characteristic dimension representing the channel cross section, for which we take the channel width, and ρ and η are the fluid viscosity and density, respectively.

With the artificial cilia switched off, the two differently coloured fluid streams remained separated and did not mix. This proved that the non-actuated cilia did not influence the flow and that diffusion was not important.[28] After switching on the cilia, all measurements showed efficient mixing. Figure 5.6 shows snapshots from a mixing visualization experiment with the low-viscosity silicone oil. The lay-out corresponds to the A–B–A design of Figure 5.5. The externally driven flow is from left to right. The flow rate is 60 μL min^{-1} which corresponds to a mean velocity of 2 mm s^{-1} and $Re = 2$. For $t < 0$ s, the two separate coloured fluid streams are clearly visible. At time $t = 0$ s all artificial cilia were switched on with a frequency of 50 Hz at an AC voltage of 100 V kHz^{-1}. A meandering flow pattern occurred within 0.04 s, it evolved, and the fluid was observed to be homogeneously mixed within 1.5 s. During this time, the fluid travelled not even 1.5 cycles through the channel, which is significantly less than what is achieved by known passive groove-type mixers.[25]

Figure 5.7 shows the mixing results for the higher-viscosity silicone oil, and the mixing pattern design C–D–C of Figure 5.5, analogous to the mixing results shown in Figure 5.6 for the lower viscosity fluid and design A–B–A. Now, the flow rate is 10 μL min^{-1} which corresponds to a mean velocity of 0.33 mm s^{-1}

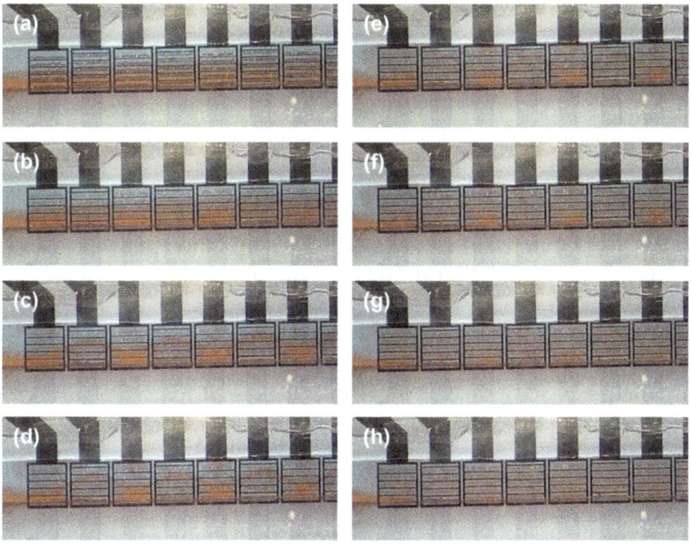

Figure 5.6 Snapshots from a mixing experiment using dyed silicone oils (viscosity 0.93 mPa·s) and mixing configuration design A–B–A from Figure 5.3. The externally driven main mean velocity (from left to right) is 2 mm s^{-1}. At time $t = 0$ s the artificial cilia are switched on with a frequency of 50 Hz at an AC voltage of 100 V kHz^{-1}. (a) Not actuated. (b) $t = 0.00$ s. (c) $t = 0.04$ s. (d) $t = 0.08$ s. (e) $t = 0.12$ s. (f) $t = 0.16$ s. (g) $t = 0.44$ s. (h) $t = 1.44$ s. A meandering flow pattern almost immediately occurs, and within 1.5 s, the fluid is homogeneously mixed. This time corresponds to less than 1.5 cycles travelling distance in the main flow direction. (Reproduced from den Toonder *et al.*[4])

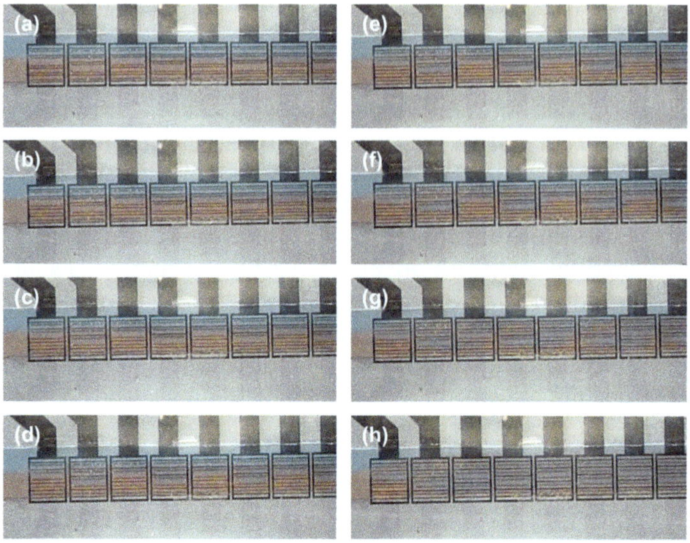

Figure 5.7 Snapshots from a mixing visualization experiment using dyed silicone oils, as in Figure 5.4, but here, the viscosity is 9.3 mPa·s and the mixing configuration is design C–D–C (Figure 5.3). The externally driven main mean velocity (from left to right) is 0.33 mm s^{-1}. At time $t=0$ s the artificial cilia are switched on with a frequency of 50 Hz at an AC voltage of 100 V kHz^{-1}. (a) Not actuated. (b) $t=0.00$ s. (c) $t=0.24$ s. (d) $t=0.48$ s. (e) $t=0.72$ s. (f) $t=0.96$ s. (g) $t=2.68$ s. (h) $t=8.76$ s. Although more slowly than in Figure 5.4, a meandering flow pattern emerges and the fluid is reasonably homogenized after about 9 s, which corresponds to about 1.5 cycles travelling distance, as for Figure 5.4.

and $Re=0.03$. Also in this case, a meandering flow pattern emerges after switching on the artificial cilia (at 50 Hz switching time, with an AC voltage of 100 V kHz^{-1}), although more slowly than in Figure 5.6. After about 9 s, the fluids are reasonably homogenized. In terms of number of cycles travelled through the channel, this time corresponds to less than 1.5 cycles, just as for the experiment of Figure 5.6. Again, this is faster than expected on the basis of passive groove-type mixers.

The artificial cilia create very efficient mixing in the micro-channel. However, it seems that, for mixing to occur by our artificial cilia, it is not required that the fluid travels through the channel and passes through a multiple number of repetitive mixing cycles. On the contrary, mixing seems to happen quite locally over just one cycle. The reason may be that the cilia do not generate a single, homogeneous vortex with one rotation rate, but a more complex flow involving more than one length- and time-scale. This idea is supported by examining the dimensionless Stokes number $St = \rho L^2/\eta t_c$, in which t_c is a characteristic (instationary) timescale. The Stokes number represents the ratio of unsteady effects over viscous effects. If $St \ll 1$, then unsteady effects can be neglected and the flow can be considered to be (quasi-) stationary. As a characteristic timescale t_c we choose the reciprocal switching frequency, which is 1/50 s in

Figure 5.6. As the characteristic length L we take the channel width, *i.e.* $L = 1$ mm. A value of $St = 50$ is then found for the low viscosity silicone oil used in Figure 5.6. For the high viscosity oil (Figure 5.7) the value is $St = 5$. Both are significantly larger than 1, indicating that unsteady effects can be substantial. This suggests that the flow generated by the artificial cilia is composed of a combination of a global vortex with a superposed secondary oscillating flow. We attribute the observed effective mixing to this combination of time- and length-scales.[29]

5.2.4 Optical Coherence Tomography Experiments

Optical coherence tomography makes it possible to observe flow patterns in cross sections of the micro-channel, perpendicular to the main flow direction. These experiments were done with the Y-shaped channel in which, similar to the mixing visualization experiments, the main flow was driven by two syringe pumps connected to the two inlets. A clear silicone oil (AK10, $\eta = 9.3$ mPa·s and $\rho = 930$ kg m^{-3}) was entered in one inlet, and the same liquid with dispersed TiO_2 particles was entered in the other inlet.

Figure 5.8 shows the results. The main flow rate is 6 μL min^{-1}, corresponding to a mean velocity of 0.2 mm s^{-1} and $Re = 0.02$. The artificial cilia are actuated

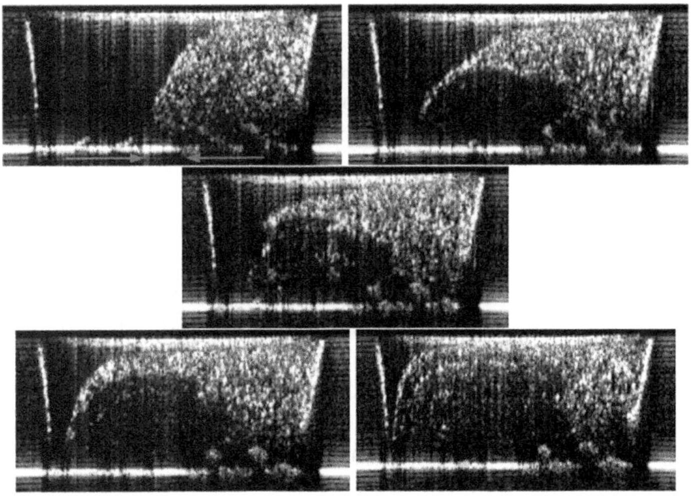

Figure 5.8 Snapshots from the OCT experiment. The images show a cross section halfway across the channel, with five cilia at the bottom. Three cilia are moving in one direction, the two other opposite to that as indicated by the arrows in the upper left image (design C–D–C of Figure 5.3). A mean velocity of 0.2 mm s^{-1} is generated by an external pump (perpendicular to the images). The fluid is silicone oil ($\eta = 9.3$ mPa·s and $\rho = 930$ kg m^{-3}) and the seeding particles are TiO_2. The cilia are actuated with a switching frequency of 80 Hz and a driving AC voltage of 90 V kHz^{-1}. Top: left: MEMS switched off; right: image at $t = 3$ s; middle: at $t = 4.5$ s; bottom: left: image at $t = 7.5$ s; right: at $t = 16$ s.

with a switching frequency of 80 Hz and a driving AC voltage of $90\,\mathrm{V\,kHz^{-1}}$. The cilia design C–D–C (Figure 5.5) is used. The cross section is positioned about halfway across the channel. Various snapshots, taken at different times after switching on the actuation voltage at time zero, are shown in the figure. The five cilia, present at the bottom of the channel, can be partly seen. As indicated in the upper left picture (time zero), three of the cilia are actuated in one direction and the two others are actuated in the opposite direction. After switching on the actuation, a flow pattern emerges. The TiO_2 particles, present in the right half of the cross section at time zero, are distributed over the complete cross section over time. This happens through a global anti-clockwise vortical motion; however, it is clear that, at a certain point in time, other vortices appear that are counter-rotating to the original global vortex, and smaller flow structures can be seen as well. This supports our earlier hypothesis that the flow generated by the artificial cilia is composed of a combination of a global vortex with a superposed secondary oscillating flow.

5.3 Uncovering the Mechanism of Fluid Manipulation Using Electrostatic Artificial Cilia

Clearly, our electrostatic artificial cilia are very effective in producing fluid flow. This raises the question as to the basic mechanism causing the flow. In small systems such as the present one, it is expected that Stokes flow conditions prevail, which means that inertia does not play a significant role, and the flow is dominated by viscous effects. This is expressed by the Reynolds number being very small, *i.e.* $Re \ll 1$. The flow is then completely reversible in time,[30] and in this situation the motion of the cilia must be asymmetric in order to have a net effect on the flow, *i.e.* the shape of the cilia must be different when moving in one direction than in the other one. If this would not be the case, after one cycle the fluid would end up exactly at the same location as where it started from, and the flow would just be oscillatory. Indeed, the cilia found in nature move in an asymmetric manner, producing an effective stroke and a recovery stroke. We should stress that the asymmetry should occur in the *shape* and not in the *time* of the motion. In other words, moving slowly in one direction and fast in the other one will not help in producing a net velocity under Stokes flow conditions.

However, as we have seen, a close look at our artificial cilia using a high-speed camera revealed that there is hardly any asymmetry in their motion, although the times of rolling out and rolling back differ significantly. Despite their resemblance to natural cilia, the mechanism of fluid manipulation by our artificial cilia seems therefore to be different. An explanation can be obtained by considering the *local* Reynolds number $Re_C = \rho U_C L_C / \eta$ of the flow, *i.e.* at the scale and location of the cilia. A characteristic size of our cilia is $L_C = 100\,\mu\mathrm{m}$ (*i.e.* their length), and the viscosity and density of the silicone oil are $\eta = 9.3$ mPa·s and $\rho = 930\,\mathrm{kg\,m^{-3}}$. Let us now base the characteristic velocity near the cilia U_C on the cilia length (100 μm) and the rolling-out time t_{ro} (which is about 1 ms or less) *i.e.* $U_C = L_C/t_{ro} = 100 \times 10^{-6}/1 \times 10^{-3} = 0.1\,\mathrm{m\,s^{-1}}$. This leads to a local Reynolds number of $Re_C \geq 1$. This means that, *locally*, the Stokes flow condition is not obeyed and inertia may well play a role, in which case asymmetry in time will indeed produce a net effect.[31] Hence, due to inertia, the flow obtained during the rolling-out phase is not cancelled out during the rolling-back phase. This can also explain the dependency of the induced flow velocity on the magnitude of the driving voltage at equal switching frequency, since the rolling-out time decreases for higher voltages, but the rolling-back time remains unaltered, as was observed from Figure 5.3.

We used numerical simulation to test the hypothesis of inertial effects being the determining factor for the flow manipulation effectiveness of our artificial cilia, as well as to study the mixing patterns in the cross sections of the channel, and compare these to the OCT results of section 5.2.4.

5.3.1 The Numerical Model

For computational reasons a two-dimensional cross section of the channel is modelled. This implies that the cilia are assumed infinitely long in the imposed flow direction. The channel and the modelled domain are depicted in Figure 5.9. As in the experiments, the channel has a height of 0.5 mm, and a width of 1.0 mm. Each cilium has a thickness t of 2 μm, and length L of 100 μm. Although in reality the artificial cilia are layered structures, with different mechanical properties, the entire cilium is modelled as a single solid with the properties of the thickest layer, which is polyimide. The initial curled shape with radius $R = 32$ μm of the cilia is shown in Figure 5.9. The cilia are attached to the channel floor at one end.

The fluid domain Ω_f is modelled as an incompressible Newtonian fluid, and the solid domain Ω_s as an incompressible, inertia less, neo-Hookean solid. The fluid is modelled in an Eulerian manner, the solid in a Lagrangian

Figure 5.9 The modelled cross section of the micro-fluidic channel, containing three cilia moving to the right and two moving to the left when actuated. (Reproduced from Baltussen *et al.*[5])

manner. This leads to the following conservation equations of mass and momentum:

$$\rho\left(\frac{\partial \vec{u}}{\partial t}+\vec{u}\cdot\vec{\nabla}\vec{u}\right)=\vec{\nabla}\cdot 2\eta D-\vec{\nabla}p_{\mathrm{f}} \quad \text{in } \Omega_{\mathrm{f}}$$

$$\vec{\nabla}\cdot\vec{u}=0 \quad \text{in } \Omega_{\mathrm{f}}$$

$$-G\vec{\nabla}\cdot(B-I)=-\vec{\nabla}p_{\mathrm{s}}+\vec{f}_{\mathrm{app}} \quad \text{in } \Omega_{\mathrm{s}}$$

$$J-1=0 \quad \text{in } \Omega_{\mathrm{s}} \tag{5.1}$$

where ρ is the fluid density, $\vec{u}$ the fluid velocity, η the viscosity, $D=1/2\left(\vec{\nabla}\vec{u}^{T}+\vec{\nabla}\vec{u}\right)$ the rate of deformation tensor, p_{f} the fluid pressure, G the modulus, $B=F\cdot F^{T}$ the Finger tensor, I is the unit tensor, p_{s} the solid pressure, $\vec{f}_{\mathrm{app}}$ the applied force which drives the system, and $J=\det F$. The deformation gradient tensor is defined as: $F=\vec{\nabla}\vec{d}$, where $\vec{d}$ is the solid displacement.

No-slip boundary conditions are taken on the channel walls and the cilia are attached to the wall at the lowest vertical boundary, *i.e.* the channel floor. On the fluid/solid boundary Γ_{fsi} the following kinematic and dynamic conditions hold:

$$\vec{u}-\dot{\vec{d}}=\vec{0} \quad \text{at } \Gamma_{\mathrm{fsi}}$$

$$\vec{f}_{\mathrm{f}}-\vec{f}_{\mathrm{s}}=\vec{0} \quad \text{at } \Gamma_{\mathrm{fsi}} \tag{5.2}$$

where $\dot{\vec{d}}$ is the time derivative of the displacement vector, $\vec{f}_{\mathrm{f}}$ is the traction force on the fluid, and $\vec{f}_{\mathrm{s}}$ is the traction exerted on the solid. By applying the kinematic interface condition as a constraint on both the fluid as solid momentum equation, the dynamic condition is automatically satisfied, since the introduced Lagrangian multiplier represents the force on either continuum.

The cilium is actuated electrostatically. The electrostatic surface force is modelled by assuming a parallel plate capacitor geometry, where the upper plate with potential ΔV is the cilium and the lower, grounded plate the channel floor. An electric field $\vec{E}=-(\Delta V/h)\vec{e}_{y}$ exists between the upper and lower plate, with h the distance between the upper and lower plate. This electric field causes a traction on both conducting plates, which attracts the cilium towards the floor so that it rolls out onto the floor. By making use of the Gauss divergence theorem and the Maxwell stress tensor this traction on the upper plate is given by:[32]

$$\vec{t}=\vec{n}\cdot\sigma_{M}=\frac{\vec{n}\varepsilon}{2\pi}\left(\vec{E}\vec{E}-\frac{\vec{E}\cdot\vec{E}}{2}I\right)=\frac{-\varepsilon}{8\pi}\left(\frac{\Delta V}{h}\right)^{2}\vec{e}_{y} \tag{5.3}$$

where ε is the dielectric constant of the fluid between the two plates, $\vec{n} = -\vec{e}_y$ the vector normal to the upper plate. For numerical reasons, the traction on the boundary is transformed into a body force by assuming a homogeneous distribution over the cilia thickness t, leading to:

$$\vec{f}_{\text{app}} = \frac{-\varepsilon}{8\pi t}\left(\frac{\Delta V}{h}\right)^2 \vec{e}_y \tag{5.4}$$

One full cycle of the cilium motion will be simulated. The actuation force will be applied until the cilia are fully unrolled, so flat against the lower wall. Then the force is released and the cilia recover elastically to their initial shape. The cycle is finished when the fluid has come to a rest. The velocity field obtained from this cycle is used as input for particle tracking for 50 cycles.

5.3.1.1 Scaling

The problem is governed by two dimensionless groups, namely the local Reynolds number (Re_C, defined by the local cilia properties as explained earlier) and R defined as the ratio between elastic and applied forces in the cilium $R = G/Lf_0$, where $f_0 = 1.75 \times 10^9$ N m^{-3} is the typical force computed from the following parameters: $\varepsilon = 8.854 \times 10^{-12}$A^2 s^4 kg^{-1}m^{-3}, $\Delta V = 100$ V, $h = 1$ µm, $t = 2$ µm. The typical length scale h for the applied force is taken equal to the dielectric layer thickness of the experimental device. The fluid has a viscosity of 9.3 mPa s. In the numerical simulations, the Reynolds number is varied from 0 to 10 by changing the density between 0 and 9300 kg m^{-3}. The solid has a modulus of 1 GPa, hence $R = 5.7 \times 10^3$ throughout all calculations.

5.3.1.2 Mixing Efficiency

The distributive mixing is modelled in a qualitative manner by convecting particles. In order to measure the mixing performance a more quantitative measure has to be used. Here the Shannon entropy increase rate κ is used,[33] which is a measure for the loss of information in the system with respect to the initial information. The entire domain is subdivided into n_c cells and the number of particles in each cell n_i is counted. From this information the mixing entropy S (eqn (5.5))[34] is computed. If all particles are homogeneously distributed the entropy is maximal, being $S_{\max} = \log n_c$:

$$S = \frac{-1}{n_{\text{part}}}\sum_{i=1}^{n_c} n_i \log \frac{n_i}{n_{\text{part}}}$$

$$\kappa = \frac{S - S_0}{S_{\max} - S_0} \tag{5.5}$$

where n_{part} is the total number of particles, n_i is the number of particles in cell i, and S_0 is the entropy of the initial state. By choosing a typical cell size, the entropy of mixing is computed for that specific length-scale, so mixing effectiveness can be computed at different length-scales with great ease. Since the device should create fluid layers which mix diffusively in a typical timescale t_{typ} of 1 s, and the diffusion coefficient D of ions in water is in the order of 10^{-9} m^2 s^{-1} the diffusion length scale is $l_d = \sqrt{Dt_{typ}} = 32$ μm, resulting in 30 cells in the x-direction and 15 in the y-direction.

5.3.2 Results

Net fluid motion for Stokes flow only occurs if the cilia motion is asymmetric as explained earlier. The asymmetry of the cilia movement can be visualized by plotting the path of the very tip of the cilium over one cycle. If the rolling path and rolling-back path do not coincide, the motion is asymmetric. The size of the area enclosed by both paths, the swept area, is a measure for extent of the asymmetry. The direction of fluid motion can also deducted from this plot. For inertial flow this is not the case, since also the time scales of rolling and rolling back play a role. The simulated tip position of the left most cilium within the computational domain is given in Figure 5.10 for $Re_C = 0$ (left) and $Re_C = 10$ (right).

From Figure 5.10 it is observed that for $Re_C = 0$ the net flow will be to the left for this cilium, since the path during elastic recovery lies above the path during unrolling. The amount of fluid which is displaced is relatively small compared to the entire swept area. For $Re_C = 10$, the net swept area increases but not substantially. Furthermore the flow direction is unknown, since also the timescales of unrolling and recovering play a role.

The distributive mixing properties for different Reynolds numbers are studied next. A rectangular block containing 2500 particles is placed in the right half of the domain, see Figure 5.11a. These particles are undergo convection by the flow for 50 cycles of cilia actuation and their final positions are plotted in Figure 5.11 for various local Reynolds numbers.

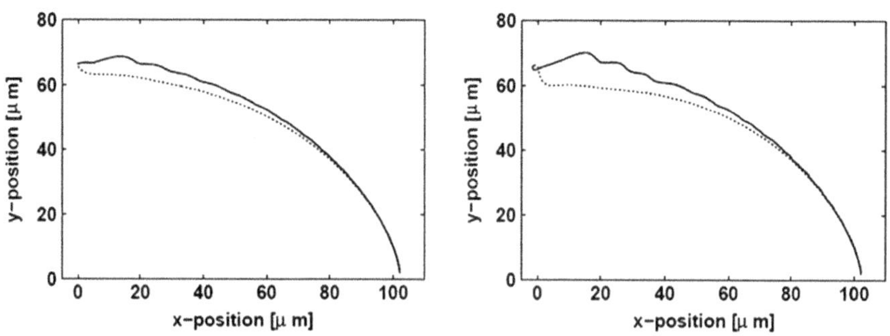

Figure 5.10 Simulated tip position of the leftmost cilium for $Re = 0$ (left) and $Re = 10$ (right). The solid line is the unrolling path when the electric force is switched on, the dashed line is the rolling path during which the force is switched off. The left most position is the starting position. (Reproduced from Baltussen *et al.*[5])

The distribution of particles over the domain becomes better for higher Reynolds number, indicating better distributive mixing. The motion of particles for $Re_C = 0$, 0.1 is in the opposite direction to that of the $Re_C = 1$, 5, 10 simulations. Therefore a Reynolds number exists where there is no global flow, and hence mixing only occurs on the local level. This is the critical Re_C where the flow changes direction due to the moving cilia. Two flow domains can be seen, one for the three left cilia and one for the two right cilia. The distinct change in flow around $Re_C = 1$ is also observed in nature.[35] In order to quantify the mixing effectiveness, the entropy increase rate κ, as defined in eqn (5.5), is computed for a typical domain size of 30×15 cells, and given in Figure 5.12. From this figure it is observed that for $Re_C = 0$, 0.1, κ is negative, indicating structure formation, so we have de-mixing. For $Re_C = 1$, κ remains about

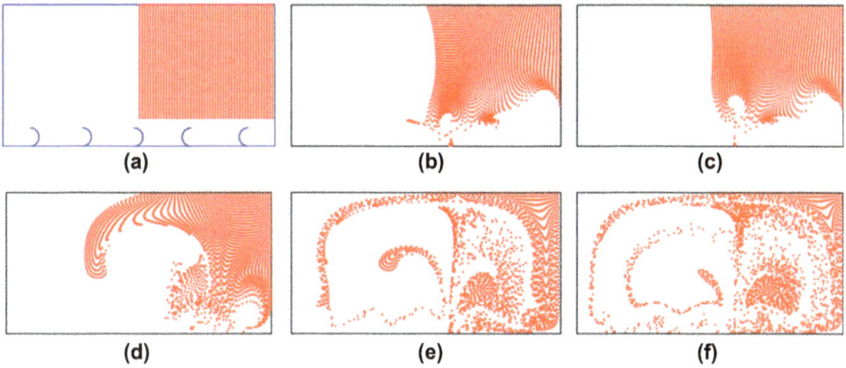

Figure 5.11 Original particle distribution (a). Particle distributions after 50 cycles for $Re = 0$ (b), $Re = 0.1$ (c), $Re = 1$ (d), $Re = 5$ (e) and $Re = 10$ (f). The Reynolds number is based on the cilium length.
(Reproduced from Baltussen *et al.*[5])

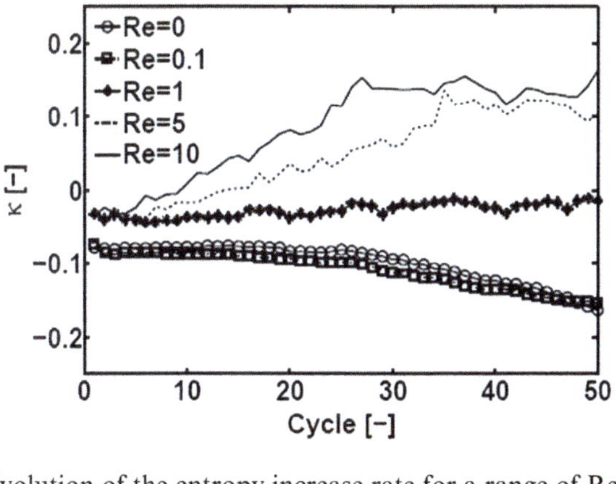

Figure 5.12 Evolution of the entropy increase rate for a range of Reynolds numbers.
(Reproduced from Baltussen *et al.*[5])

constant, so although the particles move, the distribution remains about the same. For $Re_C = 5$, 10, κ increases, indicating good mixing.

The particle distributions for $Re_C = 1$ at 50 cycles, $Re_C = 5$ at 12 cycles and $Re_C = 10$ at five cycles are almost identical, see Figure 5.13. This indicates that the net flow per cycle increases with the Reynolds number, while the basic flow structure remains the same. This means however the same mixing state can be reached for all Re_C above the critical Re_C by just increasing the number of cycles.

Thus far the numerical simulations show that Stokes flow and inertial flow give different flow patterns and that particles are better distributed for higher Reynolds numbers. As mentioned at the beginning of section 5.3, in the experiments, the local Reynolds number Re_C is estimated to be of the order 1, and it was hypothesized that local inertial effects may play an important role. This is indeed verified by comparing the flow structure as measured by OCT with the computed patterns, shown in Figure 5.14. The experimental results resemble the numerical results for high Reynolds numbers, when inertia is

Figure 5.13 Simulated particle distributions for $Re = 1$, 50 cycles (left), $Re = 5$, 12 cycles (middle) and $Re = 10$, 8 cycles (right).
(Reproduced from Baltussen *et al.*[5])

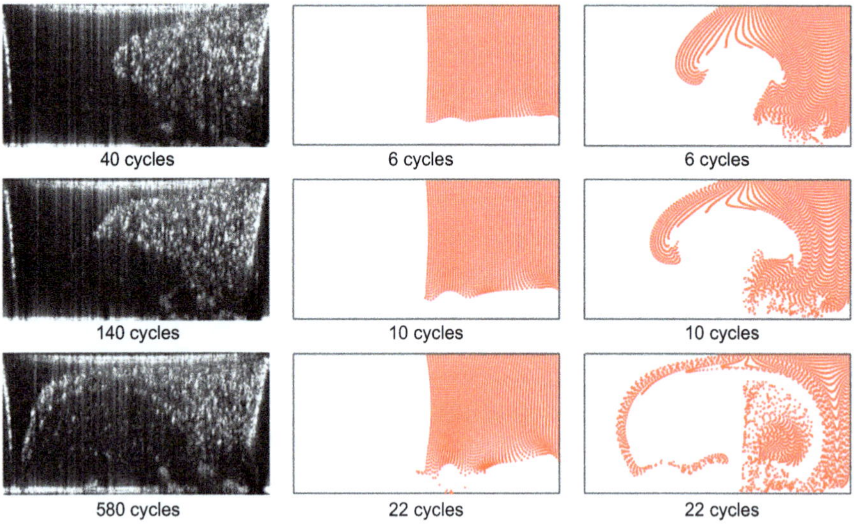

Figure 5.14 Left: images from OCT experiments. Middle: simulated particle distributions for $Re = 0$. Right: simulated particle distributions for $Re = 10$.
(Reproduced from Baltussen *et al.*[5])

important, whereas the simulated patterns for the low Reynolds numbers are completely different. The number of simulated and observed cycles differs, however. Following the previous paragraph, the same mixing state can be obtained for any Re_C above the critical Reynolds number. This indicates that $Re_C = 10$ is higher than the actual Re in experiments, but still the computed particle distributions can be used for comparison with experiments, as long as the comparison is made at a different number of cycles, as is done in Figure 5.14.

Both experiments and simulations show that the initial block of particles is stretched. Stretching alone will not give good mixing however. From simulations it is clear that a separatrix exists between the two cilia groups. Hence fluid in one domain cannot get into the other domain, and good mixing cannot be obtained. Therefore symmetry breaking is necessary,[36] which can be obtained by placing segments with mirrored alternating layouts in the channel, as shown in Figure 5.5. The alternating use of 3–2 and 2–3 configurations shifts the separatrix within the channel, breaking symmetry.

5.4 Fabrication and Experimental and Numerical Methods

5.4.1 Manufacturing of the Artificial Cilia

The artificial cilia are manufactured using a number of process steps, of which three are illustrated in Figure 5.15. The substrate is glass with a 70 nm layer of

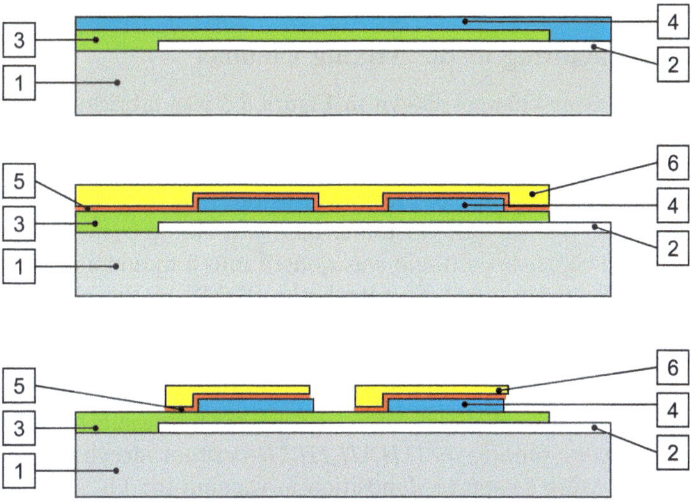

Figure 5.15 Schematic overview of three processing steps in the manufacturing of artificial cilia. Top: Al layer deposition. Middle: PI layer deposition. Bottom: Cr etching. (1) glass substrate, (2) ITO layer, (3) dielectric layer, (4) Al layer, (5) Cr layer, (6) PI layer. (The dimensions are not to scale). (Reproduced from den Toonder *et al.*[4])

indium tin oxide (ITO). The ITO is structured, by wet etching, to obtain the desired electrode pattern. A 1 μm dielectric stack is subsequently deposited. The stack consists of a 400 nm silicon oxide film, a 200 nm silicon nitride film, and another 400 nm silicon oxide film (a so-called ONO-stack). These films are deposited by plasma enhanced chemical vapour deposition (PECVD). A 300 nm aluminium (Al) layer is sputtered on top of the dielectric layer (see Figure 5.15), and subsequently patterned using wet etching. The Al layer will, in a later stage, be removed to free the micro-actuators, *i.e.* it is a sacrificial layer. On the patterned Al, a 20 nm chromium (Cr) layer is sputtered under conditions that lead to a high tensile stress in the Cr. After a UV–ozone surface treatment, a polyimide (PI, Durimide 9005, manufacturer Arch Chemicals Inc.) film is spin-coated on the Cr with a rotation speed of 850 rpm (see Figure 5.15). The PI is photosensitive and can thus be directly patterned in the shape of the micro-actuators using lithography. First, a hotplate pre-bake at 120 °C is applied for 3 min. The PI is then illuminated through a mask using UV (15 s at 9.3 mW cm^{-2}) and developed (TMA238WA developer + H_2O, 50 s). After rinsing with water, the patterned layer is dried (5 min at 110 °C) and cured (1 h at 350 °C). The Cr is then etched (using cerium ammonium nitrate) at the places where it is uncovered by PI, *i.e.* the PI acts as an etching mask (see Figure 5.15). The last step in the process is to etch the Al films from underneath the micro-beam patterns using PES etching, which leaves the Cr unaffected. During this process step, the artificial cilia are released from the surface, and curl upwards due to the internal stress present in the structure, which is caused by the different properties and thicknesses of the PI and the Cr layers.

5.4.2 Manufacturing of the Mixing Channel

The Y-shaped mixing channel shown in Figure 5.5 was fabricated as follows. The artificial cilia were manufactured on a glass substrate as described earlier. A polydimethylsiloxane (PDMS) cover, containing the Y-shaped channel structure, was mounted on top of the glass substrate. The PDMS cap was made from a mixture of 10:1 silicone elastomer base and curing agent (Sylgard 184, Dow Corning). The degassed liquid was poured into a mould and cured in an oven at 80 °C for at least 6 h. To attach the PDMS to the glass substrate, an O_2 plasma treatment (0.48 mbar, 300 W, 1 min) was applied to the PDMS, after which the PDMS cap was manually adjusted onto the glass substrate. To avoid the swelling of the PDMS by the silicone oil (working fluid), the activated PDMS was, immediately after attachment to the substrate, modified with a perfluorinated monolayer (1*H*,1*H*,2*H*,2*H*-perfluorodecyltrichlorosilane; ABCR GmbH), using gas-phase deposition in vacuum for 1 h.

5.4.3 Driving the Artificial Cilia

The actuation of our artificial cilia is done electrostatically by applying an (AC) voltage difference between the ITO electrode and the Cr layer. We used an AC voltage with a frequency of 1 kHz, which was switched on and off with typical switching frequencies between 1 and 200 Hz. The electrical signal was generated by a waveform generator (Agilent, 33220A) in combination with an amplifier (Kroh–Hite Corporation, 7602M). In the experiments reported here, all artificial cilia were addressed simultaneously. An actuation schema as shown in Figure 5.16 is generated by the waveform generator, and subsequently amplified by the amplifier.

The actuation scheme displayed in Figure 5.16, illustrating the 1 kHz AC voltage, switched on and off with switching frequency *f*. The use of an AC voltage should avoid electrolysis and should cause the ions in the liquid not to respond to the electrical field, because they do not have a sufficiently high mobility to follow the field. At the same time, the force acting on the artificial cilia is unaffected by the AC nature, because it only depends on the amplitude and not on the polarity of the voltage.

The AC frequency is independent of the frequency with which the artificial cilia are actuated, which is the second important frequency in the system. This switching frequency ranges from 1 to 200 Hz. In Figure 5.16 the switching frequency is indicated by *f*.

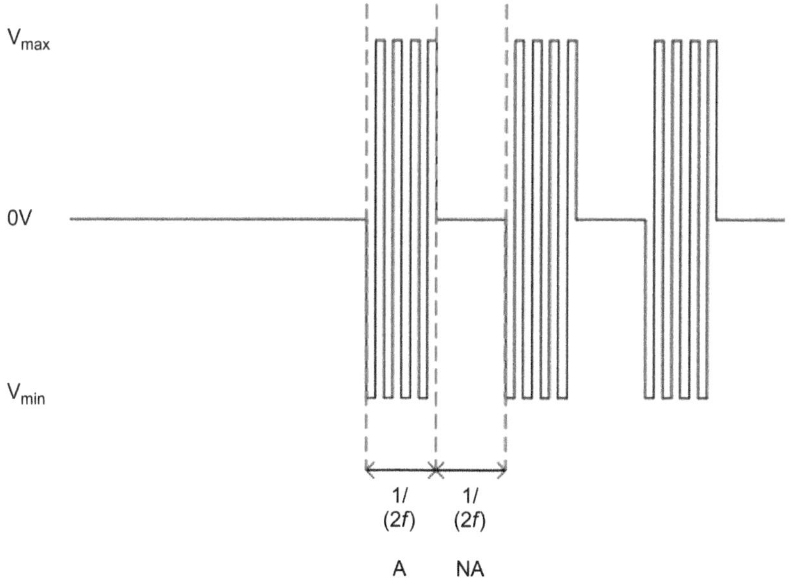

Figure 5.16 Waveform (not to scale) used for artificial cilia actuation. During the time that the cilia are in an actuated state (indicated by 'A'), an AC voltage with a high frequency (1 kHz) is applied. During the time that the artificial cilia are in a non-actuated state (indicated by 'NA'), no voltage is applied. *f* indicates the frequency with which the artificial cilia are actuated, the 'switching frequency'.

5.4.4 High-speed Camera Measurements

To determine the speed of the rolling out and rolling back of the artificial cilia both in air and in silicone oil we used a high-speed camera (a Princeton intensified camera, PI-MAX (chip THM 512×512, with a ST133 controller) mounted on an optical microscope. We focused on the cilia, obtaining a top view as in Figure 5.17. During a rolling-out and rolling-back event, images were recorded at a frame time of 25 μs in silicone oil and 500 ns in air. These images were analysed using a MatLab-based software routine to obtain the rolling times: the position of the tip of one cilium was registered in relation to the pixels making up the image, where the yellow box indicated in the figure contains 300 pixels. The starting position for the tip is approximately halfway along the yellow box. When the cilium unrolls it increasingly fills the yellow box, which is registered by the shift of the black (cilium) to grey (background) transition.

5.4.5 Particle Tracking Measurements

We carried out particle tracking experiments, of which the results are shown in Figure 5.8. The cilia were arranged on the surface in square segments of 1 mm^2, as shown in a top view in Figure 5.4a. The segment contains five columns of 20 cilia, visible in the figure as black rectangles since they are in the curled state. The surface was covered with a 0.5 mm thick silicone oil film (Wacker AK10 silicone oil, viscosity 9.3 mPa·s and density 930 kg m^{-3}), so that the cilia were completely immersed. Two types of particle, dispersed in the fluid, were used to visualize the flow, namely titanium dioxide (TiO_2) particles with a mean

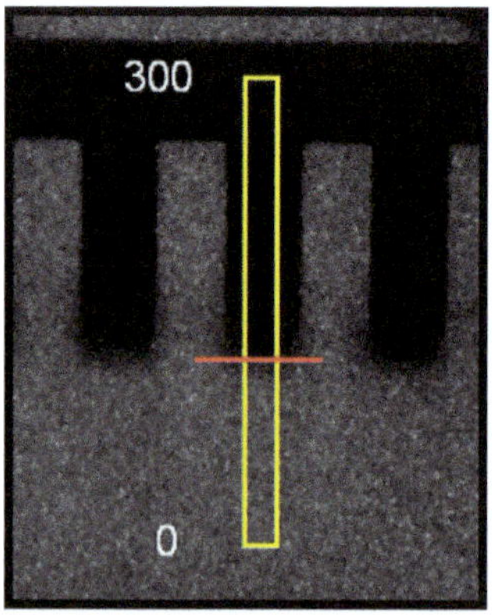

Figure 5.17 Artificial cilia movement registration in silicone oil using a high speed camera. A top view of artificial cilia (black) is visible, in the non-actuated state. The yellow box in the figure has a length of 300 pixels. The starting position for the cilium tip is approximately halfway up the yellow box. When the cilium is actuated, it unrolls and increasingly fills the yellow box. This is analysed to determined the projected position of the cilium tip over time.

diameter of 0.5 μm and hollow glass spheres with an average diameter of 12 μm. The former have a density of 4.5×10^3 kg m^{-3}, and the latter were matched to the density of the silicone oil, *i.e.* 930 kg m^{-3}. Other types of particle were tried as well: silica microspheres, polystyrene, kaolinite. All of these, however, responded substantially to the electrical field, applied to actuate the artificial cilia, by electrophoresis and/or dielectrophoresis, and they were therefore unsuitable for studying the fluid flow velocity since their movement would not represent the actual fluid movement. This effect was minimal for the used TiO_2 and hollow glass particles. Observations were done using an optical microscope. After switching on the actuation voltage, images were taken at 30 frames per second, particle tracking was done manually from the obtained movies, and the induced flow velocities were estimated.

5.4.6 Mixing Visualization Experiments

Mixing visualization experiments were carried out using the Y-shaped mixing channel device described earlier. Two different silicone oils were used in different experiments, *i.e.* Wacker AK10 silicone oil, with a viscosity of 9.3 mPa·s and a density of 930 kg m^{-3}, and Wacker AK1 silicone oil, with a viscosity of 0.93 mPa·s and a density of 930 kg m^{-3}. The fluid, entering through the two inlets, was coloured with a red and a blue dye, respectively, namely Oil Red EGN and Oil Blue N (both Sigma–Aldrich). The dies (in powder form) were added to the silicone oil up to the saturation value; the mixture was subsequently heated at 80 °C for 5 min and filtered using a Teflon filter with a pore size of 0.45 μm. The main flow rate in the channel was driven by two Harvard apparatus Model 11-plus syringe pumps connected to the two inlets, with 2 mL plastic syringes (BD Discardit II). The flow was varied between 10 μL min^{-1} and 6 μL min^{-1} for the higher-viscosity fluid, corresponding to mean velocities between 0.33 mm s^{-1} and 0.2 mm s^{-1}. For the low-viscosity fluid, a flow rate of 60 μL min^{-1} or mean velocity of 2 mm s^{-1} was used. The flow was observed with an optical microscope (Leica MZ6), and images and movies were recorded with a colour camera (CV-S2500, Jai Corporation, Japan) in combination with image analysis software (Pinnacle Studio, version 9.02, Pinnacle Systems).

5.4.7 Optical Coherence Tomography Measurements

The developing flow patterns within a cross section of the channel, perpendicular to the main flow, were visualized by optical coherence tomography (OCT).[37,38] In this procedure particles are illuminated by a low coherent light source. From combining the reflected light beam and a reference beam, an interference pattern is obtained which shows the particle positions. The main advantage of this technique is that it can measure particle positions very deep in translucent objects. OCT-generated time-dependent images of the flow pattern evolution within the same cross-sectional plane as the simulations are recorded.

The Y-shaped channel described earlier was used, with the design C–D–C shown in Figure 5.5. Both of the entries were connected to a syringe pump (Harvard Apparatus, Model 11-plus). One syringe contained silicone oil (Wacker AK10) with a viscosity of 9.3 mPa s and a density of 930 kg m^{-3}, the other syringe contained the same oil seeded with TiO_2 particles.

The Spectral Radar Optical Coherence Tomography System of Thorlabs GmbH was used, which has a depth resolution of 3.5 μm, an in-plane resolution of 10 μm, and which can capture 8 frames per second. Although three-dimensional visualisations are possible, only the cross-sectional flow at a single axial position in the channel was captured.

The experiments were conducted as follows. First the two pumps and the OCT set-up were switched on. The main flow rate was 6 μL min^{-1}, corresponding to a mean velocity of 0.2 mm s^{-1} and $Re = 0.02$. After two distinct separate flows were observed, the artificial cilia were actuated at a frequency of 80 Hz and a voltage of 90 V kHz^{-1}, and a recording was started. After about 15 min the heavier TiO_2 particles sedimented on the cilia, resulting in cilia malfunction and the experiment was stopped. Since the OCT sampling frequency is only 8 Hz, while the cilia actuation frequency is 80 Hz, individual cycles cannot be distinguished.

5.4.8 Numerical Implementation

The continuity, momentum and coupling equations described in section 5.3.1 are solved fully coupled with a finite element discretization within a fictitious domain framework.[18] The nonlinear equations are linearized via the Newton–Rhapson method and time integration of the Navier–Stokes equation is performed with a second-order backward differencing scheme with an initial backward Euler step. The constraint equation (eqn (5.2)) is integrated with the Euler backward scheme. An adaptive time stepping scheme is used. Quadratic test and trial functions are used for both fluid velocity and the solid displacement, linear functions for the pressures as well as for the Lagrangian multiplier. The fluid mesh consists of 12 000 $(N_x, N_y) = (200, 60)$ quadrilateral elements, where N_x is the number of elements in the x-direction and N_y is the number of elements in the y-direction. The fluid mesh is refined near the lower wall. The solid mesh for each cilium consists of 120 $(N_a, N_t) = (30, 4)$ quadrilateral elements, where N_a is the number of elements in the axial direction of the cilium and N_t is the number of elements over the thickness of the cilium.

During the actuation cycle of the cilia, they roll out, eventually lying flat against the lower wall. The cilia are not allowed to cross this lower wall, which can occur during the numerical solution process, however. Therefore the nodes which cross the wall are placed back on the wall. This displacement automatically generates an internal compressive force. When the applied force is released, the cilia detach from the wall automatically due to this compressive force. Due to the cilia placement, no inter cilium contact occurs.

5.5 Conclusions

We have demonstrated that our electrostatic integrated artificial cilia can generate substantial flow velocities in open-faced micro-channels, as well as efficient mixing flows in micro-fluidic channels. The concept, therefore, appears to be quite versatile. The mixing efficiency compares favourably with other passive and active micro-mixers and has a number of key advantages. The micro-mixing concept using artificial cilia has been proven to work in a wide range of Reynolds numbers, and at very high Péclet numbers. We have shown that the mixing length is shorter than for known passive mixing configurations such as the herringbone mixer. The mixing efficiency, expressed in mixing time per volume of mixed liquid, is better than for existing active micro-mixers. With the use of patterned electrodes, the effect can be switched on or off on demand, at locations of choice. These properties make artificial cilia attractive and promising for future integrated micro-fluidic devices in which active fluidic control is required, particularly in lab-on-a-chip devices for (bio-)chemical analysis in which reagents are added or stored either in a liquid or a dry state, and must be mixed quickly in micro-reaction chambers.

To explain the effectiveness of the electrostatic artificial cilia, numerical simulations and experimental visualization of the flow have been carried out. For five different local, cilia-level, Reynolds numbers Re_C, based on the cilium length and speed, a direct numerical simulation of one cilium motion cycle has been performed. The simulation results were compared with optical coherence tomography flow visualizations, and good qualitative comparison was found for the case where $Re_C = 10$. This confirms that the observed flow behaviour in the mixer based on artificial cilia is caused by inertial, rather than viscous phenomena. Although inspired by natural cilia, the working mechanism behind the artificial cilia thus was found to be different from that of the natural cilia that relies on an asymmetric motion.

Acknowledgement

Most of the research summarized in this chapter was carried out by Judith de Goede, Femke Bos and Michiel Baltussen, with substantial contributions from Patrick Anderson, Dick Broer, Laura Filippini, Murray Gillies, Martien Hulsen, Titie Mol, Mireille Reijme, Wim Talen and Hans Wilderbeek. This work was financially supported by the Dutch Polymer Institute (DPI), and the European Union (European project 'Artic', Framework 6, STRP 033274).

References

1. H. Machemer, *J. Comp. Physiol.*, 1974, **92**, 293.
2. M. A. Sleigh, *J. Exp. Biol.*, 1968, **50**, 733–743.
3. M. A. Sleigh, *Cilia and Flagella*, Academic Press, London, 1974.
4. J. den Toonder, F. Bos, D. Broer, L. Filippini, M. Gillies, J. de Goede, T. Mol, M. Reijme, W. Talen, H. Wilderbeek, V. Khatavkar and P. Anderson, *Lab Chip*, 2008, **8**, 533.

5. M. G. H. M. Baltussen, P. D. Anderson, F. M. Bos and J. M. J. den Toonder, Inertial flow effects in a micro-mixer based on artificial cilia, *Lab Chip*, 2009, **9**, 2326.
6. B. A. Evans, A. R. Shields, R. L. Carroll, S. Washburn, M. R. Falvo and R. Superfine, *Nano. Lett.*, 2007, **7**, 1428.
7. A. R. Shields, B. L. Fiser, B. A. Evans, M. R. Falvo, S. Washburn and R. Superfine, *Proc. Natl. Acad. Sci. U.S.A.*, 2010, **107**, 15670.
8. M. Vilfan, A. Potocnik, B. Kavcic, N. Osterman, I. Poberaj, A. Vilfan and D. Babic, *Proc. Natl. Acad. Sci. U.S.A.*, 2010, **107**, 1844.
9. F. Fahrni, M. W. J. Prins and L. J. van Ijzendoorn, *Lab Chip*, 2009, **9**, 3413.
10. J. Belardi, N. Schorr, O. Prucker and J. Rühe, *Adv. Funct. Mater.*, 2011, **21**, 3314.
11. S. N. Khaderi, C. B. Craus, J. Hussong, N. Schorr, J. Belardi, J. Westerweel, O. Prucker, J. Ruehe, J. M. J. den Toonder and P. R. Onck, *Lab Chip*, 2011, **11**, 2002.
12. J. Hussong, N. Schorr, J. Belardi, O. Prucker, J. Ruehe and J. Westerweel, *Lab Chip*, 2011, **11**, 2017.
13. C. E. Sing, L. Schmid, M. F. Schneider, T. Franke and A. Alexander-Katz, *Proc. Natl. Acad. Sci. U.S.A.*, 2010, **107**, 535.
14. J. V. I. Timonen, C. Johans, K. Kontturi, A. Walther, O. Ikkala and R. H. A. Ras, *Appl. Mater. Interface*, 2010, **2**, 2226.
15. C. L. van Oosten, C. W. M. Bastiaansen and D. J. Broer, *Nat. Mater.*, 2009, **8**, 677.
16. L. D. Zarzar, P. Kim and J. Aizenberg, *Adv. Mater.*, 2011, **23**, 1442.
17. B. Pokroy, A. K. Epstein, M. C. M. Persson-Gulda and J. Aizenberg, *Adv. Mater.*, 2009, **21**, 463.
18. K. Oh, B. Smith, S. Devasia, J. J. Riley and J.-H. Chung, *Microfluid Nanofluid*, 2010, **9**, 645.
19. V. Khatavkar, P. D. Anderson, J. M. J. den Toonder and H. E. H. Meijer, *Phys. Fluids*, 2007, **19**, 083605.
20. E. M. Gauger, M. T. Downton and H. Stark, *Eur. Phys. J. E*, 2009, **28**, 231.
21. M. T. Downton and H. Stark, *EPL*, 2009, **85**, 44002.
22. S. N. Khaderi, M. G. H. M. Baltussen, P. D. Anderson, D. Ioan, J. M. J. den Toonder and P. R. Onck, *Phys. Rev. E*, 2009, **79**, 046304.
23. S. N. Khaderi, J. M. J. den Toonder and P. R. Onck, *Langmuir*, 2012, **28**, 7921.
24. S. N. Khaderi, J. M. J. den Toonder and P. R. Onck, *J. Fluid Mech.*, 2011, **688**, 44.
25. R. Ghosh, G. A. Buxton, O. B. Usta, A. C. Balazs and A. Alexeev, *Langmuir*, 2010, **26**, 2963.
26. A. D. Stroock, S. K. W. Dertinger, A. Ajdari, I. Mezic, H. A. Stone and G. M. Whitesides, *Science*, 2002, **295**, 647.
27. The definition of the nature of 'complete mixing' is not straightforward. One could say that a completely mixed system consists of a combination of two or more substances that are homogeneously distributed over the whole

system, down to the molecular scale. For practical applications, it is often sufficient to reach a certain level of homogenization that is coarser, which can be quantified using various measures for mixing. In this chapter, we look at mixing only qualitatively, based on observations of microscopic images. A clear discussion of the subject is given in: C. L. Tucker III, in *Mixing in Polymer Processing*, ed. C. Rauwendaal and M. Dekker, New York, 1991, pp. 101–127.

28. The diffusion coefficient of the dyes in the silicone oil with a viscosity of $\eta = 9.3$ mPa s can be estimated as $D \approx 1 \times 10^{-11}\,\mathrm{m^2\,s^{-1}}$, in the silicone oil with $\eta = 0.93$ mPa s this is $D \approx 1 \times 10^{-10}\,\mathrm{m^2\,s^{-1}}$. For L we take the channel width, and for U the mean axial velocity. This gives Péclet numbers in the range of 20 000 to 33 000. Clearly, diffusion is not expected to play any role.
29. In fact, the mixing efficiency in turbulent flows is due to the existence of a broad, continuous range of length- and time-scales within the flow. Although not turbulent, the mixing in our mixing channel experiments may be caused by a similar generation of multiple length- and time-scales.
30. G. I. Taylor, *Proc. R. Soc. London Ser. A*, 1951, **209**, 447.
31. Applying a similar analysis to the natural cilia shows that the local Reynolds numbers are much lower in that case (namely, much smaller than 1) and inertia is not playing any role.
32. L. D. Landau, E. M. Lifschitz and L. P. Pitaevskii, *Electrodynamics of Continuous Media*, Butterworth Heinemann, Oxford, 1984.
33. T. G. Kang, M. A. Hulsen, P. D. Anderson, J. M. J. den Toonder and H. E. H. Meijer, *Chem. Eng. Sci.*, 2007, **62**, 6677.
34. C. E. Shannon, *The Bell System Tech. J.*, 1948, **27**, 379–423, 623–656.
35. S. Childress and R. Dudley, *J. Fluid Mech.*, 2004, **498**, 257.
36. J. G. Franjione, C. Leong and J. M. Ottino, *Phys. Fluids A*, 1989, **1**, 1772.
37. C. Xi, D. L. Marks, D. S. Parikh, L. Raskin and A. Boppart, *Proc. Natl. Acad. Sci. U.S.A.*, 2004, **101**, 7516.
38. Y. C. Ahn, W. Jung and Z. Chen, *Lab Chip*, 2008, **8**, 125.

CHAPTER 6

Ferromagnetic Cilia

FRANCIS FAHRNI*[a,b] AND LEO J. VAN IJZENDOORN[a]

[a] Eindhoven University of Technology, P.O. Box 513, 5600 MB Eindhoven, The Netherlands; [b] Dutch Polymer Institute (DPI), P.O. Box 902, 5600 AX Eindhoven, The Netherlands
*Email: francis.fahrni@gmail.com

6.1 Introduction

Magnetic actuation is often considered not to be efficient on small scales because magnetic forces are a volume effect, in contrast to electrostatic forces, for example, which scale with the area. This (partially true) fact, together with the readily available manufacturing technologies and materials from the microelectronics industry, gave rise to electrostatic micro-actuation, commonly known as the research field of MEMS since the late 1980s. Magnetic layers only started to be routinely applied in magnetic sensors during the last decade and therefore magnetic micro-systems have not been widely explored.[1,2] But, as we shall see later in this chapter, this scaling argument which disfavours magnetic actuation on small scales is over-simplified (see also arguments by Cugat *et al.*[2]).

Magnetic micro-actuators based on MEMS fabrication technologies are hence not abundant in literature but have been proposed by several authors since the early 1990s, based on actuation by magnetic torque[3–12] and magnetic gradient force.[13–15] The major drawback of such magnetic MEMS-like actuators is that the fabrication concepts are very complicated, prohibiting their wide use for providing micro-fluidic functions. In an effort to apply magnetic micro-actuators to micro-fluidics and make them more cost-effective, several authors have proposed devices with miniature magnets embedded in soft elastomers.[16–21] The development of magnetic polymer composites

RSC Nanoscience & Nanotechnology No. 30
Artificial Cilia
Edited by Jaap M J den Toonder and Patrick R Onck

Published by the Royal Society of Chemistry, www.rsc.org

together with their micro-structuring has only been explored, however, in the last couple of years. Recently, magnetic actuation of artificial cilia made of polymer–ferrofluid composites have been reported.[22–24]

Polymers are a natural choice for artificial cilia, with advantages that (1) they offer a broad range of processing conditions, making them for example suited for cost-effective mass-patterning methods such as micro-imprinting or inkjet printing; (2) they can be modified chemically to provide new functionalities; and (3) they generally have a lower elastic modulus than inorganic materials, providing facilitated deflection of artificial cilia. An important advantage in the use of magnetic actuation for artificial cilia intended for fluid manipulations is that magnetic fields with well-defined magnitude and orientation can be applied without any perturbation by the (bio-)chemical fluid.[25] The latter is mostly not true for other stimuli such as electrostatic, thermal or optical.

Commonly, for artificial cilia or other applications, the choice is to disperse super-paramagnetic particles in polymers, and not ferromagnetic particles, because the synthesis and stabilisation of those particles with surfactants is well known in literature.[26] Stabilised super-paramagnetic particles are even widely available commercially.[27] They have a (time-averaged) magnetization of zero in the absence of magnetic fields and therefore magnetic inter-particle interactions do not hinder uniform dispersion in fluids or polymers, while other (molecular) interactions can be overcome by using suitable surfactants for creating steric repulsions. A successful attempt at creating a nanoscale homogeneous super-paramagnetic polymer is described by Evans *et al.*[28] Because ferromagnetic particles possess a remanent magnetization they are inherently more difficult to disperse homogeneously. Few ferromagnetic polymers have been reported in literature and as soon as the particle loading is substantial (2–5 vol%) agglomerates of 10 μm or larger are present.[29–32] Those ferromagnetic polymers have, however, not (yet) been applied as micro-actuators or artificial cilia. Our own material also suffers from agglomerates, which currently limits the miniaturization and reproducibility of devices. A different approach has been to grow artificial ferromagnetic cilia from cobalt particles under influence of a magnetic field, with only few polymer molecules to bind the particles. This study[33] resulted in elastic moduli being largely influences by the particles (~2.6 GPa), several orders of magnitude larger than for our case or for other reported magnetic elastomers.

In this chapter we first introduce the actuation by magnetic torque and magnetic gradient force, and then derive the scaling relations related to both actuation mechanisms. We compare both mechanisms and highlight the differences in terms of actuation efficiency between ferromagnetic and super-paramagnetic materials. Finally, we demonstrate ferromagnetic polymer artificial cilia with large amplitudes of motion. Their fabrication is presented as well as their actuation in fluid up to beating frequencies of 50 Hz. Some advantages of their ferromagnetic nature are discussed, namely an asymmetric movement involving torsion—being tuneable by choosing the initial magnetization—and a selective addressing being dependent on the direction of rotation of the magnetic field. The artificial cilia have been tested in a micro-fluidic chamber and induce vortices as well as linear fluid transport on the scale of hundreds of micrometres.

6.2 (Ferro-)magnetic Actuation and Scaling Behaviours

An artificial magnetic cilium can be set in motion with either a magnetic gradient force or with a magnetic torque. The scaling behaviour of these two types of actuation is very different.

Considering a body of volume V over which magnetic actuation is exerted, the magnetic force is given by the following integral over the volume of the body:[34]

$$\vec{F} = \mu_0 \int \left(\vec{M} \cdot \vec{\nabla}\right) \vec{H}_0 \, \mathrm{d}V \tag{6.1}$$

and the magnetic torque similarly by:

$$\vec{\tau} = \mu_0 \int \left(\vec{M} \times \vec{H}_0\right) \mathrm{d}V \tag{6.2}$$

with μ_0 the permeability of vacuum, M the magnetization of the body and H_0 the magnetic field in the absence of the body. The magnetic force reads as the gradient of the applied magnetic field integrated over the volume of the body and is maximum if the magnetization is aligned with the applied magnetic field. It is therefore called a gradient force and attracts a magnetic body towards regions of higher field intensities if the magnetization of the body is in the same direction as the magnetic field, this always being the case for (super-)paramagnetic materials. If the magnetization is in a direction opposed to the magnetic field, then the body is repulsed from regions of high field intensities. Note that a ferromagnetic material is needed for this latter situation to occur, and there would generally also be a large magnetic torque involved. The magnetic torque reads as the cross-product of the magnetization with the applied magnetic field, integrated over the volume. The torque aligns the remanent magnetization of a ferromagnetic material with the direction of the applied magnetic field, and is maximal when the magnetization is perpendicular to the applied magnetic field. For (super-)paramagnetic materials, the torque aligns the long direction of the body with the direction of the applied magnetic field, which is a result of shape anisotropy inducing a larger magnetization in the long direction than in the direction perpendicular to it (and therefore the cross-product of eqn (6.2) is not zero). The total force acting on a magnetic body in an applied magnetic field is the sum of the gradient force and the torque. In most cases however, either the gradient force or the torque is dominating.

6.2.1 Torque

It can be shown that when integrating a constant (magnetic) torque density over the length of a cantilever, the sum can be viewed as a force acting on the end:[35]

$$F_{\mathrm{magn}} = \frac{\tau_{\mathrm{magn}}}{L} \tag{6.3}$$

The mechanical reaction force of the cantilever is given by:[35]

$$F_{\text{mech}} = \frac{\delta\, EWT^3}{4\, L^3} \tag{6.4}$$

with δ the deflection of the cantilever at its tip, E its elastic modulus and W, T and L its geometrical dimensions according to Figure 6.3. For a *ferromagnetic cilium* the ideal actuation case occurs when the remanent magnetization M_r is perpendicular to the applied magnetic field H_0, and eqn (6.2) to eqn (6.4) can be written as:

$$F_{\text{magn}} = \frac{\mu_0 M_r H_0 V}{L} \tag{6.5}$$

Both the mechanical and the magnetic force scale with the area. An important consequence is that the efficiency of actuation, characterised by the relative deflection:

$$\left.\frac{\delta}{L}\right|_{\text{ferromagn torque}} = \frac{4\mu_0 M_r H_0}{E}\frac{L^2}{T^2} \tag{6.6}$$

does not scale with size. Hence the same macroscopic applied magnetic field can be used for actuation, irrespective of how small the manufactured cilia. The same holds true for magnetic actuation by torque on a (*super*)-*paramagnetic cilium*. The magnetization of the cilium will become depend on the applied magnetic field however, but efficiency will still remain independent of the size. A quantitative comparison will be made in section 6.2.4.

6.2.2 Gradient Force

The actuation by magnetic gradient force does not have a neutral scaling behaviour, as was shown to be the case for actuation by torque in section 6.2.1. The magnetic gradient force in the case of a *ferromagnetic material* can be calculated (in a maximized situation where the remanent magnetization, the applied magnetic field and magnetic field gradient are all three perpendicular to the cilium) from eqn (6.1) as being:

$$F_{\text{ferromagn}} = \mu_0 M_r V \frac{\partial H_0}{\partial z} \tag{6.7}$$

with z the axis perpendicular to the cilium. For a (*super-*)*paramagnetic material*, this force becomes:

$$F_{\text{paramagn}} = \frac{1}{2}\mu_0 \chi V \frac{\partial\left(H_0^2\right)}{\partial z} \tag{6.8}$$

with χ the magnetic susceptibility of the material. In the case of a macroscopically applied magnetic field gradient (*i.e.* constant over the volume of the cilium), these forces act uniformly on the volume of the cilia and the mechanical reaction force is given by:[35]

$$F_{\text{mech}} = \frac{2\delta\, EWT^3}{3\, L^3}. \tag{6.9}$$

The efficiencies of actuation can thus be characterized by:

$$\left.\frac{\delta}{L}\right|_{\text{ferro gradient}} = \frac{3\mu_0 M_r}{2E}\frac{L^3}{T^2}\frac{\partial H_0}{\partial z} \quad \text{and} \quad \left.\frac{\delta}{L}\right|_{\text{para gradient}} = \frac{3\mu_0 \chi}{4E}\frac{L^3}{T^2}\frac{\partial\left(H_0^2\right)}{\partial z} \tag{6.10}$$

for ferromagnetic respectively (super-)paramagnetic materials. These expressions scale with the first power of the dimension of the cilium. Consequently, for a same external stimulus, the magnetic actuation will become linearly less efficient for smaller cilia. It will be shown quantitatively in section 6.2.3 that—for a same cilium and with external actuation—actuation by magnetic torque is favoured over actuation by magnetic gradient force on the micro-scale.

Conservation of a high actuation efficiency for small sizes would require the magnetic field gradient to scale inversely proportional to the dimension of the cilia, which would be the case if the means of generating the magnetic field—commonly an electromagnet or a moving permanent magnet—is scaling together with the cilia (when reducing the size of a (electro)magnet the field gradient increases at the same rate). But miniaturizing electromagnets or miniaturizing moving permanent magnets is mostly unpractical. However, a practical way of achieving high magnetic field gradients on small scales is by running an electrical current through a micro-fabricated wire. The magnetic field of a current wire is given by:

$$H_0 = \frac{I}{2\pi r} \tag{6.11}$$

with I the current and r the distance from the wire to the cilium. The magnetic gradient is very localized and the current wire is ideally placed close to the tip of the cilium. In such a situation the actuation is exerted predominantly on the tip of the cilium. We will consider that the gradient is constant over the tip, and that only the tip is magnetic over a length equal to the thickness T. This assumption yields a result comparable to finite element simulations[36] for aspect ratios of cilia between 10 (actuation over-estimation by ~30%) and 50 (actuation under-estimation by ~30%). The equivalent magnetic volume is then equal to $V = W\,T^2$ and replacing H_0 in eqn (6.10) one obtains:

$$F_{\text{ferro}} = \mu_0 M_r WT^2 \frac{I}{2\pi r^2} \quad \text{and} \quad F_{\text{para}} = \frac{1}{4}\mu_0 \chi WT^2 \frac{I^2}{\pi^2 r^3} \tag{6.12}$$

for ferromagnetic respectively (super)-paramagnetic materials. For a wire located at a distance r from the tip, measured perpendicular to the cilium, the actuation efficiencies can now be given by:[36]

$$\left.\frac{\delta}{L}\right|_{\text{ferro wire gradient}} = \frac{\mu_0 M_\mathrm{r}}{\pi E}\frac{L^2}{Tr^2}I \quad \text{and} \quad \left.\frac{\delta}{L}\right|_{\text{para wire gradient}} = \frac{\mu_0 \chi}{2\pi^2 E}\frac{L^2}{Tr^3}I^2 \qquad (6.13)$$

For a constant current, these relations scale inversely with the first, respectively second power of the length scale and both favour small scale actuation. Paramagnetic actuation reaches the same efficiency as ferromagnetic actuation when:

$$\frac{\chi I}{2\pi r} = M_\mathrm{r} \quad \text{or} \quad \chi H_0 = M_\mathrm{r}. \qquad (6.14)$$

Typical bulk magnetic materials have $\chi = 10$ for an assembly of super-paramagnetic particles and $M_\mathrm{r} = 6\times10^5\,\mathrm{A\,m^{-1}}$ for a ferromagnetic material,[37] and the above condition would be satisfied for a magnetic field of $6\times10^4\,\mathrm{A\,m^{-1}}$, *i.e.* a magnetic induction of 75 mT. For this induction, super-paramagnetic materials are close to saturation, meaning that paramagnetic linear laws no longer apply and ferromagnetic laws will apply instead, with $M_\mathrm{r} = M_\mathrm{sat}$. Hence one can conclude that, in practice, magnetic actuation on small scales with a current wire is limited by the scaling of the ferromagnetic law. A similar conclusion can be made for eqn (6.10). In other words, paramagnetic actuation is always lower or equal in deflection amplitude to ferromagnetic actuation.

6.2.3 Scaling

The actuation efficiencies derived above in sections 6.2.1 and 6.2.2 can be summarized for *ferromagnetic actuation* in a single graph (*cf.* Figure 6.1), where the grey area symbolizes the region without useful actuation. Several assumptions were made to bring these relations into the picture. The elastic modulus was taken to be 1 MPa and the magnetic loading of the polymer 5 vol%, representing typical magnetic polymer cilia design values. The susceptibility of super-paramagnetic particles was taken to be 10 per particle which relates to a susceptibility $\chi = 0.5$ for the cilia material (assuming the loading of 5 vol%). The remanent magnetization of ferromagnetic particles was taken to be $6\times10^5\,\mathrm{A\,m^{-1}}$, which is the saturation magnetization of iron, even if our particles could so far only achieve 15% of this value (*cf.* section 6.4.1). At a concentration of 5 vol% this relates to $M_r = 3\times10^4\,\mathrm{A\ m^{-1}}$ for the ideal ferromagnetic cilia material. The magnetic fields and field gradients were approximated by the values at a distance of 5 mm on the axis of a permanent Nd–Fe–B rod magnet (length = 2 cm, diameter = 5 mm). Hence $\mu_0 H_0 = 50\,\mathrm{mT}$, $\mu_0 \partial H_0/\partial z = 10\ \mathrm{T\ m^{-1}}$ and $\mu_0{}^2 \partial H_0{}^2/\partial z = 1\ \mathrm{T^2\ m^{-1}}$. The maximum current in a wire was assumed to be 10 A, which is the maximum current that a common source can supply. The scale s relates to the cilia according to Figure 6.1 with

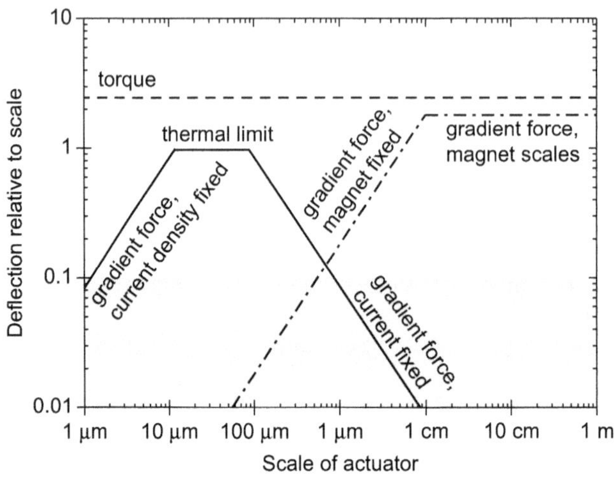

Figure 6.1 Scaling behaviour for different actuation concepts with ferromagnetic cilia. The scale of the actuator is here given by its length ($s=L$). The different curves and assumptions are discussed in the text. Results presented in section 6.4 follow the scaling of the dashed line.

$L=s$, $T=s/20$, $r=s/5$. The aspect ratio of the actuator (ratio L/T) is thus 20 and also represents the upper limit for samples we could fabricate.

As noted previously already, actuation by magnetic torque (dashed line in Figure 6.1 or eqn (6.6)) does not depend on the scale of the cilia and can be exerted with external means for any scale. On the contrary, actuation by magnetic gradient force reduces linearly with size (dash–dotted line in Figure 6.1 or eqn (6.10)). Above the scale of the magnet we considered (centimetre sized), the actuation by gradient force does not improve above actuation by torque because in a real situation the magnet should remain at least the size of individual cilia or larger. For smaller scales the actuation by gradient force with an external centimetre-sized magnet ceases to be effective below the millimetre. Actuation by gradient force with a current wire (plain line in Figure 6.1 or eqn (6.13)) begins to yield interesting deflections below the millimetre, which is typically the size of interest for artificial cilia. It can be shown that a thermal limit is then reached at a scale of 100 µm, related to the heat dissipation of the 10 A current in a micro-wire. For even smaller scales the current cannot be maintained when electromigration occurs, limiting the current density to a maximum of $10^{12}\,\mathrm{A\,m^{-2}}$.[38] For a current of 10 A electromigration therefore occurs for a wire with a cross-section smaller than $1\times1\,\mu\mathrm{m}^2$, hence for a scale s of $\sim10\,\mu\mathrm{m}$. A typical cilium of length between 10 and 100 µm it therefore best suited for actuation by gradient force, which happens to be the desirable size for most cilia with regards to their integration in microdevices. Note that with our considerations, and for cilia lengths below 100 microns, the magnetic induction induced by the current is larger than 100 mT at the cilium location. Super-paramagnetic cilia would therefore be mostly saturated and will behave the same as ferromagnetic ones. Given both behave the same, it would be advantageous to use super-paramagnetic materials since they (1) can more easily be dispersed homogeneously in polymers and (2) do not present a remanent magnetization which might interfere with the system when the actuation is not on-going.

The configurations discussed above are not an exhaustive list of possibilities. For example, a magnetic torque might also be exerted on a cilium with the magnetic field generated by a current wire. The main asset of a current wire is, however, that it can generate locally very high gradients but not high fields, or very inhomogeneous ones. Therefore this option does not seem very promising, even if it might lead to a feasible device on small scales. Furthermore, integrated actuation might also be achieved with integrated coils. An integrated planar coil can generate higher magnetic fields and might be better suited for generating a magnetic torque than an integrated current wire. However, planar coils are more complicated to fabricate than simple current wires.

6.2.4 Ferromagnetic Versus Super-paramagnetic

It was discussed at the end of section 6.2.3 that—in the case of *actuation by magnetic gradient force*—the material considerations and assumptions on cilia geometries meant that ferromagnetic cilia do not offer advantages on small scales. The magnetic fields involved would anyway almost saturate the super-paramagnetic cilia and these would therefore yield the same deflection as for ferromagnetic cilia.

For *magnetic actuation by torque* the comparison is, however, much in favour of ferromagnetic cilia. The relation for ferromagnetic materials actuated by torque was already derived in eqn (6.6). For (super-)paramagnetic materials, the torque originates from shape anisotropy. In the case of ideal shape anisotropy (*i.e.* demagnetization factors of 1 in the perpendicular direction and 0 in the parallel direction, which is fair for aspect ratios above 20), the magnetization parallel and perpendicular to the long axis of the cilium is:

$$M_{\text{parall}} = \chi H_0 \cos(\alpha) \quad \text{and} \quad M_{\text{perp}} = \frac{\chi}{1+\chi} H_0 \sin(\alpha) \tag{6.15}$$

with α the angle between the applied magnetic field and the long axis of the cilium. The magnetic torque as derived from eqn (6.2) becomes:

$$\vec{\tau} = \mu_0 \left[\left(\vec{M}_{\text{parall}} \times \vec{H}_0 \right) + \left(\vec{M}_{\text{perp}} \times \vec{H}_0 \right) \right] V. \tag{6.16}$$

with an amplitude:

$$\begin{aligned} \tau &= \mu_0 \left[\left(M_{\text{parall}} H_0 \sin(\alpha) \right) - \left(M_{\text{perp}} H_0 \cos(\alpha) \right) \right] V \\ &= \frac{1}{2} \mu_0 H_0^2 \left(\chi - \frac{\chi}{1+\chi} \right) \sin(2\alpha) V \end{aligned} \tag{6.17}$$

which is maximal for $\alpha = 45°$. Combining eqns (6.17), (6.3) and (6.4), the relative deflection for magnetic actuation by torque on paramagnetic materials becomes:

$$\left.\frac{\delta}{L}\right|_{\text{para torque}} = \frac{2\mu_0\chi^2 H_0^2}{E(1+\chi)}\frac{L^2}{T^2} \tag{6.18}$$

Both eqn (6.6) and eqn (6.18) do not scale with length scale and actuation remains efficient with the same external stimulus irrespectively of how small the cilia will get. But the efficiency of actuation for ferromagnetic cilia can potentially by one or several orders of magnitude higher than for super-paramagnetic ones. Figure 6.2 is a plot of eqn (6.6) and (6.18) for the same material properties as chosen in section 6.2.3 (thin plain line and thin dashed line respectively in Figure 6.2). Our own developed ferromagnetic polymer has up to now been limited to a bulk remanent magnetization $M_r = 90\,\mathrm{kA\,m^{-1}}$, or $M_r = 4.5\,\mathrm{kA\,m^{-1}}$ at 5 vol% (fat plain line in Figure 6.2).

The deflection obtained with a *super-paramagnetic torque* increases quadratically with the induction of the applied magnetic field and useful deflection are reached above 20 mT (dashed line in Figure 6.2a). Around that value, the quadratic increase is attenuated by the saturation of the super-paramagnetic particles. The curve taking into account this saturation is calculated with a field dependent susceptibility defined as the ratio M/H in the measured magnetization curve of typical composites with 5 vol% particles.[36] As can be seen in the graph, a super-paramagnetic torque does not benefit much from magnetic inductions above 100 mT. The deflection reached with a *ferromagnetic torque* increases only linearly with the induction of the applied magnetic field, but no saturation effects are present. For the lower remanent magnetization (fat plain line in Figure 6.2) useful deflections are reached above 20 mT, whereas for the maximum remanent magnetization (thin plain line in Figure 6.2) useful deflections would be reached with inductions as low as 1 mT. Both

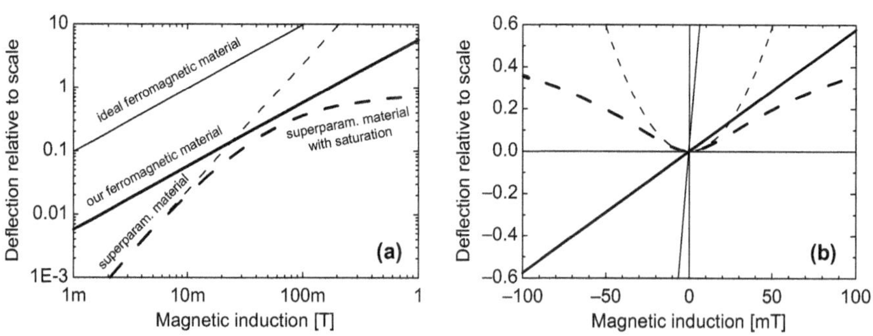

Figure 6.2 Maximum relative deflection of a cilium with respect to the induction of the homogeneous magnetic field inducing a torque. Logarithmic (a) and linear (b) representations. The different curves, as well as the assumptions about cilia materials and geometries, are discussed in the text.

ferromagnetic curves (plain lines in Figure 6.2) keep increasing linearly and the only limitation factor is the coercivity of the ferromagnetic material. As can be seen from Figure 6.2b, the super-paramagnetic torque is not depending on the sign of the magnetic field, in contrary to the ferromagnetic torque. This can be understood as the remanent magnetization aligns with the *direction* of the magnetic field for a ferromagnetic torque, whereas the long orientation of the actuator aligns only with the *axis* of the magnetic field. Upon a rotation of the magnetic field over 360°, a super-paramagnetic cilium is thus experiencing two strokes, while a ferromagnetic cilium is experiencing only one.

In conclusion to the discussion on actuation and scaling, ferromagnetic and super-paramagnetic cilia offer similar performance in case of actuation by magnetic gradient force, but gradient forces cannot efficiently be exerted with external means. External means of actuation are mostly only efficient in the case of actuation by magnetic torque. And for actuation by torque, ferromagnetic materials offer much improved efficiency over super-paramagnetic materials. Furthermore, as will be demonstrated in section 6.4.2.4, the ability to tune the remanent magnetization of ferromagnetic materials can also be advantageous.

6.3 Fabrication and Experimental Methods

6.3.1 Ferromagnetic Polydimethylsiloxane Composite

To create artificial cilia we use a ferromagnetic elastomer composite created by mixing commercially available ferromagnetic nanoparticles into a polydimethylsiloxane (PDMS) copolymer. The ferromagnetic particles are obtained from MK Impex Canada and are 70 nm iron particles including a carbon shell of 2–5 nm thickness (MKN–Fe/C-070, 99.8% purity), hereafter denoted as Fe–C particles. The carbon shell is intended to protect the particles from oxidation which would alter their magnetic characteristics. The deflection of artificial cilia is proportional to their magnetization (eqn (6.6)) and the Fe–C particle concentration should be maximized. A concentration of 2 vol% is chosen. Above 3 vol% macroscopic chunks of particles were noticed and the material is therefore not suited for micro-fabrication. We first dispersed the Fe–C particles in chloroform[39] by sonication in a weight ratio 1:100. Subsequently this mixture was added to a PDMS copolymer, shaken and sonicated again. After evaporating the chloroform overnight on a hotplate at 90 °C, the appropriate curing agent for the PDMS was added and the ferromagnetic PDMS composite was cured into a functional elastomer. Particle clusters ranged up to a size of tens of micrometres for the maximum loading of 3 vol% of magnetic material.

The ferromagnetic PDMS composite can be made starting from several commercially available forms of PDMS with different curing mechanisms. For the magnetic characterisation, macroscopic samples are produced with standard PDMS (Sylgard 184, Dow Corning) which is mixed in a ratio of 10:1 with its curing agent. Samples are spin-coated onto a small round substrate which results in samples with a diameter of 5 mm and a thickness up to 0.5 mm. Then the samples are cured thermally at 60 °C during at least 4 h. For the fabrication of artificial cilia, a procedure has been chosen in which the ferromagnetic PDMS composite is structured by UV–lithography, hence an appropriate PDMS had to be chosen. RMS-033 (Gelest, supplied by ABCR as (2–4% methacryloxypropyl-methylsiloxane)–dimethylsiloxane copolymer) is a PDMS with acrylate side groups that can be cross-linked with the common photo-initiator 2,2-Dimethoxy-2-phenylacetophenone (DMPAP, 98% purity, ABCR).[40] This photo-initiator is used in a weight ratio of 1% and is first dissolved in xylene (*p*-xylene, Merck) in a ratio of 1:2, before being mixed with the PDMS. For the elastic modulus measurements, this same PDMS has been cured with the thermal-initiator 2,2′-azobisisobutyronitrile (AIBN, Akzo) at 100 °C for 1 h, because the Fe–C particles would totally absorb the UV light for samples thicker than a couple of tens of micrometres and hinder normal cross-linking with a photo-initiator. For both the photo-initiator and the thermal initiator, the cross-linking has to be done in a nitrogen atmosphere since oxidation inhibits the cross-linking reaction. Without particles the elastic modulus is around 1 MPa, for both initiators. Up to 2 vol% this modulus would not significantly change (~1.3 MPa), as measured on macroscopic samples (~20×5×0.5 mm) with a mechanical analyser (Diamond TMA, Perkin Elmer).[41] During our actuation experiments (*cf.* section 6.3.4) the artificial cilia will be immersed in isopropanol which swells them to a certain extent. It was verified that the elastic modulus was only slightly affected (~1.1 MPa, for a sample with 2 vol% magnetic content being immersed in isopropanol).

6.3.2 Fabrication of High Aspect Ratio Lying Artificial Cilia

High aspect ratio lying artificial cilia are fabricated with a common lithographic structuring and sacrificial layer etch to release the cilia. Polyvinyl alcohol (PVA, 87–89% hydrolysed, Mw 31k–50k, Aldrich) was chosen as a sacrificial layer because it can be dissolved in water and will not be affected by common solvents used in the processing of the ferromagnetic PDMS composite. Because PVA is not photosensitive in itself, a lift-off photo-resist is used to structure it. The process steps are illustrated in Figure 6.3.[41] Typical dimensions of our cilia are $W \sim 100\,\mu m$, $L \sim 300\,\mu m$ and $T \sim 15\,\mu m$ which corresponds to structures with an aspect ratio of ~20.

First a glass substrate (Thermo Scientific microscope glass slide, Menzel) is cleaned with an alkaline solution and rinsed with ethanol. Then it is exposed to a UV–ozone treatment for 10 min (PSD-UV, Novascan Technologies Inc.). A layer of photo-resist (ma-N 1420, micro-resist technology) is spin-coated at 3000 rpm for 30 s, and baked for 5 min on a hotplate at 120 °C. The photo-resist is exposed through a mask with a UV lamp (Omnicure series 1000, filter 320–500 nm, EXFO Canada). Contact lithography is used and the sample is exposed 45 s with an irradiance estimated at $10\,mW\,cm^{-2}$. Subsequently, the sample is developed with mild manual agitation during 90 s in the adequate

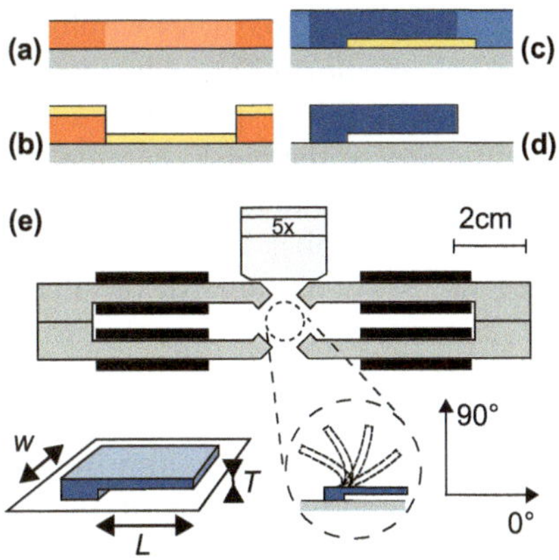

Figure 6.3 Schematic view of the processing steps for artificial ferromagnetic cilia and the quadrupole set-up used for actuation and microscope inspection. (a) Spin coating of lift-off photo-resist and UV exposure through a mask. (b) Development of lift-off photo-resist and spin coating of PVA sacrificial layer. (c) Lift-off of the sacrificial layer, spin coating of the ferromagnetic PDMS composite and exposure through a mask. (d) Development of the ferromagnetic PDMS and dissolution of the PVA sacrificial layer in water. (e) Cross-sectional view of the quadrupole with soft iron core (grey) and four coils (black) that creates a rotating magnetic field in the centre region where the artificial cilia are placed. A core connects the left coils to the right coils in the plane behind the drawing in order to increase the flux guiding; for clarity this is not drawn in the schematic. The dimension of the quadrupole in the plane perpendicular to the plane of the sketch is 1 cm. Microscopy inspection with a long range working distance lens (×5) is possible from above the quadrupole. The reference angles for rotation of the magnetic field and orientation of the cilia are indicated.

developer (ma-D 533 S, micro-resist technology), rinsed in demi-water and blow-dried with nitrogen. All the parameters, *i.e.* the baking time and temperature, the exposure dose and the developing conditions, are of crucial importance to reach a good undercut[42] which will ensure clean lift-off in the following steps. A 5% solution of PVA in water is spin-coated at 1000 rpm for 30 s on the lift-off photo-resist, baked for 2 min on a hotplate at 90 °C, and lifted off by immersing 4 min in an ultrasonic bath of acetone. The sample is then rinsed with isopropanol and blow-dried with nitrogen. The resulting thickness of the PVA layer is less than 1 μm.

The ferromagnetic PDMS composite is prepared as indicated in the previous paragraph (RMS-033, 1% DMPAP and 2 vol% of Fe–C particles) and spin-coated at 5000 rpm for 30 s. The resulting layer is ~16 μm thick. For longer spin-coating times, the layer thickness converged to ~11 μm but the amount of Fe–C particle clusters remaining in the PDMS layer kept decreasing after reaching terminal thickness. This is most probably due to the fact that the Fe–C particle clusters are only loosely fixed in the (not yet cross-linked) PDMS matrix and the centrifugal force continues to exert on the clusters. Thus, to ensure that the initial amount of Fe–C particles would remain in the artificial cilia, the spin-coating time is kept short (30 s). The ferromagnetic PDMS layer is then exposed through a mask with a spacing of a couple hundred micrometres to avoid the layer sticking to the mask. Manual alignment of the mask with the sacrificial layer is performed under a microscope after removing part of the ferromagnetic PDMS with a rubber blade to reveal the sacrificial layer. Note that the presence of the Fe–C particles in the PDMS was not found to change significantly the required exposure dose due to the absorption of the UV light by the particles. First, an exposure of 20 s with an irradiance of 10 mW cm^{-1} is performed in air; then, in a second step, an exposure of 80 s with the same irradiance is performed in a nitrogen atmosphere. We found that the first exposure would ensure a good adhesion of the ferromagnetic PDMS composite on the glass substrate whereas the second exposure would actually cross-link the layer, since the cross-linking reaction is inhibited in presence of oxygen during the first step. The ferromagnetic PDMS layer was then developed in xylene for 15 s without agitation and dipped for 15 s in isopropanol without agitation. Thereafter, the sample was very gently blow-dried with nitrogen. It is to be noted that during the development, the layer is swelling because xylene is not only dissolving the unexposed PDMS but also swelling the exposed PDMS, and some structures might shift due to the swelling (if the first exposure step of 20 s in air is omitted, most structures would actually swell off the substrate). As a last processing step, the PVA sacrificial layer can be dissolved in water and the artificial cilia are released. The cilia do not stick to the glass substrate as long as the sample is kept in fluid (if dried, the artificial cilia stick irreversibly to the substrate). In the case of a lab-on-a-chip device, one can easily envisage that this last processing step is performed as the first step when using the device. When a micro-fluidic channel is filled with an aqueous solution, the PVA sacrificial layer will dissolve and release the cilia.

6.3.3 Quadrupole Set-up for a Rotating Magnetic Field

A quadrupole electromagnet set-up is used to create a rotating magnetic field for actuation by magnetic torque of artificial cilia (see Figure 6.3e). The magnetic core of the quadrupole is made of soft iron (ARMCO pure iron, AK Steel) and the four pole tips are aligned under an angle of 45° with the vertical to allow microscopy inspection with a 5× objective having a long range working distance (11 mm). The four coils consist each of 800 windings of insulated copper wire with a diameter of 0.25 mm and resistance of 10 Ω each. The coils are addressed individually through function generators that drive power transistors in a push–pull configuration. A power supply (System DC Power Supply 6622A, Agilent) provides the necessary currents in the coils (up to 20 V and 3 A). The magnetic quadrupole has a linear current–field characteristic in all orientations, up to a magnetic induction of 50 mT. For this magnetic induction the current is ~700 mA in each of two coils for a diagonal orientation or ~500 mA for each of the four coils for a horizontal or vertical orientation. In the horizontal direction, a magnetic induction up to 100 mT can be obtained ($\sim$1.5 A in each coil). For generating a rotating magnetic field, the currents in the four coils follow a programmed sinusoidal shape. At high frequencies some losses in amplitude of the magnetic induction are observed, caused by eddy currents in the core. Up to 100 Hz, and for a desired induction of 50 mT, the measured losses are approximately 0.5% of magnetic induction per hertz. The magnetic induction is hence equal to $\sim$25 mT at 100 Hz instead of 50 mT as in a static case. Magnetic field gradients of our quadrupole are low and will not play any significant role in actuation. The maximal magnetic field gradient within a sphere of radius 2.5 mm located at the centre of the quadrupole was simulated to be smaller than 4.5 T m^{-1} for an induction of 50 mT. Calculations with these values following eqn (6.10) show that the angular deflection due to gradient force of an artificial cilium not being centred in the quadrupole would be insignificant (below 0.1°).

The artificial cilia manufactured on a glass substrate were placed in isopropanol in an open container that fits in the centre region of the quadrupole. Isopropanol was chosen because of its very low surface tension, which allows us to move the samples between different containers without the cilia drying and getting stuck onto the substrate. The movements of the artificial cilia were visualized with a high-speed camera (MotionPro® HS-3, Redlake) and the angle of deflection of the artificial cilia was determined by projecting the tip of the cilia in the horizontal plane and observing the apparent rotation point of the cilia for large deflections. Dark-field microscopy was used in order to obtain a good contrast of the moving PDMS cilia in fluid.

6.3.4 Closed Micro-fluidic Channel Assembly and Particle Tracking Experiments

A closed micro-fluidic channel was assembled to visualize the fluid movement induced by our artificial cilia. A PDMS cover with a micro-fluidic channel of length 2 cm, width 2 mm and height 400 μm was placed on the glass substrate with the artificial cilia and sealed by clamping mechanically. The PDMS cover was fabricated by pouring Sylgard 184 (Dow Corning) and its curing agent over a mould, degassing it under vacuum, and curing it at 60 °C for at least 4 h. A drop of water containing 3.15 μm polystyrene particles (PGP-30-5, Spherotec) was spread in the open channel on the PDMS cover, before closing the device with the glass substrate containing the artificial cilia. As soon as the sacrificial PVA layer was put into contact with this drop, the cilia were released and could operate.

To visualize the fluid movement induced by the artificial cilia, the device was placed in the centre of the quadrupole. Bright-field microscopy enabled the visualization of the movement of the cilia whereas a dark-field view with a powerful external lamp enabled the visualization of the fluid movement through the tracer particles. Images were recorded with the high-speed camera at a frequency equal to the actuation frequency of the cilia.

6.4 Results and Discussion

First, we present results on the measurements of the magnetization of our ferromagnetic PDMS composite. Then we analyse and interpret the movements of our artificial cilia in several configurations. Finally, we present our observations of fluid manipulation by artificial cilia inside a micro-fluidic chamber.

6.4.1 Magnetization of the Ferromagnetic PDMS Composite

The magnetization behaviour of our PDMS composite was measured with a vibrating sample magnetometer (VSM 10, DMS Magnetics). The results of a small and a large hysteresis loop are plotted in Figure 6.4. The data are normalized to the measured saturation magnetization of the Fe–C particles at 2 T. The magnetization behaviour was found to be the same for volume concentrations of particles with a range between 0.05% and 3% (*i.e.* the normalized magnetization curve has the same shape, unlike for our previous studies with super-paramagnetic particles).[36] The saturation magnetization M_{sat} of the Fe–C particles was measured to be equal to $6 \cdot 10^5\,\mathrm{A\,m^{-1}}$, such that the remanent magnetization was $8.8 \times 10^4\,\mathrm{A\,m^{-1}}$ (14.7% of M_{sat}) for a 2 T hysteresis loop and $6.8 \times 10^4\,\mathrm{A\,m^{-1}}$ (11.3% of M_{sat}) for a 100 mT hysteresis loop. Note that to obtain the net magnetization of the ferromagnetic PDMS composite, the values have to be multiplied by the particle volume concentration. The coercive field H_c was measured to be −18.9 mT for a 2 T hysteresis

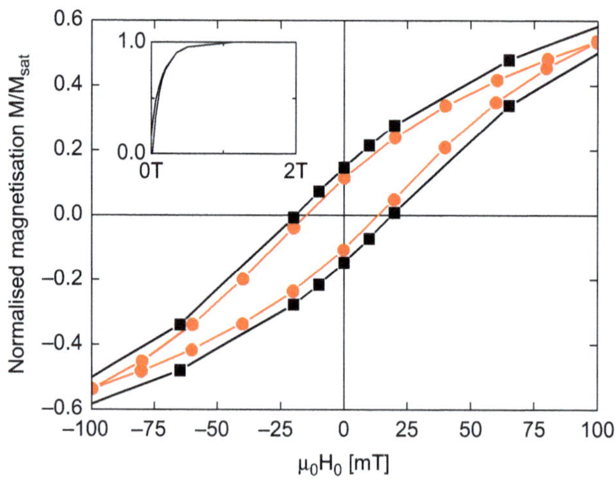

Figure 6.4 Magnetization curve of Fe–C 70 nm iron particles dispersed in PDMS with a concentration of 3 vol%. The magnetization is normalized to the saturated magnetisation measured for Fe–C particles at 2 T. (■) Hysteresis loop for a sweep of ± 2 T. The full positive side of the curve is shown in the inset. (●) Hysteresis loop for a sweep of ± 100 mT.

loop and −14.6 mT for a 100 mT hysteresis loop. These coercive fields are rather low (and certainly lower than the fields we use for actuation) but as the next results will indicate, this situation turns out to be favourable for creating asymmetry in the movement of the artificial cilia.

6.4.2 Actuation of Artificial Cilia in a Homogeneous Rotating Magnetic Field

The geometrical dimensions and magnetic content of adjacent cilia on a same sample vary somewhat due to the fabrication process, which is illustrated by the rough edges of the cilium shown in Figure 6.7 (higher resolution images are provided in our paper).[41] The consequence is that cilia have slightly different behaviours, particularly in their compliance to the magnetic torque. The behaviour of cilia on different samples are however comparable and the actuation for each cilium is reproducible in time.

6.4.2.1 Magnetic Field Rotating Anti-clockwise

In this experiment the cilia were magnetized in their long direction with an induction of 100 mT, in order to create a remanent magnetization that tends to align the cilia with the magnetic field during the actuation phase. Actuation experiments were performed with a magnetic field rotating anti-clockwise (*i.e.* from 0° to 360°, *cf.* Figure 6.3) for increasing field strengths. Results are shown in Figure 6.5 for two individual cilia that are representative of cilia with a small (cilium C1) and large (cilium C2) deflection. The experiments were done with a rotation frequency of 0.1 Hz, which is equivalent to a quasi-static situation.

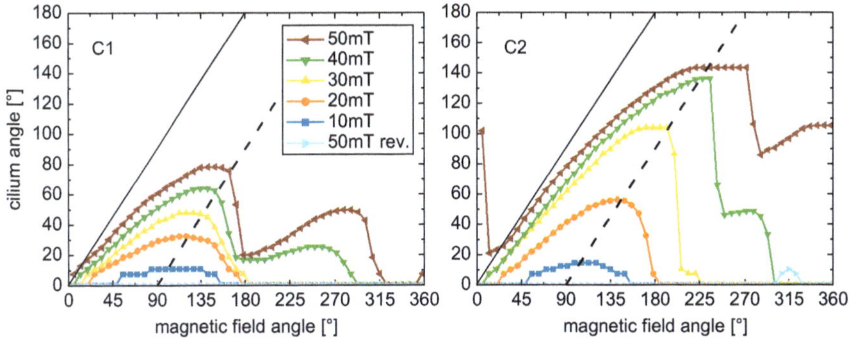

Figure 6.5 Angular deflections of a stiff (C1) and a more compliant (C2) cilium depending on the angle of rotation of the magnetic field, measured for increasing magnetic inductions. The magnetic field is rotated anti-clockwise (*i.e.* from 0° to 360°), except for one series (▶) where it is rotated in the clockwise direction. The plain line indicates the angle of the field itself and the dashed line indicates the position of maximum torque for the cilium (*i.e.* field angle minus 90°).

Initially, a cilium has a deflection angle of 0°. As the magnetic field is rotated to positive angles, the cilium follows the orientation of the magnetic field. Due to a finite elasticity of the cilium, its deflection is always less than the angle of the magnetic field. As the angle between the cilium and the field increases, so does the torque on the cilium. A maximum torque is reached when the angle between field and cilium reaches 90° (cross-product of eqn (6.2) is maximal). The expected maximum deflection angle of the cilium is indicated in Figure 6.5 by the dashed line.[43] As the field continues to rotate, the torque collapses and the cilium comes back to its resting position. For magnetic inductions lower than the coercive field of the Fe–C particles, the cilium stays in its resting position because the torque it experiences as the magnetic field rotates from 180° to 360° induces a movement towards the glass substrate. The situation is different for magnetic fields that exceed the coercive field of the Fe–C particles. After the torque has collapsed and the cilium has come back to its resting position, the remanent magnetization reverses by the applied field and, as the magnetic field continues to rotate anti-clockwise from 180° to 360°, the cilium again experiences a torque that induces a deflection to positive angles of the cilium.[44] This second stroke of the cilium in the same revolution of the magnetic field can be observed in Figure 6.5 for magnetic inductions of 40 and 50 mT. The amplitude of the second stroke is somewhat smaller than the first stroke due to the fact that the cilium was initially magnetized with a higher magnetic induction (100 mT) than the induction used in the actuation experiments (up to 50 mT). Due to the hysteresis of the Fe–C particles, the remanent magnetization is lower in the reversed direction than in the original direction (*cf.* Figure 6.4). For a more compliant cilium (*cf.* Figure 6.5, panel C2), the first stroke is much longer such that when the movement collapses, the magnetic field is already oriented towards angles of ~270°. In this case the cilium does not return to its resting position but experiences the second stroke already and eventually returns to the rest position after the second stroke (*cf.* Figure 6.5, panel C2). Note that for an ideally compliant cilium, or for higher magnetic fields, the torque would be high enough for the cilium to stay constantly in its most deflected position, analogously to the situation of a cilium in a magnetic field rotating clockwise (*cf.* section 6.4.2.3). In such a situation there would not be any more actuation.

6.4.2.2 *Amplitude of Deflection*

The maximum deflection of cilia increases approximately linearly with increasing magnetic fields (*cf.* Figure 6.5). As a consequence, we assume that linear beam deflection formulas (eqn (6.4)) apply as a first approximation for our artificial cilia, even for large deflections. We verify the consistency of our model by comparing the apparent thickness determined by optical microscopy inspection (vertical distance between two focal planes) to the value obtained with the model. Eqn (6.6) applies when the torque and hence deflection are maximal, *i.e.* when the angle between field and remanent magnetization is 90°.

The equivalent thickness is then given by:

$$T = \sqrt{\frac{4\,\mu_0\,M_r\,H\,L^3}{E\,\delta}} \tag{6.19}$$

with $M_r = 1.36\,\text{kA}\,\text{m}^{-1}$ the assumed remanent magnetization of the cilium, $L = 300\,\mu\text{m}$ the length of the cilium, and δ its deflection.[46] In a magnetic induction of 30 mT, cilia C1 and C2 have a maximum deflection of 47.5° and 103.8°, respectively, which implies an equivalent thickness of 4.2 and 2.8 μm. The apparent thickness of cilia C1 and C2 found by optical microscopy is $16 \pm 2\,\mu\text{m}$. Several factors, related to the inhomogeneous character of the cilia, can explain the observed difference. The Fe–C particle concentration is likely to be higher than 2 vol%, because of the clusters of particles which remained attached to the cilia, and will effectively increase the value of M_r. Also, the model does not describe defects in the geometry of a single cilium, which could make it more compliant, and does not account for deformation of the strip to which cilia are attached.

6.4.2.3 *Magnetic Field Rotating Clockwise*

When the magnetic field is rotating in a clockwise direction (*i.e.* from 360° to 0°) and for magnetic fields below the coercive field of the Fe–C particles, the cilia respond to the field identically as the case of anti-clockwise rotation. This is expected because the movement is quasi-static and the remanent magnetization is locked. For magnetic fields above the coercive field of the particles, the situation is dramatically different. For increasing magnetic inductions up to 50 mT, the movement of the cilia is progressively diminished and almost completely reduces to zero as shown in Figure 6.6 for cilium C3. This can be understood from the direction of the remanent magnetization. In the initial position of the magnetic field and the cilium, the remanent magnetization is oriented in the 0° direction. When the magnetic field is rotated clockwise, the torque exerted on the cilium pulls it towards the substrate. Once the field has reached an angle of 180° the remanent magnetization reverses and, as the field continues to rotate clockwise from 180° to 0°, the cilium again experiences a torque that moves it towards the substrate. Only for very compliant cilia we do see a little stroke when the magnetic field approaches 0° in a clockwise direction, since in that situation the remanent magnetization has already partially been reversed to the original direction and the angle between field and cilium is still big enough to create a small torque (*cf.* Figure 6.6 for 50 mT). Note that cilium C3 is one of the most compliant in the population we analysed. For cilia such as C1 and C2, the movement for clockwise rotation of the magnetic field at 50 mT is almost non-existent (*cf.* Figure 6.5, series ▶).

The actuation of our cilia with high magnetic inductions (50 mT) is therefore selective to the rotation direction of the magnetic field. Cilia pointing in opposite directions can be actuated fully independently, since the external magnetic field is rotating effectively in an opposed direction for opposite cilia.

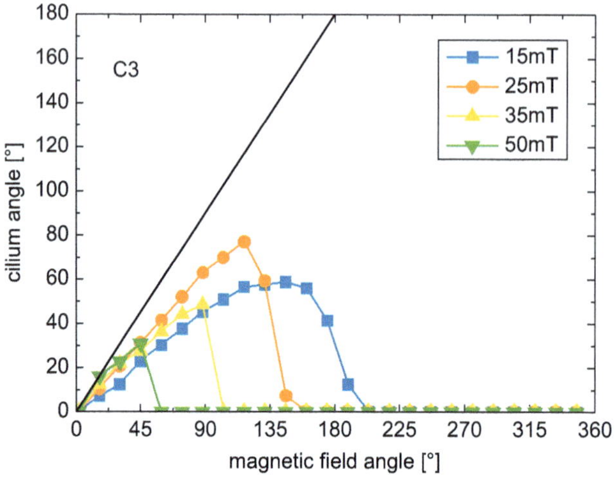

Figure 6.6 Angular deflections of cilium C3 depending on the angle of rotation of the magnetic field, measured for increasing magnetic inductions. The magnetic field is rotated clockwise (*i.e.* from 360° to 0°). The plain line indicates the angle of the field itself.

6.4.2.4 Asymmetric Movement of Artificial Cilia

The movements of artificial cilia described so far do not a priori contain asymmetry. But provided with adequate initial conditions, we were able to induce a large asymmetry in the movement of our artificial cilia by superposition of a torsion to their bending movement. We first magnetized the artificial cilia in the transverse direction (rather than in the longitudinal direction) with a magnetic induction of 100 mT. By transverse direction we mean the direction along the width W of the artificial cilia as depicted in Figure 6.3. When the magnetic field is rotated as in previous experiments, the movement of the cilia is not constrained to the plane of rotation of the field anymore. A torsion movement occurs due to the angle between the transverse remanent magnetization and the field. The movement can clearly be seen in Figure 6.7 for cilium C2 whose initial remanent magnetization is induced either (a) longitudinally, (b) transversely or (c) in the opposite transverse direction. When the remanent magnetization is longitudinal in the cilium, its movement is constrained to the plane of rotation of the magnetic field (Figure 6.7a). When the initial remanent magnetization is transverse, it induces a torsion movement of the cilium to one or the other side (Figure 6.7b and c). The torsion of the cilium for angles of the magnetic field between 0° and 180° is always opposite to the torsion for angles between 180° and 360 and is due to the initial transverse remanent magnetization. We attribute the bending movement to the existence of a longitudinal magnetization component, generated by the field used for actuation. It is to be noted that for a cilium with limited movement (deflection < 90°) the torsion is on one side for the first stroke and on the other

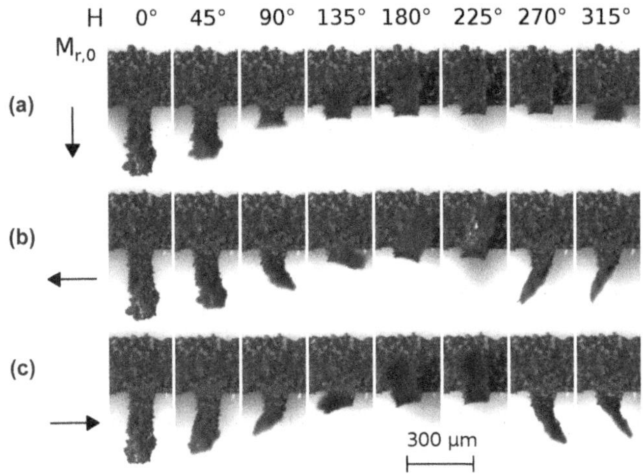

Figure 6.7 (a–c) Top microscope view of the movement of cilium C2 in a rotating magnetic field with an induction of 50 mT. The angle of the magnetic field H is given according to Figure 6.3. The initial remanent magnetization $M_{r,0}$ is longitudinal for situation (a) as indicated by the arrow. For situations (b–c) the initial remanent magnetization is transverse.

side for the second stroke. Since the cilium comes almost back to its resting position between these two strokes, there is actually no asymmetry in the movement. Only for compliant cilia with movements up to 140° does the torsion lead to an asymmetric movement. In that case the cilia do not return to their resting positions between the strokes, but only at the end of both strokes.

6.4.2.5 Movement of Cilia at Higher Frequencies

In all previous sections, the artificial cilia were actuated in a quasi-static way with a magnetic field rotating at a frequency of 0.1 Hz. At higher actuation frequencies the viscous drag of the cilia in the fluid becomes visible. Figure 6.8 shows the amplitude of movement of cilium C2 in a magnetic induction of 50 mT for increasing field frequencies. As stated in the experimental section, the quadrupole experiences eddy current losses that decrease the magnetic induction with increasing frequencies. However, the magnetic induction at 100 Hz remains at least 25 mT and the eddy currents alone cannot explain the strong drop in amplitude observed in Figure 6.8. The cilium stops moving effectively above a couple of tens of hertz and eventually stops moving at all at 100 Hz. For a significant movement to be preserved above 50 Hz, the magnetic field should be increased in order to increase the torque on the cilia and hence apply a force greater than the viscous drag. A much higher magnetic field used for actuation would however entirely re-magnetize the cilia in the longitudinal direction, making them lose their initial transverse remanent magnetization and an asymmetric movement would no longer be possible.

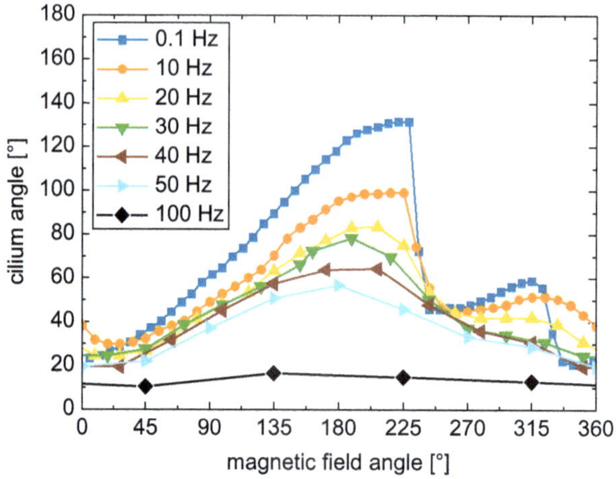

Figure 6.8 Angular deflections of cilium C2 depending on the angle of rotation of the magnetic field, measured for increasing frequencies of the rotating magnetic field. The magnetic induction drops with increasing frequencies from 50 mT to ~25 mT at 100 Hz.

Our cilia are still quite large for typical applications one could have in mind for artificial cilia. Scaling down is not possible at the moment because of the limitation in resolution of the lithographic process for our ferromagnetic PDMS composite and because of the large size of particle clusters. The scaling down of the system is, however, possible from the magnetic actuation point of view (eqn (6.6)) and from the point of view of the viscous drag. The drag force and the force related to the magnetic torque applied on the cilia are both proportional to the area, and therefore the scaling is neutral.

6.4.3 Fluid Manipulations by Ferromagnetic Cilia

Particle tracking experiments reveal the fluid manipulations induced by our artificial cilia. We performed experiments for cilia with symmetric and asymmetric movements. In both cases we observed an oscillatory fluid movement and a net fluid flow (movies are available in the supplementary material of our paper).[41] A visualization of the net fluid flow is provided in Figure 6.9. By changing the focus plane, the movement of particles at several depths in the shallow micro-channel could be observed and no large differences in particle movement were found.

For a cilium moving with torsional asymmetry, a net induced vortex is clearly visible at its location (*cf.* Figure 6.9a). The rotation direction of the vortex depends on the direction of the initial transverse remanent magnetization, which is expected because the magnetization determines the direction of the torsion. Since the actuation of the cilia was shown to be sensitive to the rotation direction of the magnetic field, two cilia being in the same plane but pointing into opposite directions can be actuated fully independently in the

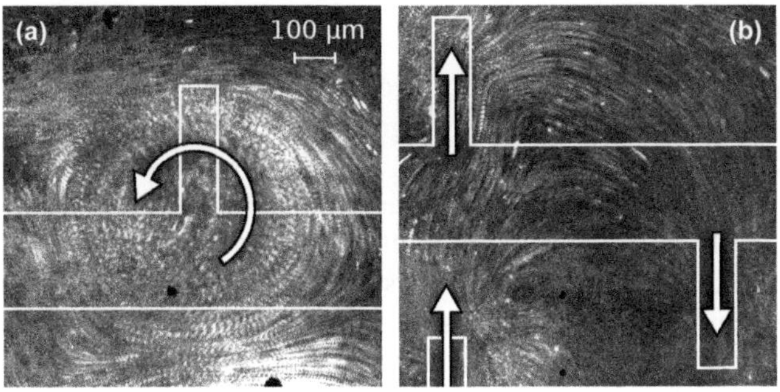

Figure 6.9 Top view visualization of the net fluid flow induced in a micro-fluidic chamber for cilia with a torsional asymmetry (a) and without a torsional asymmetry (b). The positions of the cilia are indicated with the white lines. The actuation frequency was 5 Hz and the magnetic induction of the rotating field was 30 mT. The visualization images were obtained by subtracting consecutive frames, for the first 20 frames of the movies, and plotting for each pixel the maximum absolute value obtained from these subtractions. The arrows indicate the direction of the fluid flow that is observed in the movies.

same external magnetic stimulus, depending on the direction of rotation of the magnetic field. Such an independent actuation means that two different sets of vortices can be started and stopped individually in a micro-fluidic device, providing for example the opportunity to mix fluids with a blinking vortex protocol.[47–49] The mixing could be done in a micro-fluidic chamber and would not need any applied fluid flow like all passive mixers require.

For a cilium moving symmetrically, the induced net fluid movement is locally translational instead of rotational (*cf.* Figure 6.9b). Note that the global flow is only rotational because of recirculation in the closed cavity and that pumping of fluids through a micro-fluidic channel would be possible with numerous cilia oriented in the same direction. If the bending movement of the cilium was fully symmetric, there should be no net fluid flow due to the low Reynolds number prevailing in the micro-channel.[50] We conclude that the movement must be asymmetric, probably because the shape of the cilium is slightly different during the movement leaving the rest position and during the movement coming back to it. This explanation may be supported by the simulations of an artificial cilium by Khaderi *et al.*[51] in which buckling of a ferromagnetic cilium leads to asymmetry. A buckling asymmetry would indeed induce a translational fluid displacement as we observed and would be analogous to the fluid manipulation of a natural cilium.

Manual particle tracking indicates that the maximum fluid velocity observed in the vicinity of a cilium with a torsional asymmetric movement at a frequency of 5 Hz was $\sim$200 μm s^{-1} whereas it was $\sim$70 μm s^{-1} in the case of a cilium moving (a priori) symmetrically. Induced fluid velocities could be reached up to $\sim$500 μm s^{-1} for actuation frequencies of 50 Hz and cilia moving asymmetrically.

6.5 Conclusion

In this chapter we have discussed the various ways magnetic actuation can be applied to cilia, namely by torque or gradient force. Scaling behaviours have been derived and discussed, the main conclusions being that—with external means of generating magnetic fields—actuation by magnetic gradient force is getting gradually less efficient for smaller scales while actuation by magnetic torque was remaining constant for all scales. Actuation by magnetic gradient force would tend to require means of generating local (integrated) magnetic fields in order to reach acceptable efficiencies. Ferromagnetic polymers would perform equally when compared to super-paramagnetic polymers for this latter type of actuation. In the case of actuation by magnetic torque, ferromagnetic polymers were shown to have a more efficient deflection. The advantage in deflection can potentially be up to one or two orders of magnitude compared to super-paramagnetic polymers.

Fabrication of ferromagnetic cilia with a length of 300 μm and an aspect ratio of 20 were demonstrated. While this dimension is still rather large for artificial cilia, the scaling behaviour of the system is neutral and scaling down is only limited by the fabrication process. The largest challenge lies in the homogeneous dispersion of ferromagnetic particles in polymers. Since all materials used in the fabrication are polymeric, methods such as micro-imprinting and inkjet printing could be used that will enable low cost artificial cilia.

Our studies have also shown that ferromagnetic cilia offer the possibility to tune the remanent magnetization and thereby controlling asymmetric or selective movements, which cannot be achieve in the same way for super-paramagnetic cilia (due to the lack of magnetic hysteresis).

Acknowledgements

This research forms part of the research programme of the Dutch Polymer Institute (DPI), Project #532. Figures 6.3–6.9 and part of the text are reproduced from the journal *Lab on a Chip*, published by the Royal Society of Chemistry.[41]

References

1. T.-S. Chin, *J. Magn. Magn. Mater.*, 2000, **209**, 1–3.
2. O. Cugat, J. Delamare and G. Reyne, *IEEE Trans. Magn.*, 2003, **39**, 6.
3. L. H. Lu, K. S. Ryu and C. Liu, *J. Microelectromech. Syst.*, 2002, **11**, 462.
4. J. W. Judy, R. S. Muller and H. H. Zappe, *J. Microelectromech. Syst.*, 1995, **4**, 162.
5. C. Liu and Y. W. Yi, *IEEE Trans. Magn.*, 1999, **35**, 1976.
6. Y.-H. Jang and Y.-K. Kim, *J. Micromech. Microeng.*, 2003, **13**, 853.
7. C. T. Pan and S. C. Shen, *J. Magn. Magn. Mater.*, 2005, **285**, 422.
8. J. W. Judy and R. S. Muller, *J. Microelectromech. Syst.*, 1997, **6**, 249.
9. Y.-P. Hsieh, *J. Magn. Magn. Mater.*, 2012, **324**, 939.

10. C. Liu, *Mechatronics*, 1998, **8**, 5.
11. C.-W. Chang and W. Hsu, *J. Micromech. Microeng.*, 2009, **19**, 105026.
12. S. Yao, S. Goto, J. Sakurai, T. Shinshi, M. Uehara and H. Yamamoto, *IEEE Proceedings on Nano/Micro Engineered and Molecular Systems,* 5–8 January 2009, Shenzhen, China, ed. N.F. de Rooij, IEEE Conference Publications, New York, 2009.
13. R. L. Smith, R. W. Bower and S. D. Collins, *Sens. Actuators, A*, 1990, **24**, 47.
14. H. J. Cho and C. H. Ahn, *Proc. IEEE MEMS*, 2000, **686**, 78–84.
15. K. L. Tsai, M. Ziaei-Moayyed, R. N. Candler, W. Hu, V. Brand, N. Klejwa, S. X. Wang and R. T. Howe, *J. Microelectromech. Syst.*, 2011, **20**, 1.
16. S. Santra, P. Holloway and C. D. Batich, *Sens. Actuators, B*, 2002, **87**, 358.
17. N. J. Sniadecki, A. Anguelouch, M. T. Yang, C. M. Lamb, Z. Liu, S. B. Kirschner, Y. Liu, D. H. Reich and C. S. Chen, *Proc. Natl Acad. Sci. U.S.A.*, 2007, **104**, 14553.
18. M. Shen, C. Yamahata and M. A. M. Gijs, *J. Micromech. Microeng.*, 2008, **18**, 025031.
19. T. Pan, S. J. McDonald, E. M. Kai and B. Ziaie, *J. Micromech. Microeng.*, 2005, **15**, 1021.
20. M. Khoo and C. Liu, *Sens. Actuators, A*, 2001, **89**, 259.
21. C.-Y. Lee, Z.-H. Chen, H.-T. Chang, C.-Y. Wen and C.-H. Cheng, *Microsyst. Technol.*, 2009, **15**, 1171.
22. B. A. Evans, A. R. Shields, R. L. Carroll, S. Washburn, M. R. Falvo and R. Superfine, *Nano Lett.*, 2007, **7**, 1428.
23. J. Belardi, N. Schorr, O. Prucker and J. Rühe, *Adv. Funct. Mater.*, 2011, **21**, 3314.
24. See Chapter 11.
25. The magnetic induction in a medium is given by $B = \mu_0\ (1+\chi)\ H$, with μ_0 the permeability of vacuum, χ the susceptibility of the medium, and H the magnetic field applied to the medium. A measure for the perturbation of the magnetic field by the medium is obtained by comparing χ to 1. The susceptibility of aqueous solutions or solvents is typically $\chi \approx -10^{-5}$.
26. A. H. Lu, E. L. Salabas and F. Schüth, *Angew. Chem., Int. Ed.*, 2007, **46**, 8.
27. N. Pamme, *Lab Chip*, 2006, **6**, 24–38.
28. B. A. Evans, B. L. Fiser, W. J. Prins, D. J. Rapp, A. R. Shields, D. R. Glass and R. Superfine, *J. Magn. Magn. Mater.*, 2012, **324**, 501.
29. A. Boczkowska, S. F. Awietjan, S. Pietrzko and K. J. Kurzydłowski, *Composites, Part B*, 2012, **43**, 636.
30. K. L. Tsai, M. Ziaei-Moayyed, R. N. Candler, W. Hu, V. Brand, N. Klejwa, S. X. Wang and R. T. Howe, *J. Microelectromech. Syst.*, 2011, **20**, 1.
31. H. Denver, T. Heiman, E. Martin, A. Gupta and D.-A. Borca-Tasciuc, *J. Appl. Phys.*, 2009, **106**, 064909.

32. J. Li, M. Zhang, L. Wang, W. Li, P. Sheng and W. Wen, *Microfluid. Nanofluid.*, 2011, **10**, 919.
33. J. V. I. Timonen, C. Johans, K. Kontturi, A. Walther, O. Ikkala and R. H. A. Ras, *ACS Appl. Mater. Interface*, 2010, **2**, 8.
34. A. Engel and R. Friedrichs, *Am. J. Phys.*, 2002, **70**, 4.
35. R. J. Roark and W. C. Young, *Formulas for Stress and Strain*, McGraw-Hill, New York, 1975.
36. F. Fahrni, M. W. J. Prins and L. J. van IJzendoorn, *J. Magn. Magn. Mater.*, 2009, **321**, 12.
37. Both χ and M_r would have to be multiplied by the magnetic volume loading percentage to represent the polymer composite characteristics.
38. J. R. Black, *IEEE Trans. Electron Device*, 1969, **16**, 4.
39. Among common solvents, chloroform was found to produce the most stable colloid. When mixed with polydimethylsiloxane (PDMS), a more stable colloid results in a magnetic PDMS with smaller particle clusters. Processing with toluene, xylene or tetrahydrofuran produced larger clusters in the PDMS than processing with chloroform. The solvents investigated were selected for being able to swell cross-linked PDMS (according to Lee *et al.*[52]), hence having chemical affinity with PDMS monomers.
40. J. C. Lötters, W. Olthuis, P. H. Veltink and P. Bergveld, *J. Micromech. Microeng.*, 1997, **7**, 3.
41. F. Fahrni, M. W. J. Prins and L. J. van IJzendoorn, *Lab Chip*, 2009, **9**, 3413.
42. A. Voigt, M. Heinrich, K. Hauck, R. Mientus, G. Gruetzner, M. Topper and O. Ehrmann, *Microelectron. Eng.*, 2005, **78–79**, 503.
43. An exception can be seen in Figure 6.5, panel C2, for a magnetic induction of 50 mT. In this situation the cilium has already reached the maximum mechanical deflection at around 140° and thus cannot move further.
44. We assume here that the magnetization is either aligned with the long axis of the cilium, pointing to its tip, or that it is anti-parallel to that direction. Indeed, the hysteresis curve of Figure 6.4 was recorded for a re-magnetization in the anti-parallel direction. It is reasonable to assume that the magnetization is mostly in the longitudinal plane of the cilium because an out-of-plane magnetization requires much higher fields than an in-plane magnetization.[45]
45. B. Edwards and D. I. Paul, *J. Magn. Magn. Mater.*, 1995, **147**, 331.
46. The deflection is here assumed to be the distance covered by the tip of the cilium on a circular path.
47. H. Aref, *J. Fluid Mech.*, 1984, **143**, 1.
48. V. V. Meleshko and H. Aref, *Phys. Fluids*, 1996, **8**, 3215.
49. It is important to note that the use of a mixing protocol (such as a blinking vortex protocol) is crucial to obtain efficient mixing. A single vortex created by one cilium is not likely to produce efficient mixing in itself. The blinking vortex mixing protocol consists of a chamber with two inter-spaced vortices that can be switched on and off. By sequentially activating both

vortices, an optimal mixing regime can be created, *i.e.* maximal material exchange between both vortices.

50. At the fastest point in an actuation cycle at 5 Hz, a cilia with a width of 100 μm effects a stroke over a distance of ~200 μm in 20 ms. Therefore we can estimate the Reynolds number to be a maximum of 1 in our experiment. At such low Reynolds numbers, viscous effects are expected to dominate over inertial effects and temporal asymmetry in the movement will play a minor role in net flow movement.
51. S. N. Khaderi, M. G. H. M. Baltussen, P. D. Anderson, D. Ioan, J. M. J. den Toonder and P. R. Onck, *Phys. Rev. E*, 2009, **79**, 046304.
52. J. N. Lee, C. Park and G. M. Whitesides, *Anal. Chem.*, 2003, **75**, 23.

CHAPTER 7

Light-actuated Artificial Cilia Based on Liquid Crystal Networks

CASPER L. VAN OOSTEN,[a] CEES W. M. BASTIAANSEN[b] AND DIRK J. BROER*[b]

[a] Peer + , P.O. Box 2374, 5600 CJ Eindhoven, The Netherlands;
[b] Department of Chemistry and Chemical Engineering, Eindhoven University of Technology, P.O. Box 513, 5600 MB Eindhoven, The Netherlands
*Email: D.Broer@tue.nl

7.1 Introduction

Polymer networks with controlled molecular organization possess unusual, but accurately adjustable and addressable optical, electrical and mechanical properties. A presently widely applied method to form three-dimensional ordered polymer networks is photo-initiated polymerization of monomers that possess a liquid crystal (LC) phase.[1–4] The principle of the process is shown in Figure 7.1. The monomer is brought to a temperature at which the LC phase manifests itself. At this temperature a short exposure with UV light activates the photo-initiator present in a low concentration, and the liquid crystal order of the LC monomer is frozen in permanently in a cross-linked polymer network. The variety in LC phases of the LC monomers provides a diversity in molecular order, all self-organizing and preserved by photopolymerization.[5–8] Techniques known for LCs to establish monolithic molecular order, such as rubbed polymer coatings (typically 30 to 60 nm thick polyimide coatings, as

RSC Nanoscience & Nanotechnology No. 30
Artificial Cilia
Edited by Jaap M J den Toonder and Patrick R Onck

Published by the Royal Society of Chemistry, www.rsc.org

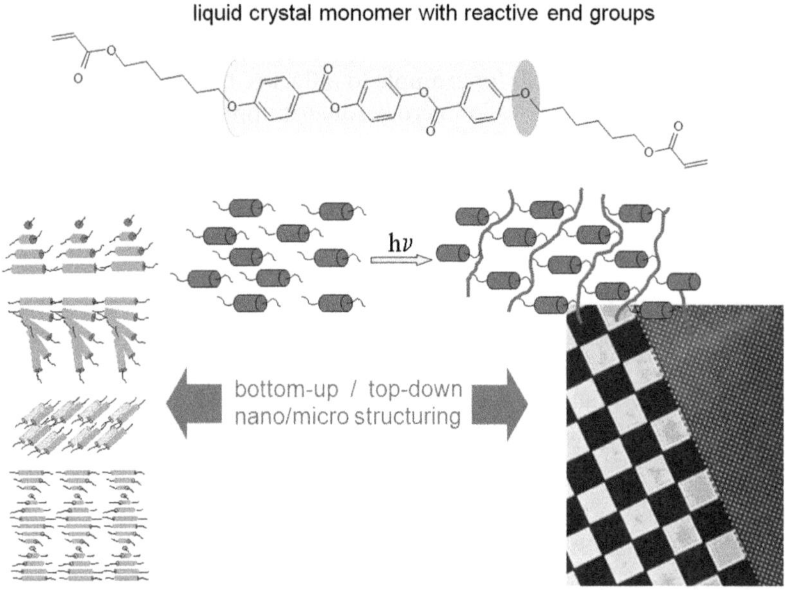

Figure 7.1 Schematic representation of the formation of a liquid crystal network. The example molecule has a nematic phase at elevated temperatures which can be aligned. In the aligned state the molecule is polymerized by UV light thus forming a polymer network with a frozen-in molecular order. The molecular alignment can be varied by various means, as some examples show on the left. By using photomasks micrometre-sized structures can be formed as shown by the polarization microscope insert showing elements of uniaxial order (bright) alternating with elements that are isotropic (black between the crossed polarizers). The combination of self-organization and lithography is an example of bottom-up, top-down structuring technology.

used in the display industry, which are unidirectionally buffed with a velvet-like polyester fabric), surfactant treated surfaces or external electric or magnetic fields, can be applied or even can be combined.[9] The use of light to initiate the polymerization process enables structure formation on a length scale of micrometres or higher. The molecular composition of the monomers can be tailored, *e.g.* to optimize on the mechanical and optical properties of the films. The mechanical properties of the LC networks (LCNs) are, apart from their anisotropic nature, of the same order as those of the isotropic acrylate and epoxide networks. This means that the modulus and strength are similar and depend strongly on the molecular parameters like cross-link density and the ratio between stiff and flexible molecular units in the LC monomer.[10]

Polymer materials in general are of interest for the fabrication of artificial cilia because they offer a low-cost alternative to the traditional inorganic materials while increasing the range of potential applications.[11–14] Processing tools are available for producing micrometre-sized features, including embossing, lithographic processing and printing. An easy adjustable chemical composition provides a wide control over the properties and polymers are capable of deformations which are orders of magnitude larger than those of inorganic actuators. In many cilia or MEMS applications a motion perpendicular to the plane is desired, for example to influence flow structures in microfluidics or for cantilever beams in micro-resonator applications. In these cases, out-of-plane bending actuators offer many advantages as they can be manufactured in-plane and small, in-plane strains are sufficient to create large, out-of-plane motion. The LCN actuators as discussed in this chapter are fully compatible with these requirements.

In this chapter we will first explain the basic principle of shape deformation in LC networks using the most straightforward case of a thermal actuator. Using the thermal actuator as an example, effects of various alignments such as twisted nematic (TN), splay and cholesteric are shown. LCN actuators can be made to respond to light as trigger. The chapter finishes with an example of miniaturization of actuators for a micro-fluidic application, where actuators are structured using inkjet printing.

7.2 Anisotropy in Mechanical Properties and Thermal Expansion

Monolithic (single domain), densely cross-linked polymer networks like the LC acrylate-based systems used throughout this chapter display a large difference between the properties in the directions parallel and perpendicular to the common average direction of the monomeric units (director). The polymer network systems indicated in Figure 7.2 are by far the best studied glassy liquid

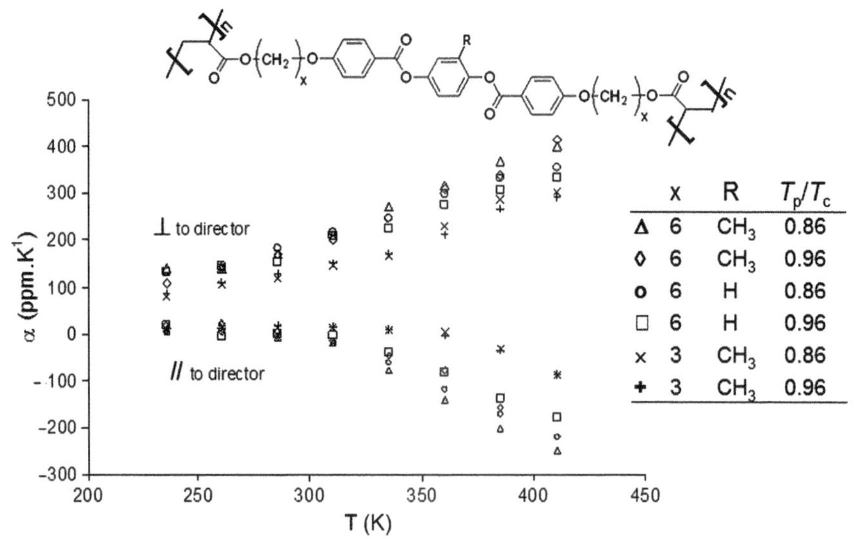

Figure 7.2 Linear thermal expansion coefficients α for uniaxially aligned networks. Networks were made by polymerization at various reduced temperatures (T_p/T_c) and some structural variations with respect to spacer length x and core substituents R.

crystalline system in relation to MEMS applications.[4,15] These polymers exhibit a glass transition temperature (T_g) above room temperature, typically around 60 °C. At room temperature the modulus is of the order of 1–3 GPa.[10] Due to the anisotropy of the system, the compliance of the polymer network is roughly three times higher perpendicular than parallel to the director. The properties are influenced by the length of the flexible spacer between the aromatic core and the acrylate moieties. A longer spacer introduces more mobility into the network, reducing the glass transition temperature with an increasing spacer length, also resulting in a lower room-temperature modulus. Attaching a methyl group to the central aromatic ring has several effects on the properties of the monomer and polymer. In the monomer phase, a methyl side group suppresses the presence of a smectic LC phase of the monomer and thus helps processing. In the polymer, the methyl group does not affect the T_g, but it does decrease the order of the network because of steric effects. Upon heating, the steric hindrance of the methyl group also causes the network to show more pronounced expansion behaviour than the non-substituted analogues.

Like any polymer, the volume of LCNs increases with heating. For aligned LCNs, however, this deformation is not uniform but has a strong anisotropy in their thermal expansion coefficients.[4,15] Figure 7.2 shows this behaviour for the CnR systems. Below T_g, the thermal expansion coefficient in the direction parallel to the director is close to zero, but above that the thermal expansion becomes negative. Orthogonal to the director, the thermal expansion rapidly increases above T_g. As can be expected, the systems with a longer aliphatic spacer have a larger temperature response. Furthermore, the systems that were cured close to the nematic–isotropic transition (clearing temperature, T_c) of the monomer show a smaller response than the systems cured further below this LC transition. The reduced temperature, the curing temperature divided by the clearing temperature of the monomers, is listed for comparison reasons for the systems in Figure 7.2. A decrease in the curing temperature only slightly affects the order of the polymer network, but leads to a significant increase in the temperature response of the system. For example, for C6M (the monomer with a spacer length of six methylene units and a methyl side group) changing the reduced curing temperature from 0.96 to 0.86 increases the order parameter of the network from 0.71 to 0.76, but changes the strain parallel to the director upon heating from −50 °C to 150 °C from −1.3% to −1.7%.

At temperatures below T_g the system expands with temperature due to an increasing molar volume. The preferential expansion direction is perpendicular to the long axis of the molecule, as the expansion is dominated by increasing intramolecular distances. Around and above T_g there is a small and reversible loss of molecular order causing additional deformation. The reduction of order is favourable for entropic reasons but is limited by the polymer network. The measured change in order parameter is small, only of the order of a few per cent as concluded from birefringence measurements, but geometrical arguments show that this indeed explains for the observed dimensional changes. Figure 7.3 illustrates this principle. The loss of order leads to an increasing average tilt of

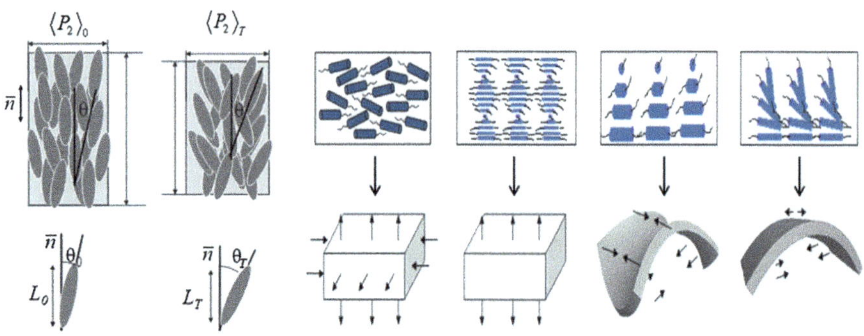

Figure 7.3 Left: Representation of the network order with director $\bar{n}$. The projected length on the orientation axis changes from L_0 to L_T when the average angle of the mesogenic units with the director increases from θ_0 to θ_T. Right: various types of molecular order, planar uniaxial, cholesteric, twisted nematic (TN) and splay, with their corresponding deformations with decreasing the order parameter.

the mesogenic unit θ which decreases the projection of the end-to-end length of the monomeric units. The order parameter is given by:

$$\langle P_2 \rangle = {}^1\!/_2\left[3\left(\cos^2\theta\right) - 1\right]$$

and the length change parallel to the director can be then estimated using:

$$\left(\frac{\Delta L}{L_0}\right)^{\prime\prime}_{\langle P_2 \rangle} = \left(\frac{2\langle P_2 \rangle_T - \langle P_2 \rangle_0}{2\langle P_2 \rangle_0 + 1} + 1\right)^{1/2} - 1$$

The total length change is then given by the sum of the length change due to molar volume increase and the change due to order loss:

$$\left(\frac{\Delta L}{L_0}\right)^{\prime\prime} = \left(\frac{\Delta L}{L_0}\right)^{\prime\prime}_{V} + \left(\frac{\Delta L}{L_0}\right)^{\prime\prime}_{\langle P_2 \rangle}$$

For LC networks the length changes estimated from order parameter changes are consistent with the experimentally obtained length changes.

7.3 Molecular Alignment Configurations

The linear expansion of a planar uniaxially aligned LCN is relatively small and the corresponding strain, typically <2%, is therefore of limited use for actuation. During processing in the monomeric phase, molecular alignments like splay, twist (TN) or helical (cholesteric) ordering can be generated. After photopolymerization, the LCNs with these alignments will have properties that vary in one or more directions of the system. The techniques available for creating this molecular alignment are mostly taken from the display manufacturing industry and range from rubbed polyimide alignment layers to electrical and magnetic fields. Helical order can be induced by the addition of chiral (liquid crystal) molecules or chiral (liquid crystal) monomers.

Figure 7.3 shows a number of alignment configurations and their corresponding deformations upon heating as consequence of small reductions of the order parameter. The TN and cholesteric alignments only differ in the degree of director rotation, but the two alignments show distinct deformation behaviour. The gradient in director orientation in the TN film results in a gradient in thermal expansions through the thickness of the film. As a result, a free-standing film will deform upon heating into a saddle like structure. Figure 7.4 shows images of the heat induced deformation of a clamped film. The film is clamped such that at the top of the film the molecular director is parallel to the length axis of the film. Two competing bending axes are present: one in the length direction of the film and one perpendicular to it. Where the film is clamped, one bending axis is suppressed and the film curls up trying to satisfy the saddle-like deformation. When the pitch of the molecular rotation becomes small relative to the film thickness, as is the case for the cholesteric orientation, the in-plane deformations will average each other out and the main deformation takes place orthogonal to the film surface.

Bending of thin films is of practical interest for application as actuators in MEMS. Most bending actuators operate similar to bimetals and need two or more layers to bend.[16,17] Using LCNs, it is possible to create a gradient in the

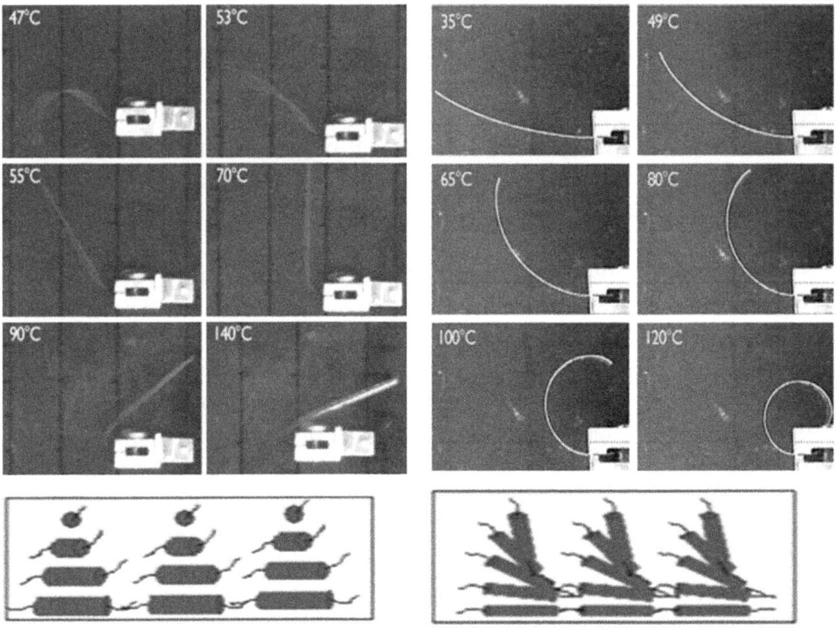

Figure 7.4 Temperature triggered deformation in LC network films with a gradient in molecular alignment over the cross section of the film. Left: twisted nematic alignment. Right: splayed alignment.

degree of expansion through a single material and bending can be realized in a single monolithic layer. In the twisted-nematic alignment, this behaviour was already visible. However, the competing bending axes make the behaviour unpredictable and therefore it is not the most suitable molecular configuration. A configuration that does not exhibit this conflict in strain is the splay configuration, where the molecular director rotates from in-plane alignment to homeotropic alignment over the thickness of the film as shown in Figure 7.3. Figure 7.4 shows the bending of a splayed actuator with temperature. The bending is smooth and no clamps are required to enforce bending over a single axis. In the in-plane direction perpendicular to the cross section shown, the thermal expansion of the film is constant over the thickness of the film. Therefore, no bending is observed into that direction.

7.4 Light-induced Deformation

For MEMS applications heat is often not a practical stimulus. Light triggered actuation is an attractive alternative because it in general shows a fast response and a light source can be used remotely, and sometimes even patterned by means of a mask or interference, to drive the actuator. Azobenzene molecules are well known to induce photomechanical effects.[18–24] The azobenzene unit undergoes a *trans–cis* isomerization upon illumination with UV light as illustrated in Figure 7.5a. Finkelmann *et al.*[18] showed that this conformation change of the azobenzene moiety leads to an amplified response when embedded in an LC polymer matrix. The shape changes are reversible and the dichroic nature of the azo moiety makes it possible to control the deformation direction by changing the polarization state of the light.[21,22,25] In glassy LC network actuators modified with azobenzene moieties a similar effect can be achieved but here the control over the molecular alignment of the LC network, as indicated in Figure 7.3, offers an additional handle on the deformation direction.

To minimize gradients in light intensity over the cross section of the films we often work with relatively low concentrations of azo molecules, *e.g.* <10 wt%. In these systems the dimensional change is mainly caused by a reduction of the order parameter of the LCN, similar to the thermal systems. The bending azo group affects the average orientation of the neighbouring molecular units which by cooperative interactions affect the whole molecular system. At room temperature and under ambient light conditions, the azobenzene unit is predominantly in the elongated *trans* form. Upon illumination with light that addresses the *trans* absorption band the bended *cis* state will be formed until a photostationary state is formed corresponding to the equilibrium between the forming *cis* molecules and the thermally induced (or visible-light induced) reversely forming *trans* molecules. Depending on the intensity and spectrum of the light source, thermal deformation may contribute significantly to the deformation. Due to the light-absorption of the film, the film will heat up and deform like a thermal actuator. Because the host LC material of the network is

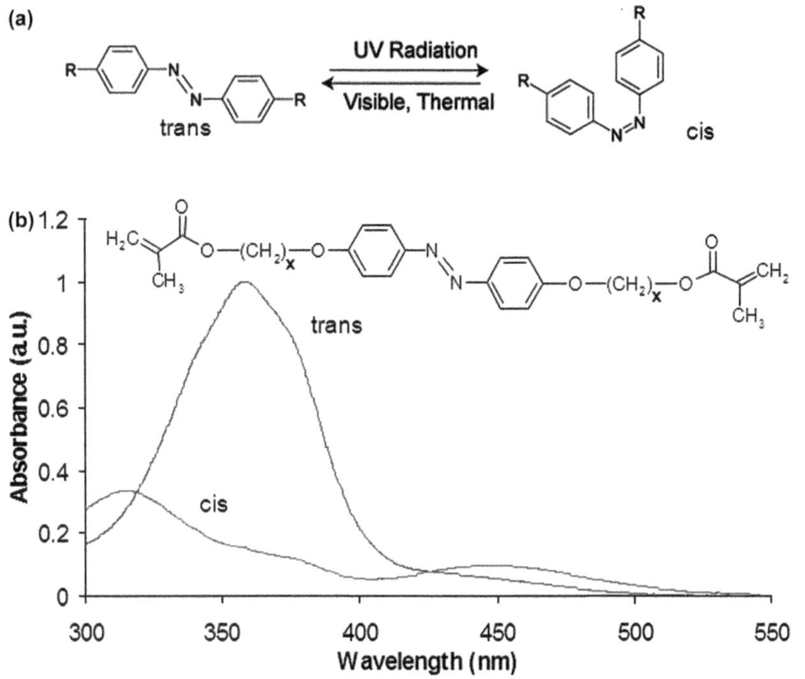

Figure 7.5 Conformation change upon exposure of azobenzene compounds (a) and the change in absorption spectra of the azobenzene monomer in solution after exposure with visible light corresponding to a preferentially *trans*-state of the monomer, and after exposure with UV light corresponding to a preferentially *cis*-state of the monomer.

the same as that of a thermal actuator discussed in the foregoing section, the responses to temperature are very similar.

The azobenzene monomer we have chosen to copolymerize with our LC monomers is shown in Figure 7.5b. The absorption peaks are around 360 nm for the *trans* state and at 320 and 440 nm for the *cis* state. LC networks incorporating this azobenzene monomer show strain responses that are in the order of a few per cent. Figure 7.6a shows the response of a uniaxially aligned film thermostatted in a water bath, parallel and perpendicular to the director. In this case, the film is cycled with UV and visible light, with a power of 100 mW cm^{-2} for the UV and 10 mW cm^{-2} for the visible light. For reference, Figure 7.6 shows also the heating of the LC network without azobenzene dye under the same illumination cycles. In addition to an overall drift due to some heating of the water bath, the LC network without dye shows a small response to the cycling of the light. When this temperature effect is subtracted from the measurements, the maximum strains display a linear dependence on the azobenzene concentration for dye loads up to 10 wt% (Figure 7.6b). Similarly, the deformation speed scales linearly with the light intensity.[25]

Typically the pure azobenzene monomer, and therefore also a network which is based on solely this material, will absorb 90% of the light within the first

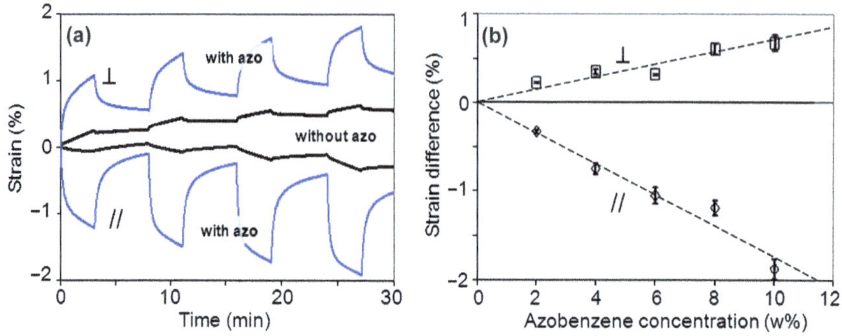

Figure 7.6 Response to cycled UV illumination of 100 mW cm^{-2} in water perpendicular and parallel to the molecular alignment of a uniaxial planar LC network film containing 0 and 8 wt% azobenzene monomer (a). Maximum strain difference between the UV-illuminated state and the visible-illuminated state, for increasing azobenzene monomer content, perpendicular and parallel (b).

micrometre. Also in the diluted case with azobenzene concentrations <10 wt% the presence of azo-dye in the film causes the light to attenuate through the thickness of the film, creating a gradient in light intensity. When illuminated from the top or bottom, a planar uniaxially aligned film with azo-dye therefore bends towards the light source. If splayed films are used, the deformation gradient will be much larger and the film can be made to bend to or from the light source as dictated by the molecular alignment in the material. At the bottom of the film the splayed film will expand in the plane of the film, whereas the planar film will contract. For the same concentration of azobenzene, a splayed film will therefore display sharper bending than a planar film. In addition to a steeper strain gradient, the splayed system also dictates the deformation direction and the bending axis thus becomes independent from the location of the light source. When the film is positioned with the homeotropic side towards the light source, the film will bend away from the source.

Similar to the thermal actuators, these systems are glassy at room temperature, with a room temperature modulus of 1.3 GPa parallel to the director and 0.6 GPa perpendicular to the director. The responses of the splayed and the TN systems are fast, reaching 70% of its final deformation in about 0.25 s in response to 100 mW cm^{-2} UV.[26] The reverse relaxation under ambient conditions is slower, in the order of 15 s. The speed of this relaxation can be increased by exposure with visible light to address the absorption band of the *cis*-state, or by increasing the temperature. The TN and splayed systems show similar responses in speed as well as in final bending radius. When not restricted by the clamp, the TN system displays a second bending axis similar to what has been reported in the thermal systems. In terms of light-input, the splayed system presents a more efficient bending mechanism than the planar system. The planar–uniaxial system loses part of its energy in a length contraction that does not contribute to the bending of the beam. The splayed system will achieve the same stationary bending state upon illumination at a lower concentration of azobenzene dye.

7.5 Inkjet Printing Actuators: Towards Polymer Cilia

The capability of liquid crystal polymers to undergo reversible shape deformations in response to heat or light make them attractive candidate materials for making micro-actuators. Early reports already mention the potential of light-driven actuators for MEMS devices[21] and lab-on-a-chip applications.[27] Numerous MEMS components can be fabricated from polymers; however, the large-scale manufacturing of polymer micro-actuators using, for instance, roll-to-roll processing on a single substrate or integrated in a device still faces a few challenges. The polymer actuators reported so far are hardly suitable for microstructuring:[28,29] for instance, the multilayer structure of conjugated polymer micro-actuators complicates the production.[30] A further challenge is that the actuators are often electrically addressed, requiring inorganic (metal) electrodes, thus leading to less favourable or costly manufacturing conditions, that is, processing occurs at high temperatures and/or in acidic or basic environments.[31] Liquid crystal network actuators can overcome these challenges: all-polymer microdevices can be fabricated using inkjet printing technology in combination with the self-organizing properties of liquid crystal networks. This self-assembling of the liquid crystal makes it possible to create large strain gradients. Light-driven actuation is chosen to enable simple and remote addressing. A simple actuator design is adopted, mimicking cilia in nature to act as active pumps or mixers in micro-fluidic systems.[32] By using multiple inks in the inkjet printing process, micro-actuators with different subunits are created that can be selectively addressed by changing the wavelength of the light.[20] Large deformations of the actuator structure are achieved, and it is shown that the different substructures can be independently addressed using different colours of light.

As we have seen above, the most common way of obtaining light-driven order change in LC polymers is through the inclusion of molecules incorporating the azobenzene unit using their ability to undergo the *trans–cis* isomerization upon exposure with light. There are a number of azobenzene dyes that can cause shape deformations when included in an LC network or LC rubber.[33–37] Depending on their chemical structure, they have absorption bands for the *trans*-state that vary from UV light to visible (green) light. For the purpose of inkjet printed cilia we used two different azo dyes: A3MA and DR1A (Figure 7.7). A3MA has an absorption maximum at 358 nm for the *trans*-state; whereas the *trans*-absorption band of DR1A is at 490 nm. The optimal dye concentration depends on the molar absorption of the dye and the thickness of the film.[38] An azobenzene concentration of 4 wt% for the A3MA dye and 1 wt% for the DR1A dye was chosen. These concentrations are high enough to give a significant response, but low enough so that in a 10 μm thick film the light penetrates deeply enough to affect more than just the top layer. The host matrix of the actuator is based on a combination of liquid crystals

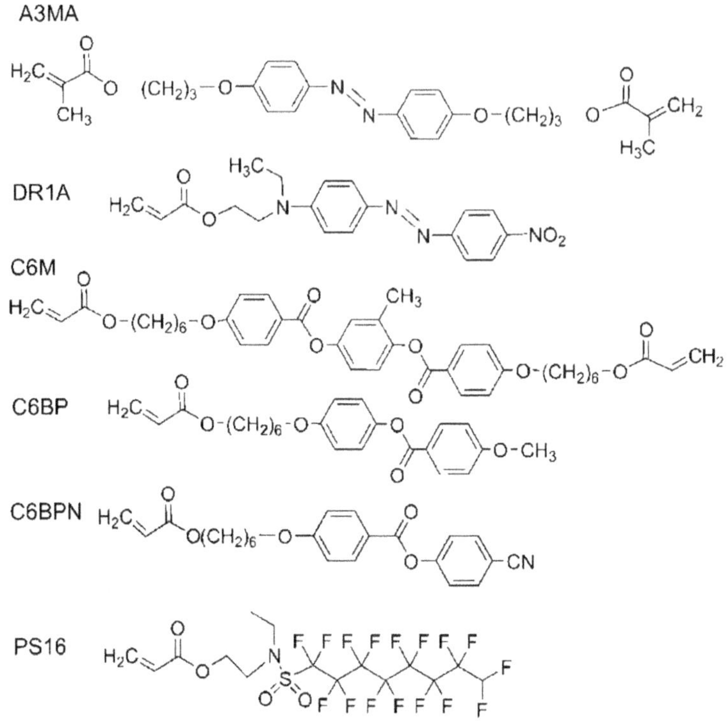

Figure 7.7 The monomers used to form the inkjet printed actuators. The two photoactive monomers are A3MA and DR1A. The host matrix is built from C6M, C6BP and C6BPN. The surfactant PS16 is used to obtain the desired alignment at the air interface.

functionalized with acrylates or methacrylates (Figure 7.7). We use a one-step manufacturing route, generating the optimized molecular monomer alignment in the polymer network for the desired motion of the actuator.[20] In many actuators, bending is achieved by creating a bi-layer structure, using two materials with a different degree of responsiveness. Here, the self-organizing and anisotropic properties of the liquid crystal were exploited to create an internal gradient in a single layer. By changing the director orientation through the thickness of the film in a splay-bend configuration, significantly stronger bending can be achieved than with a uniaxially aligned film of the same chemical composition. A splay-bend configuration is presented in Figure 7.8a: the liquid crystalline units undergo a gradual orientational change through the thickness of the film, from perpendicular to the substrate at the bottom, to parallel at the top. In the polymerized state the splayed molecular alignment dictates both the bending direction and the bend axis of the film, making the direction of the response independent from the direction of the incident light. To facilitate manufacturing the mix of reactive liquid crystals presented in Figure 7.7 has a balanced ratio of LC elastic constants so that in the monomeric state a splay-bend configuration is energetically permitted using surface

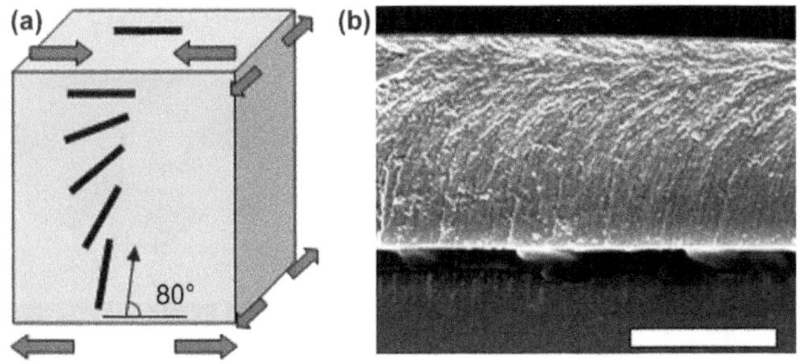

Figure 7.8 The splay-bend molecular organization through the thickness of the film. (a) Schematic representation of a splay-bend molecular orientation through the thickness of the film. The arrows indicate the direction of the material response upon actuation. (b) The molecular alignment is visible in a cross section of a fractured film. The scale bar indicates 5 μm.

alignment techniques on a single substrate. This mixture of monomers serves as a host material for the dyes: the resulting polymer will be referred to as A3MA polymer or DR1A polymer, depending on the dye that is included.

An essential step towards achieving large amplitude actuators is controlling the self-assembly of the liquid crystals so that an upward-bending (rather than a downward-bending) splayed actuator is created. To achieve this, the molecular director should be pointing perpendicular to the substrate at the bottom and parallel to the substrate at the top of the film, causing the bottom of the film to expand and the top to contract (Figure 7.8a). Under unforced conditions, the mix of liquid crystals monomers will align perpendicular to the air interface, minimizing its surface free energy. To overcome this, in-plane alignment at the air interface is obtained by adding a small amount of surfactant PS16 (Figure 7.7) to the reactive monomeric mix. The directionality of the alignment at the air interface is introduced by giving the molecular alignment at the bottom surface a slight pre-tilt from the surface normal. A molecular anchoring of 80° with respect to the substrate is obtained by the unidirectional rubbing of a homeotropic-aligning polyimide layer, which reproduces the desired molecular alignment. The alignment becomes visible by breaking the sample and inspecting the fracture surface using SEM (Figure 7.8b). Inkjet printing was chosen as a process to apply the microstructured actuators because it allows variations of the material composition in the plane of the substrate in a single processing step, while alternative micropatterning techniques such as lithography require several more steps to achieve the same.[39] Using a commercial inkjet printer, the monomeric LC mix containing one of the two dyes was deposited on the substrate. In our case, the reactive monomeric mix is crystalline with a crystalline–nematic transition just above room temperature, and the mix was therefore dissolved to enable printing. After evaporation of the solvent, supercooling keeps the monomer in the LC state, also at room temperature.

The goal of this procedure is to produce artificial cilia that are driven by light for use as mixers or pumps in micro-fluidic systems. Artificial cilia are

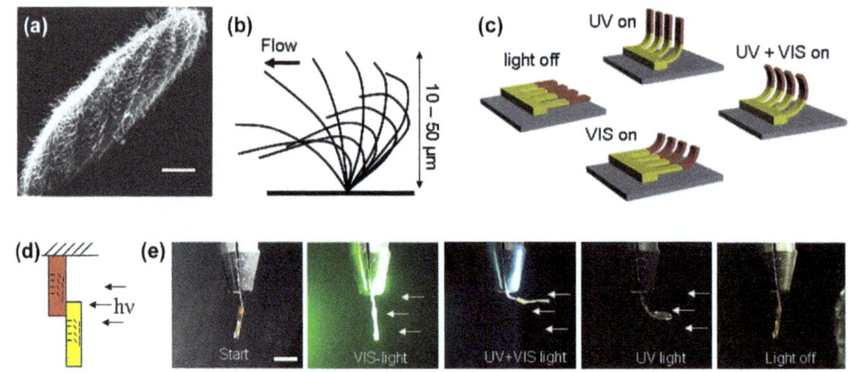

Figure 7.9 Natural and artificial cilia and their motions. (a) Cilia can be found on several micro-organisms, such as paramecia. (scale bar 20 µm) (b) A paramecium uses the beating motion of the cilia, characterized by different forward and backward strokes, for self-propulsion. (c) Artificial, light-driven cilia produce an asymmetric motion controlled by the spectral composition of the light. (d) Schematic representation of the macroscopic set-up, showing the orientation of the molecules. (e) Steady state responses of a 10 µm thick, 3 mm wide and 10 mm long modular LCN actuator to different colours of light (scale bar 5 mm).

nature-inspired structures and mimic the behaviour of cilia that are, for instance, used by paramecia to move through fluids (Figure 7.9a). An individual cilium makes a flapping, asymmetric motion, with a backward stroke different from the forward stroke, causing an effective flow in the surrounding fluid (Figure 7.9b). Artificial cilia can provide effective mixing in micro-channels.[32] Using a single dye, asymmetric motion can be introduced into LC azobenzene actuators by varying, over time, the light intensity over the actuator surface. Given the small dimensions of the actuator and the large deformations we are aiming for, this is difficult to realize in practice. Instead, using two dyes, varying the composition of the actuator in the plane, is proposed (Figure 7.9c). By selecting the colour composition of the light, the position of the actuator can be brought into four positions: in the dark, the actuators are flat. With only UV light, the yellow A3MA dyed part of the actuator bends. At the same time, any DR1A in the *cis*-state absorbs (with peak absorption at 360 nm) and is therefore actively brought to the unbent *trans*-position. When both UV and visible light are used, the flap bends over its total length. Finally, if only visible light is used, only the red DR1A part of the flap bends. Simultaneously, *cis*-A3MA (with peak absorption at 440 nm) absorbs and is therefore actively driven to the flat, *trans*-dominated state. Switching between these four positions produces a cilia-like motion with an asymmetric stroke. The weak *cis* absorption in both dyes adds to the thermally driven recovery to the flat state, speeding up the rate at which the illumination cycle can be run.

To demonstrate the effect on a macroscopic scale, two strips of A3MA polymer and DR1A polymer were cut and glued together at one end. Care was taken to orient the films so that they both had the same molecular alignment and that the strip was cut parallel to in-plane alignment of the molecules. Figure 7.9d provides a schematic representation of the set-up and the alignment, with the DR1A polymer mounted at the top and the A3MA polymer mounted at the bottom.

The strip was actuated in air using light of different colours, resulting in the steady state responses shown in Figure 7.9e. When the strip was illuminated with visible light, 455–550 nm, the top part with the DR1A polymer showed a small bend. Then, the strip was illuminated with a combination of UV and visible light, resulting in rapid and strong bending over the whole length, assuming a final state completely bent into the light. With only UV illumination, the top part of the film relaxed and only the bottom part (the A3MA polymer) of the film was bent. With the light off, the film completely relaxed back to its original position. This shows the desired motion of the actuator.

To manufacture free-standing miniature actuators, four basic process steps were taken (Figure 7.10). As a release layer, a 1 μm thick polyvinyl alcohol (PVA) layer was used. PVA was chosen because it is soluble in water but not in the solvents used for the processing of the polyimide or the LC monomers. Consequently, the sample was coated with the 'soft' homeotropic aligning polyimide and lightly buffed to create the desired molecular pre-tilt angle of 80° to the surface.

The monomeric LC mix was then deposited with the inkjet printer, using separate cartridges for the monomeric mixtures with the two different dyes. The reactive monomers can be cured after printing every separate colour, or after printing the full structure, depending on whether a distinct separation of the colours or a strong adhesion between the two parts is desired. For the remainder the first approach was followed for practical reasons: with the available printing set-up, only a single mix could be printed at one time. Polymerizing directly after printing helps to preserve the morphology when a printer cartridge is exchanged. The cilia are released from the substrate by dissolving the PVA layer in water. The polyimide layer is thin enough that it

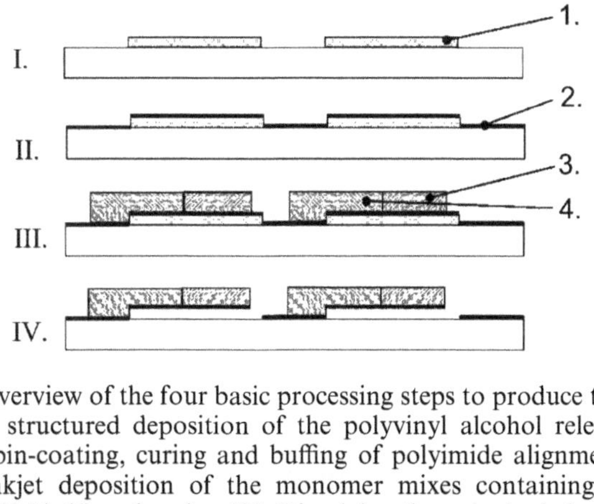

Figure 7.10 Overview of the four basic processing steps to produce the modular cilia. I. structured deposition of the polyvinyl alcohol release layer (1); II. Spin-coating, curing and buffing of polyimide alignment layer (2); III. Inkjet deposition of the monomer mixes containing DR1A (3) and A3MA (4) and curing; IV. Dissolving the polyvinyl alcohol release layer.

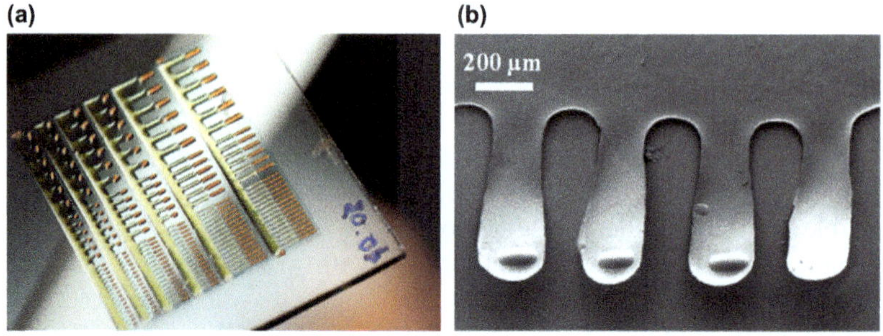

Figure 7.11 Inkjet printed artificial light-driven cilia before their partial release from the substrate (a). The polymeric, light responsive actuators, in yellow and red, are structured in different sizes on the release layer (blue) and glass substrate. SEM image of free-standing LC network actuators (b).

breaks, and in some cases it was observed that the polyimide broke upon first actuation of the cilia. The printing process is well controlled and different sizes can be printed (Figure 7.11a). Inspection in the SEM reveals that the structures are indeed partially free-standing on the substrate (Figure 7.11b).

A number of micro-actuators are shown in Figure 7.12a–d. The printed structures have a thickness of around 10 μm at the location of the free-standing flaps, being slightly thinner at the base than at the top. It is believed that the combination of the print geometry and the drop coalescence causes this height difference. At the base of the flaps, where the structures are permanently attached to the substrate, the printed layer has a thickness of 20 μm. In this thicker layer, the monomers do not form a monodomain-oriented structure, as is visible from the crossed polar images (Figure 7.12b). Because this part of the structure does not contribute to the motion, the poor alignment of the base is not expected to harm the performance of the actuators. At smaller line widths, the structures become somewhat irregular due to dewetting of the monomers on the substrate. With the current set-up, using a drop spacing of 15 μm, the minimum line width needed to produce a bending flap is 100 μm, but it is expected that this can be decreased with further optimization of the printer, substrate and monomeric mix to a few tens of micrometres.

The response of A3MA polymer cilia in water to cycled light is shown in Figure 7.12e. All flaps bend in response to illumination with UV light. The forward, UV-induced response is much faster than the thermally driven back-reaction. The actuation of the cilia was repeatable without loss of the stroke length. The experiments were carried out in water to minimize initial sticking of the cilia to the substrate and limit the thermal heating from the light source to a maximum increase of 2 °C. The difference in responses between the different cilia is mainly attributed to the sticking to the substrate and variations in the thickness of the cilia.

The response of the colour printed cilia is shown in Figure 7.12f. When illuminated with visible light, the DR1A part of the actuators show a small bending. Upon illumination with UV light, the A3MA polymer part bends

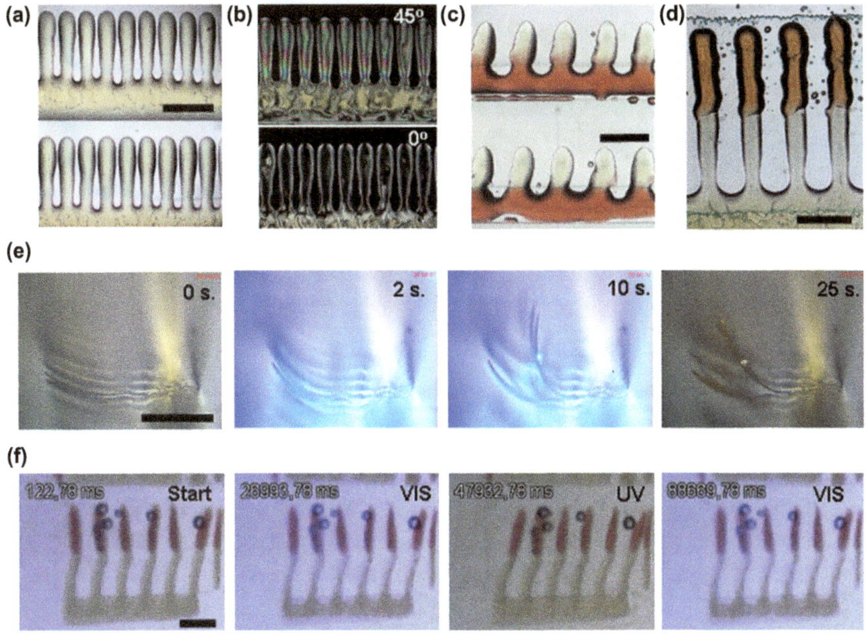

Figure 7.12 Microstructured cilia and their response to light. (a) Shows arrays of A3MA polymer flaps (yellow) on polyvinyl alcohol (blue). (b) Crossed polar microscopy with the sample at 45° and 0° shows that the splayed alignment in the flaps is parallel to the length of the flap. (c) Cilia can be manufactured with a gradient in composition from DR1A to A3MA polymer. (d) Alternatively, cilia are made with two separate parts of DR1A and A3MA polymer by separately polymerizing the two parts. (e) Side view of the actuation of A3MA polymer cilia with UV light ($1\,\mathrm{W\,cm^{-2}}$) in water. (f) Frontal view of actuation of multicolor cilia in water addressed with VIS ($4\,\mathrm{mW\,cm^{-2}}$) and UV ($9\,\mathrm{mW\,cm^{-2}}$) light. All scale bars indicate 0.5 mm.

strongly. Visible actuation drives the A3MA polymer back to its original state and bends the DR1A part again. With the current set-up and actuators, motions are still too slow to induce measurable flow. It is expected that 10 times faster motions can be achieved with thinner actuators and higher light intensities. For asymmetric cilia movements, motions of a few hertz will provide effective fluid manipulation in microsystems.[40]

To characterize the steady-state light-response, uniform flaps with only A3MA dye or only DR1A dye were produced. The incident intensity of the light was varied while a few cilia were monitored using optical microscopy. The images were analysed by fitting the bend radius of the cilia with a circle. At maximum intensity, the A3MA polymer cilia show a full bend with respect to the surface at an intensity of $2\ \mathrm{W\ cm^{-2}}$, with an average bending radius of 220 μm. To compare the performance of the inkjet-printed flaps with bulk material, the strain difference is estimated using $\Delta\varepsilon = t/r$ ADVANCE\u 5, where t is the film thickness, r the bending radius and $\Delta\varepsilon$ the strain difference

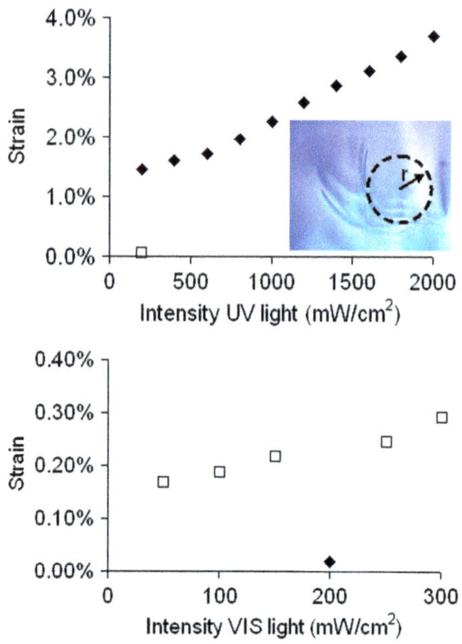

Figure 7.13 Strain response of printed cilia. Calculated strains from A3MA polymer (◆) and DR1A polymer (□), in response to UV light (250–390 nm), randomly polarized, and to visible light (488 nm), with polarization state parallel to the molecular director. The inset shows how the bending radius was measured by fitting a circle to the bending part of the flap.

between the top and the bottom of the film. Figure 7.13 shows this strain as a function of intensity UV and visible light. At an intensity of 100 mW cm^{-2}, the average bend radius for A3MA flaps is 0.7 mm, giving a strain difference of 1.4%. For bulk material, the deformation in a film of 10 μm was measured at the same intensity, and found to be −0.75% and 0.35% for the directions parallel and perpendicular to the director.[26] For a splayed orientation, the predicted in-plane strain difference is thus 1.1%, closely matching the deformation of the micro-actuator. The slightly better performance of the printed flaps than the bulk film is ascribed to an overestimation of the strain due to thickness variations.

The response of the DR1A polymer to visible light was significantly less than that of the A3MA for UV light of the same intensity. This is due, in part, to the fact that the molar concentration is 3.16 times lower for the DR1A dye than for the A3MA dye in the polymer host. Also it is speculated that the molecular structure, and, in particular, the fact that the DR1A dye acts as a pendant group on the acrylate backbone rather than as a cross-linker, results in a less efficient strain transfer to the host material.

Finally, it was verified that the A3MA and DR1A polymers can be addressed independently, by actuating the A3MA polymer flaps with visible light and the DR1A flaps with UV light (Figure 7.13). In both cases, the response of the flaps was minimal, confirming that the response is indeed wavelength specific. The simple addressing of the cilia by changing the colour of the light allows a well-controlled motion that has the potential to be more effective in mixing than the electrostatic driven counterparts that can only perform an asymmetric motion.[32] By actuation with light, this design overcomes the inherent problems related to the use of electrical fields in wet environments.

7.6 Conclusion

Liquid crystal networks provide a platform for a range of actuators for miniaturized systems. Their one-step processing from monomer to polymer is fully compatible with current MEMS manufacturing processes and allows actuators to be manufactured *in situ* using photopolymerization. Polymerization in the aligned nematic phase results in a strong anisotropy in the properties of the material, which can be tuned through the control over the molecular alignment of the monomers before polymerization. The anisotropy of the material dictates the direction of the deformation of the actuator. A change of the order parameter is the underlying mechanism that drives the deformation of the material. Here we demonstrated the use of heat and UV light to modulate the molecular order. The high cross-link density of the networks allows for only moderate in-plane strains, but due to the high modulus of the material they have a large work potential. Bending mode actuators are created using alignments such as twisted nematic or splay, where the direction of the twist or splay dictates deformation direction. Patterning of aligned submillimetre structures using photolithographic techniques was demonstrated for thermal actuators. Furthermore, in-plane patterning of the alignment of thermal actuators was demonstrated, creating in-plane variations of the response. Inkjet printing of the actuators opens new possibilities, such as colour printing in order to vary actuator responsiveness in the plane. Using reactive liquid crystals, the actuators can be made in only a few processing steps, exploiting the self-assembling capacity of the material. The processing allows fabrication of large-area and roll-to-roll active all-polymer devices and opens up possibilities for rapid prototyping of low-cost MEMS. We conclude that liquid crystal networks are a promising technique for the production of micro-actuators.

Acknowledgement

Some of the work presented here was supported by the Dutch Polymer Institute under project number 532.

References

1. D. J. Broer, J. Boven, G. N. Mol and G. Challa, *Makromol. Chem.*, 1989, **190**, 2255.
2. D. J. Broer, R. A. M. Hikmet and G. Challa, *Makromol. Chem.*, 1989, **190**, 3201.

3. D. J. Broer, G. N. Mol and G. Challa, *Makromol. Chem.*, 1991, **192**, 59.
4. D. J. Broer, in *Radiation Curing in Polymer Science and Technology*, ed. J. P. Fouassier and J. F. Rabek, Elsevier Science Publishers, London, 1993, vol. III, ch. 12, p. 383.
5. D. J. Broer and I. Heynderickx, *Macromolecules*, 1990, **23**, 2474.
6. R. A. M. Hikmet and B. H. Zwerver, *Mol. Cryst. Liq. Cryst.*, 1991, **200**, 1997.
7. H. S. Kitzerow, H. Schmid, A. Ranft, G. Heppke, R. A. M. Hikmet and J. Lub, *Liq. Cryst.*, 1993, **14**, 911–916.
8. J. Lub, D. J. Broer, R. A. M. Hikmet and K. G. J. Nierop, *Liq. Cryst.*, 1995, **18**, 319.
9. D. J. Broer, *Mol. Cryst. Liq. Cryst.*, 1995, **261**, 513.
10. R. A. M. Hikmet and D. J. Broer, *Polymer*, 1991, **32**, 1627.
11. S. R. Quake and A. Scherer, *Science*, 2000, **290**, 1536–1540.
12. Q. M. Zhang, H. F. Li, M. Poh, F. Xia, Z. Y. Cheng, H. S. Xu and C. Huang, *Nature*, 2002, **419**, 284–287.
13. R. Pelrine, R. Kornbluh, Q. B. Pei and J. Joseph, *Science*, 2000, **287**, 836–839.
14. E. W. H. Jager, E. Smela and O. Inganas, *Science*, 2000, **290**, 1540–1545.
15. D. J. Broer and G. N. Mol, *Polym. Eng. Sci.*, 1991, **31**, 625–631.
16. M. Christophersen, B. Shapiro and E. Smela, *Sens. Actuators B*, 2006, **115**, 596–609.
17. S. Timochenko, *J. Opt. Soc. Am. Rev. Sci. Inst.*, 1925, **11**, 233.
18. H. E. Finkelmann, G. Nishikawa, G. Pereira and M. Warner, *Phys. Rev. Lett.*, 2001, **87**, 015501.
19. C. L. Van Oosten, D. Corbett, D. Davies, M. Warner, C. W. M. Bastiaansen and D. J. Broer, *Macromolecules*, 2008, **41**, 8592–8596.
20. C. L. Van Oosten, C. W. M. Bastiaansen and D. J. Broer, *Nat. Mater.*, 2009, **8**, 677–682.
21. T. J. Ikeda, Y. L. Mamiya and Y. Yu, *Angew. Chem. Int. Ed.*, 2007, **46**, 506–528.
22. N. Tabiryan, S. Serak, X. M. Dai and T. J. Bunning, *Opt. Express*, 2005, **13**, 19.
23. P. Palffy-Muhoray, M. Camacho-Lopez, H. Finkelmann and M. Shelley, *Nat. Mater.*, 2004, **3**, 307–310.
24. K. Lee, M. L. Smith, H. Koerner, N. Tabiryan, R. A. Vaia, T. J. Bunning and T. J. White, *Adv. Funct. Mater.*, 2011, **21**, 2913–2918.
25. K. D. Harris, R. Ruypers, P. Scheibe, C. L. Van Oosten, C. W. M. Bastiaansen, J. Lub and D. J. Broer, *J. Mater. Chem.*, 2005, **15**, 5043–5048.
26. C. L. Van Oosten, K. D. Harris, C. W. M. Bastiaansen and D. J. Broer, *Eur. Phys. J. E*, 2007, **23**, 329–336.
27. S. J. Woltman, G. D. Jay and G. Crawford, *Nat. Mater.*, 2007, **6**, 929–938.
28. J. D. W. Madden, N. A. Vandesteeg, P. A. Anquetil, P. G. A. Madden, A. Takshi, R. Z. Pytel, S. R. Lafontaine, P. A. Wieringa and I. W. Hunter, *IEEE J. Oceanic Eng.*, 2004, **29**, 706–728.

29. T. Mirfakhrai, J. D. W. Madden and R. H. Baughman, *Mater. Today*, 2007, **10**, 30–38.
30. Q. M. Zhang, H. F. Li, M. Poh, F. Xia, Z. Y. Cheng, H. S. Xu and C. Huang, *Nature*, 2002, **419**, 284–287.
31. E. Smela, O. Inganas and I. Lundstrom, *Science*, 1995, **268**, 1735–1738.
32. J. Den Toonder, F. Bos, D. Broer, L. Filippini, M. Gillies, J. de Goede, T. Mol, M. Reijme, W. Talen and H. Wilderbeek, *Lab Chip*, 2008, **8**, 533–541.
33. Y. L. Yu, M. Nakano and T. Ikeda, *Nature*, 2003, **425**, 6954.
34. N. Tabiryan, S. Serak, X. M. Dai and T. Bunning, *Opt. Express*, 2005, **13**, 19.
35. M. Kondo, Y. L. Yu and T. Ikeda, *Angew. Chem. Int. Ed.*, 2006, **45**, 1378–1382.
36. M. Camacho-Lopez, H. Finkelmann, P. Palffy-Muhoray and M. Shelley, *Nat. Mater.*, 2004, **3**, 307–310.
37. T. J. White, N. V. Tabiryan, S. V. Serak, U. A. Hrozhyk, V. P. Tondiglia, H. Koerner, R. A. Vaia and T. J. Bunning, *Soft Matter*, 2008, **4**, 1796–1798.
38. M. Warner and L. Mahadevan, *Phys. Rev. Lett.*, 2004, **92**, 134302.
39. Z. Nie and E. Kumacheva, *Nat. Mater.*, 2008, **7**, 277–290.
40. V. V. Khatavkar, P. D. Anderson, J. M. J. Den Toonder and H. E. H. Meijer, *Phys. Fluids*, 2007, **19**, 083605.

CHAPTER 8

Hydrogel-Actuated Integrated Responsive Systems (HAIRS): Creating Cilia-like 'Hairy' Surfaces

ALISON GRINTHAL*[a] AND JOANNA AIZENBERG*[a,b,c,d]

[a] School of Engineering and Applied Sciences, Harvard University, USA; [b] Wyss Institute for Biologically Inspired Engineering, USA; [c] Department of Chemistry and Chemical Biology, Harvard University, USA; [d] Kavli Institute for Bionano Science and Technology, USA
*Email: agrinth@fas.harvard.edu; jaiz@seas.harvard.edu

8.1 Introduction

By bending and wiggling just 10 stubby fingers, humans have learned to trace out intricate works of art, write an endless variety of stories, and evoke symphonic opuses from a piano. Meanwhile, at the nanoscale, our cells and tissues have learned to talk to their environment by bending and swirling thousands of nanoscale hairs: lungs detecting an inhaled pathogen vigorously beat their cilia and drive it out;[1] hormone-triggered ciliary waves propel an egg as it leaves the ovary,[2] and, soon after, the developing embryo sprouts its own cilia and spins them in spiral patterns to direct molecular and cellular traffic as it grows.[3] Nanoscale hairs' unique combination of high aspect ratio and high surface-to-volume ratio makes them one of the most limber structures known, and with thousands of them, the potential diversity and complexity of collective

RSC Nanoscience & Nanotechnology No. 30
Artificial Cilia
Edited by Jaap M J den Toonder and Patrick R Onck

Published by the Royal Society of Chemistry, www.rsc.org

motions undoubtedly surpass anything we have ever been able to dream of at the macroscale. It is no wonder, then, that the idea of constructing surfaces covered with synthetic analogs has caught the imagination of physicists, chemists, bioengineers, and many others wishing to extend humans' creative and manipulative dexterity to the nanoscale. As natural cilia demonstrate, motions in this realm can be translated into flows and forces able to disrupt bacterial settlement behavior, intervene in cell development, set up patterns and gradients in moving fluids, steer swimming particles, modulate chemical and biomolecular processes, and more, yet they also enable us to envision manipulating flow in its broadest sense: of heat, current, even light.

Nanoscale dexterity does not come cheap in the natural world, where it takes over 650 different proteins to build one cilium. A promising biomimetic approach is to dispense with the biological bells and whistles and distill cilia down to the mechanical element most essential for moving things around: their outer shape. Various synthetic structures have been designed to generate rhythmic pumping and mixing forces and do so quite well, often under remote control by magnetic or electrostatic fields, light, or other inputs.[4–7] Nevertheless, this streamlining leaves behind another, equally ingenious dimension of their biological heritage. Cilia in the wild continuously negotiate an onslaught of signals, not only hydrodynamic but all kinds of mechanical and chemical stimuli wafting through their territory, and autonomously adapt their motions to their ever-changing environment. If a micro-organism bumps into an obstacle or encounters a noxious chemical, its cilia detect the problem, adjust their beating, and steer away.[8,9] Respiratory cilia actively increase their beat frequency if the mucus on top of them grows more viscous or accumulates nasty compounds;[1,9,10] cilia in the reproductive tract beat when hormones tell them to.[11] The result is motions that evolve in tandem with the environment and automatically remain finely tuned to the situation. Although holding artificial cilia to these standards requires braving another look at their biological complexity, in the end designing them to read and adapt to their surroundings on their own is far more efficient than trying to keep up with every local change and constantly reprogram them ourselves.

For translating signals into motion, it is what's inside each cilium that matters most. The guts consist of a microtubule skeleton that provides overall shape, interwoven with a network of muscle-like proteins. The muscle network directly senses a wide assortment of chemical and physical signals such as calcium, magnesium, nucleotides, redox state, temperature, and mechanical pressure, and transduces them into chemical events; specifically, ATP hydrolysis and atomic rearrangements within the proteins.[1,12–14] These, in turn, produce larger scale muscle reconfigurations that generate mechanical forces, and the mechanical forces bend and slide the microtubules.[15–18] As a network, this system integrates the various signals across space and time and turns them into an infinite variety of corresponding cilia motions, translating the complex environmental scene into a mechanical language of bends, twists, and swirls.

The prospect of trying to engineer such a system into every nanohair is not inspiring at all, but nature is also our inspiration when it comes to irreverent interpretation of the details. Life isn't shy about taking a good idea and changing it around to suit different contexts, and has in fact done so over and over with the idea of pairing chemo-mechanical 'muscles' with structural elements. Variations on the theme range from the filaments that coordinate cell division to skeleto-muscular systems that walk, with the specific configurations and muscle molecules different in every case.[19–21] For our synthetic version of cilia, then, we have kept the concept of coupled muscle and structural elements but turned the cilia inside out: instead of putting muscle inside each structure, we embed an array of solid nanohairs in a network of muscle-like responsive hydrogel. Like the natural muscle proteins, the networked hydrogel polymers undergo molecular rearrangements in response to a wide range of stimuli, integrate the chemical changes into mechanical forces—in this case, swelling and contracting—and bend and twist the nanohairs. This design not only makes each hair's motions responsive to the changing environment, but, by embedding all the hairs in a single muscle network, also introduces a new capacity for creating large-scale collective motions in intimate conversation with chemical and physical stimuli. In this chapter, we discuss our strategy for, first, designing and fabricating versatile surfaces of nanohairs capable of a tremendous diversity of complex, reversible motions;[22–31] and, second, integrating them with equally versatile hydrogel designs that selectively evoke these motions over multiple scales in synchrony with spatiotemporal changes in the environment.[32–38]

8.2 Creating the Lexicon of Nanohair Motions

Much of our appreciation for the many ways nanohairs can move has come from witnessing what they do in human-made systems that, though not alive, have nevertheless remained untamed wilderness for most of their history. Attempts to create synthetic arrays of nanohairs—an increasingly popular biomimetic pursuit in recent years[39]—frequently yield disorder due to the hairs' notorious tendency to buckle, clump and collapse as they dry.[40,41] Closer scrutiny, however, reveals an amazing diversity of ways to make a mess; strewn over the surface are all sorts of bundles and swirls of different shapes, sizes, even chirality and hierarchical composition. Randomly thrown together they may be worthless, but each attests to its own story of a unique set of motions that produced it. Since everything takes place under our noses in rather simple synthetic systems, we have no excuse that these motions arise from biological complexity beyond our ability to understand and control. We have therefore turned this delinquent capacity for motion into our ally by deciphering how the various motions and structures depend on the hairs' mechanical, geometric, and surface properties and on their interaction with the evaporating liquid, and then using this insight to program the arrays for specific dynamic repertoires.

8.2.1 Nanohair Gymnastic Abilities

At the most basic level, high aspect ratio nanohairs are excellent at bending. Given enough force, they can bend far enough to meet each other in small

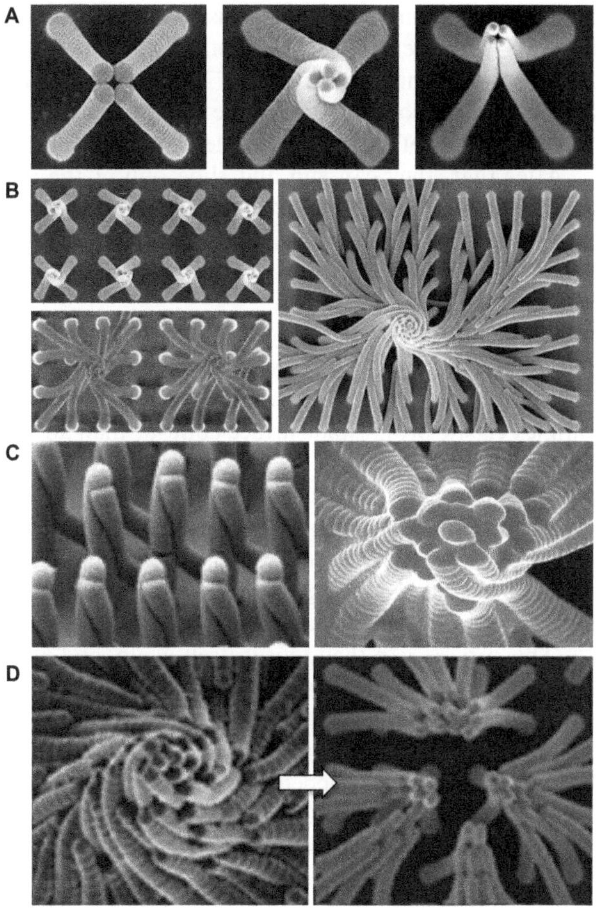

Figure 8.1 Nanohair dynamics. (A) Bending hairs can meet at their tips, twist into chiral swirls, or zip together. (B) Each cluster can combine with others in increasing levels of hierarchical complexity. (C) Chirality and symmetry can be controlled by introducing anisotropy into the hairs. (D) Clustering and swirling are reversible, with larger clusters disassembling in hierarchical stages.

clusters, coming together like grasping claws or creating bridges from one hair to the next (Figure 8.1A, left). But the fun does not end there: once they meet, they can slip, slide, and twist around each other, swirling their tips and potentially the entire hairs in a spiral (Figure 8.1A, center).[23,25] Alternatively, the hairs can zip together in straight-laced fashion, coming together to form bundles of parallel hairs (Figure 8.1A, right).[25] Interestingly, these same types of motion underlie the behavior of natural cilia, whose asymmetric movements are determined by parallel bundles of microtubules slipping, sliding, and twisting along each other inside each cilium.[15–18] Bundling and sliding are also fundamental to how stereocilia—closely related nanohairs in our ears that translate sound waves into nerve signals—distinguish pitch and frequency; bundling determines effective stiffness, and stick–slip motions mediated by surface glycans fine tune mechano-sensitivity.[42–44] In our synthetic arrays, these motions are controlled by a combination of the elastic and adhesive properties of the hairs.[23,25] Hairs must of course be flexible enough to reach each other in the first place. After that, they can slip and slide only if the adhesive forces between them are not too strong, and, if they do slip, their flexibility determines how far they can twist around each other. Zipping, on the other hand, requires being both sticky enough not to slip and flexible enough to accommodate extensive bending.

Each of these configurations can then become a moving entity in its own right. Bridged, helical, and zipped clusters develop new characteristic stiffness, symmetry, and surface properties: for example, a helical bundle may have a much more complex elastic profile and surface contour than an individual hair; a zipped cluster will be stiffer; and even simple bridged clusters will be influenced by the displacement between the hairs' bases and tips. Clusters can consequently bend and swirl in chiral or other patterns not seen with the individual hairs. If their collective elastic and adhesive properties allow, the clusters can also come together, bridge, twist around each other, or align in parallel, creating yet another level of elastic, surface and dynamic complexity.[23,25] This process can then repeat itself again, and again, creating a hierarchy of collectively moving groups of hairs (Figure 8.1B). The number of hierarchical stages can be controlled by the effective stiffness, adhesive properties, and spacing of the hairs, although it is not a simple equilibrium function of these factors but rather depends on the stepwise acquisition of new properties as the system progresses dynamically from one stage to the next.[25]

If the hairs are perfectly vertical, cylindrical and evenly spaced, their bending direction and the handedness of their helical twisting and chiral swirling is random. However, introducing just a slight but systematic anisotropy into the hairs can overcome the randomness to produce whole arrays that bend and swirl in unison, with prescribed directions and chirality. A simple way to do this is to tilt all the hairs in the same direction. If they are arranged in a square lattice, for example, tilting them slightly off the lattice direction is enough to bias them to come together, slip and wind with synchronized chirality[23,25] (Figure 8.1C, left). Orientation is likewise a critical feature of biological cilia and is believed to underlie their chiral swirling in growing embryos.[45] Instead of tilting, the same effect can also be obtained in our synthetic arrays by giving the hairs an asymmetric cross section that makes them effectively stiffer and less likely to bend along one axis than another. Biasing their bending direction in this manner also provides control over how clusters come together at higher hierarchical stages. For example, hairs with an elliptical cross section form anisotropic bundles with cross sections reflecting those of the hairs[25] (Figure 8.1C, right).

Most essentially, all of this behavior can work in reverse. As long as the hairs are elastic, they are inherently poised to spring back to their original shapes and positions as soon as the driving force is removed. Within hierarchical clusters, dissociation takes place through a series of hierarchical stages, with large clusters giving way to smaller groups (Figure 8.1D), each of these subsequently breaking into smaller groups, and so on.[25] The tendency to spring apart is counteracted by the hairs' tendency to stick to each other. Once they meet, a tug-of-war between elastic and adhesive forces determines how readily they separate at each stage. By manipulating this balance, we can specify whether clusters become essentially permanent new composite cilia, or reversibly cluster and uncluster, or, for larger bundles, some combination of the two.[25,26] Fine tuning the surface chemistry also provides a way to make reversibility responsive to the surroundings, by making their affinity for each other dependent on the pH, composition, or other aspects of their environment.

8.2.2 Programming the Elements of Motion

Although ultimately the nanohairs' motions will be shaped by the gel and environmental stimuli, a crucial part of designing the responsive system is to program how the hairs can and cannot move under different forces. Doing so requires specifying their elastic, symmetry and surface properties as discussed above and explored by others,[41,46–51] a fabrication feat we have made simple and straightforward by introducing a set of techniques for controlling each feature independently and with almost arbitrary precision. Using a variant of soft lithography, we create replicas of a master array in any material we choose, *via* an intermediate elastomeric mold that can be stretched, twisted, or bent before the materials cure to produce an assortment of new geometries.[22,24] A complementary benchtop technique, Structural Transformation by Electrodeposition on Patterned Substrates (STEPS), lets us sculpt these or the original master structures into hairs with a continuum of widths and conical, overhanging, bent or hairy shapes with arbitrary proportions.[28,29] As we describe, these methods, plus surface chemistry modification,[25,26] make it possible to pick and choose the fundamental elements of motion and assembly that can be induced and combined into complex responses by the different stimuli.

We program how far hairs can bend under force by the choice of both materials and geometry (Figure 8.2, top left). To account for these features' combined effects, we define a unified quantity, effective stiffness, which expresses the force required to deflect a hair a given distance as a function of its elastic modulus, length and radius.[22] The modulus is set by the material composition, which can be anything we pour into the elastomeric mold using the double replica technique: materials can range from very stiff, such as ceramics, to a variety of intermediate to very soft polymers. To access more or less any arbitrary point along a continuum of moduli, we mix high- and low-flexibility polymers in the appropriate proportions and apply the mixture to produce the replica. Using the STEPS technique, we can also set the radius to any arbitrary value. Starting from thin hairs, we deposit a smooth conformal film around each hair to increase its radius, which can be varied continuously by controlling the deposition conditions. This modified array can then be

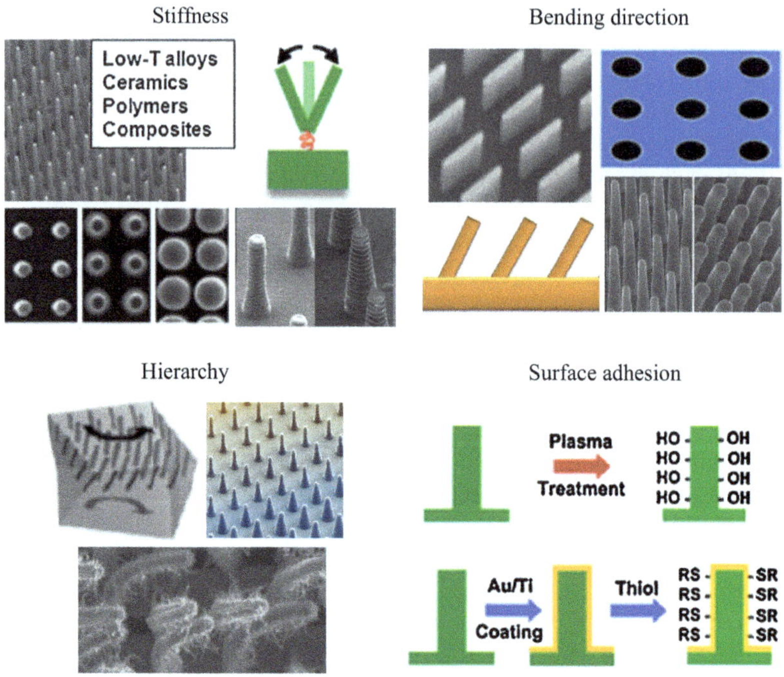

Figure 8.2 Programming mobility. Stiffness (top left) is controlled by material and geometry. The elastic modulus is varied by replicating the array in different materials and composites, and the radius is systematically modified by STEPS (see text). Either can be varied along the hair, *e.g.* to create flexible linkers or mechanically reinforced cones. Bending direction (top right) is specified by anisotropic cross section or orientation, introduced in the original master array (plates), by stretching or shearing the mold (elliptical or tilted), or by STEPS (bent). Hierarchical patterning (bottom left) can be introduced at the array level, by complex mold deformations or by gradients in STEPS conditions, as well as at the sub-hair level by depositing tiny hairs on each hair. Surface adhesion (bottom right) can be varied from weak to strong either by plasma treatment or by coating with thiol-attached organic molecules.

replicated as above, such that the radius and modulus can be chosen completely independently of each other.[27] Each parameter can also be varied along the length of the hair. If we want a hair to deflect under low force but to be sturdy along most of its length, we can create a flexible linker region at the base by sequentially adding high- and low-modulus material precursors to the elastomeric mold during replication.[24] A version of STEPS also allows us to transform cylindrical hairs into tapered shapes with radii varying from bottom to top, such as mechanically reinforced cones with wide bases and thinner tips.[29]

A hair's bending direction in response to an isotropic force is programmed by the choice of both shape and orientation (Figure 8.2, top right). As discussed above, an anisotropic cross section makes a hair proportionately easier to bend along one axis than another. We can create such structures in the original master array, as we have done with our rectangular microfins, or we can deform the elastomeric mold during replication to stretch, for example, cylindrical hairs into ellipses. Since the mold can be stretched or twisted in any direction and to any extent within the elastic range of the mold, we can create nearly any shapes and proportions we want from the same starting array. Orientation can likewise be specified arbitrarily by shearing the mold to tilt the hairs any amount in any direction. Rather than tilting, we can also create bent hairs using another version of STEPS. These approaches can readily be combined to create still more complex possibilities for translating forces into directed motions.[27]

Both the double replica and STEPS methods also allow us to program elements of multi-scale architecture (Figure 8.2, bottom left). Stretching or shearing the elastomeric mold uniformly stretches or tilts the hairs, but we can also bend, twist and/or otherwise deform the mold in countless ways to create arrays with hair orientations and cross sections varying in different patterns over the surface. In addition to these patterns, STEPS provides a way to create gradients of shapes and sizes progressing smoothly from one end of the array to the other. Both approaches, alone or in combination, enhance and control the range of options for how hairs might interact or bend collectively over the scale of the surface. Going in the other direction, we can also create multi-scale motion on the surface of each hair. Using STEPS, we deposit tiny hairs on each hair,[28] producing hairy hairs that not only have much more complex hierarchical shapes but can potentially undergo motion on both scales. Any array can be transformed into hairy hairs, enabling us to independently specify the features of hairs, arrays, and sub-hairs.

A hair's adhesion to other hairs—which governs its ability to slide or dissociate—can be programmed by the choice of materials, but to specify it independently and with finer control, we directly modify the surface chemistry after the hairs are constructed (Figure 8.2, bottom right). One approach we have used is to plasma treat polymeric hairs to generate free hydroxyl groups; adhesion increases with increasing plasma treatment time.[25] For even finer control and access to a wider range of chemical properties, we instead coat the hairs with a thin layer of gold and then with self-assembled monolayers (SAMs) of alkylthiol molecules.[26] By choosing from an assortment of functional groups that interact through van der Waals, hydrogen bonding, or disulfide interactions of varying strengths, and using a range of carbon chain lengths, we are able to program adhesion along an entire spectrum from weak to strong. This approach also gives us the flexibility to choose SAMs whose adhesive strength is sensitive to chemical changes in the surroundings.

8.2.3 Cooperation with the Solvent

Of course, even in a plain evaporating liquid, the array properties are just half the picture. Before we take on adding a hydrogel muscle, this 'simple' system offers us valuable insight into how nanohairs cooperate with the surrounding medium and how the medium can be manipulated to shape local and array-wide motion. The key feature of an evaporating liquid is that its surface contour changes continuously. If the surface is pinned to the hairs, the shrinking liquid level pulls the hairs down and the hairs resist, pulling the surface up along their sides. In other words, the surface contour shapes the hairs and the hairs shape the surface contour. In dynamic terms, this means that (1) if one hair moves, it alters the local surface contour seen by the hair next to it, such that motions are propagated among hairs *via* shape changes in the medium; and (2) we can step in and influence the process by controlling the surface contour independently.

Striking examples of the unique patterned motions that can be produced by the latter approach can be witnessed by placing a simple mask over an array of drying hairs.[30] If a mask with, say, a raised honeycomb pattern is placed face down on the liquid surface, the liquid pins to the texture and develops a honeycomb patterned surface contour. As evaporation proceeds, the liquid is also pinned to the hairs and bends them toward the honeycomb, so that the whole array moves according to the imposed pattern (Figure 8.3, left). The closer the scale of the mask pattern gets to the spacing of the hairs, the more intricately we can control the hair motions. If we just use an identical hair array as the mask, but rotate it slightly, the resulting surface contour bends all the hairs in a periodic chiral Moiré pattern (Figure 8.3, right). Tiny changes in the rotation angle allow precise tuning of the periodicity, and the chirality can be switched by rotating the mask in the opposite direction. These examples use only garden-variety arrays of cylindrical, vertical hairs; they barely begin to suggest what might be possible if surface contour manipulations are superimposed on the hair structural manipulations already known to influence their dynamic complexity.

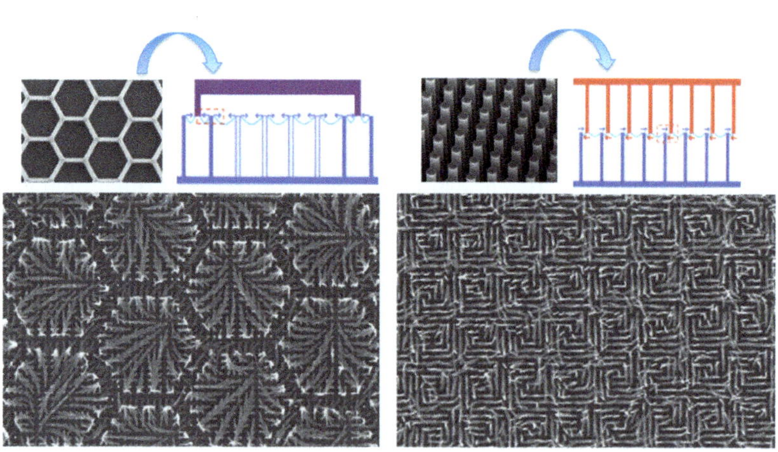

Figure 8.3 Shaping the surface contour of the evaporating liquid by pinning it to a textured mask induces hairs to bend in complex long-range patterns on the scale of the mask features.

8.3 Hydrogel as the Driving Force: Nanohairs Meet Their Match

Now it is time to explore stimuli beyond evaporating liquids and program hair motion to be selective by providing a directional muscle bundle, a hydrogel. As driving forces, hydrogels preserve some of the basic features of an evaporating liquid: their surface runs from one hair to another, their level shrinks as they dry, and their changing surface contour shapes and is shaped by embedded hairs. But at the same time, they are much more interesting. Their behavior is tuned in to nearly any feature of the environment we design it to be, whether chemical, mechanical or any other form. While a liquid is gone once it evaporates, the polymer network of a hydrogel stays in place and shrinks and swells as stimuli come and go, making the whole system directly reversible and continuously responsive. As elastic networks, hydrogels can hold a structure of their own rather than being enslaved to surface tension, so they can be designed with inherent complex shapes and surface contours that complement those of the hairs. Structure also means they can be compartmentalized, with gel properties varying over the array, and provides a multitude of options for hair–gel configurations. In incorporating hydrogels as the muscle component of artificial cilia, we aim to make use of all these qualities in their full splendor.

8.3.1 Tailoring Hydrogels as Chemo-mechanical Transducers

Chemical tinkering with the hydrogel polymer network provides numerous possibilities for choosing which stimuli it can sense and how it responds.[52–57] All hydrogels work essentially the same way: the polymer chains can either pack densely with little water inside (the contracted state), or they can loosen their interactions, absorb water, and expand (the swollen state). Their chemical functional groups determine the polymers' attraction or repulsion for each other and/or for water, and have been extensively tailored to make swelling and contracting responsive to a wide range of stimuli such as humidity, pH,[58] salts, biomolecules,[59] temperature,[60] light,[61,62] electric fields[63] or mechanical pressure[64,65] (Figure 8.4). For example, polyacrylamide-based gels containing acrylic acid side chains contract at acidic pH, but above the pK_a, the acid's protons dissociate, an osmotic imbalance is created, and the network absorbs water and swells.[58] Sensitivity to temperature is achieved by using pNIPAAm (poly(*N*-isopropylacrylamide)) gel: at low temperatures, it swells due to hydrophilic interactions between the side chains and water, but above a critical temperature (the lowest critical solution temperature, or LCST), hydrophobic interactions between the isopropyl groups dominate, water is expelled, and the gel contracts.[60] Sensitivity to either stimulus can be fine tuned over a wide range by choosing side groups with different pK_a values or transition temperatures. The swelling ratio—how much the gel can swell, and thus how much force it can exert—can also be controlled by controlling the density of cross-links between polymers.[66,67] Chemistry, cross-linking density, and other factors can also be manipulated to tune the kinetics of both swelling and contracting.[58]

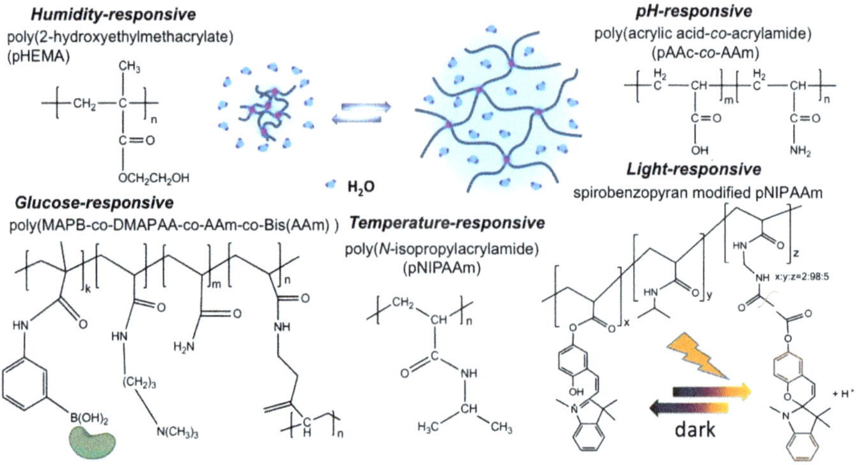

Figure 8.4 Hydrogels shrink and swell *via* chemical changes that occur in response to different stimuli.

Besides being highly tunable in terms of when, how much, and how quickly they respond, hydrogels can also be designed to swell and contract in elaborate chemo-mechanical patterns. As Turing suggested, the dynamic interplay between chemical and mechanical processes can introduce higher order complexity into a system,[21] and, as chemo-mechanical transducers, hydrogels are a prime example. Such patterns can take the form of complex propagation of chemical changes within the shape-changing gel, as has been investigated and modeled in self-oscillating gels that locally expand and contract in conjunction with an oscillatory chemical reaction taking place within them.[68–71] If a gel is attached to a rigid base or otherwise restricted in the directions it can expand, it can wrinkle or crease as it swells or contracts.[72–75] Introducing nanoscale protrusions and textures into the base has been found to generate complex, reversible three-dimensional patterns in the gel's surface contour.[74]

A variety of set-ups enable the forces generated by these chemo-mechanical processes to be translated into nano-structure motion. The gel itself can be shaped into nano-structures, with bending driven by propagation of chemical signals within and between neighboring structures, as has been demonstrated for artificial cilia made entirely from self-oscillating gels.[76,77] If a layer of gel is attached to a layer of flexible polymer, its expansion can bend or curl the polymer, with diverse possibilities for controlling the shape and extent of bending by varying the geometry of how and where the gel is placed on the polymer structure.[78] Placing the gel inside a bundle of carbon nano-tubes enables it to reconfigure the bundle as it expands and contracts, with diverse and complex options for controlling motion by the orientation and mechanical profile of fibers within the bundle.[79] Embedding an array of nanohairs in hydrogel allows us to build on elements of each of these scenarios.

8.3.2 Putting Hairs and Gels Together

To make the most of both nanohairs' and hydrogels' immense and complementary ranges of behavior and tunability, we take a modular approach that enables us to program them independently of each other and then put them together.[36] Before introducing any gel, we design and construct the nanohair arrays using the full complement of techniques described above to specify the material, shape, multi-scale and adhesive properties that govern their range of motion. In a separate step, we chemically design and synthesize the molecular components of the gel to determine the network's sensitivity to stimuli. We then combine the two and polymerize the gel on the array, creating our surface of gel-embedded nanohairs.[32,34,37,38] A critical feature of construction, however, is that the array and gel must be attached to each other; just as an evaporating liquid must be pinned to the hairs to impose force, the gel polymers need to be stably coupled to the hairs to transmit mechanical forces to them efficiently. We accomplish this by introducing a covalent chemical bonding step into the assembly.[32,34] If we use silicon hairs, we react them with poly(glycidyl methacrylate) (pGMA); for polymeric materials we incorporate GMA directly into the polymer precursor during replication. Either way, the hairs end up with glycidyl groups that are then coupled to a reactive molecule such as acrylic acid that covalently bonds to the gel's side chains during polymerization. The result is a hybrid surface we call HAIRS, for Hydrogel-Actuated Integrated Responsive Systems (Figure 8.5).

The nanohairs are now set up to move in synchrony with environmental stimuli. If a stimulus (or its absence) favors the dehydrated form of the gel, the gel contracts and pulls the hairs down with it (Figure 8.5, right). If the conditions favor hydration, the gel swells and lets the hairs stand back up. If the

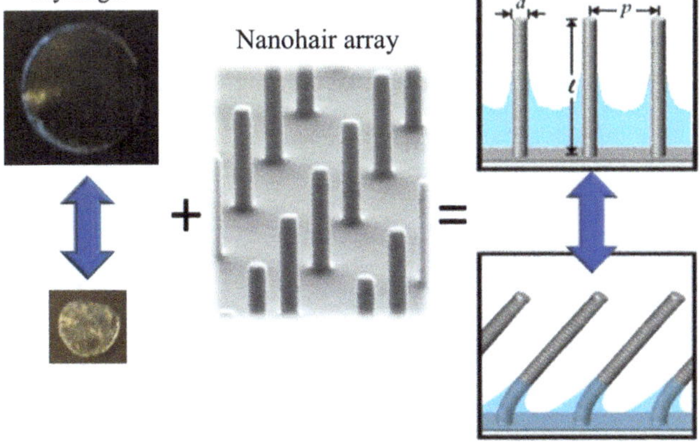

Figure 8.5 Nanohairs embedded in a layer of responsive hydrogel (Hydrogel-Actuated Integrated Responsive Systems, or HAIRS) bend and straighten as the gel contracts and swells.

gel swells and contracts isotropically around a hair, the hair will bend either randomly or in a direction programmed by its shape or orientation, but whatever it does, it will pull or push back on the surrounding gel, change the local surface contour, and propagate its motion to the adjacent hairs. Modeling of the system suggests that the interplay between hairs and gel may extend beyond mechanics and reach all the way back to the gel's chemistry: the moving hairs may influence the gel's tendency to absorb or release water and potentially modify how it responds to stimuli.[80,81] The hairs, gel and environment thus communicate through a complex, dynamic exchange between chemistry and motion.

8.3.3 Sculpting the Gel

The gel's shape, as we have mentioned, is not just a matter of expanding and contracting while being pushed and pulled around by the hairs. Like the hairs, the gel is elastic and has its own distinctive shape that can be programmed independently, remembered over cycles of swelling and de-swelling, and brought to bear on its interaction with the hairs. Finite element modeling of the hybrid system gives us an idea of how this can work.[34] All other factors being equal, if a gel's surface is perfectly flat, it will contract or swell equally on all sides of a hair and will not impose a net force in any direction. However, if the gel has an uneven surface topography that makes it thicker on one side of the hair, modeling indicates that as the gel contracts, the hair will bend toward the thicker region, where the absolute change in volume is greatest. For example, if the surface is dome-shaped, all the hairs within the dome will bend toward the center (Figure 8.6). Since gels are rarely perfectly smooth, this mechanism no doubt underlies much of the random bending directions we see in simple systems of cylindrical hairs and flat gels; a defect that makes the gel thicker in one spot creates local domains of hairs that bend toward it.

Clearly, then, structuring the gel is just as fundamental to programming the nanohair motions as designing the hairs. Although the gel precursors are liquid

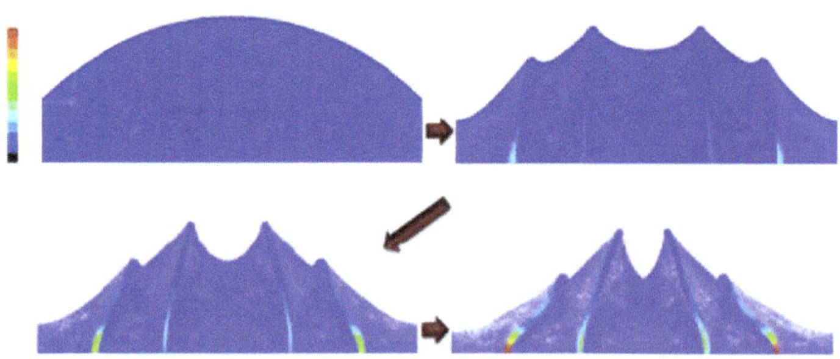

Figure 8.6 Finite element modeling indicates hairs bend toward regions where the hydrogel is thickest.

when we add them to the hair array, we can shape the surface topography independently by placing a three-dimensional patterned mask over the liquid so that the surface takes the shape of the pattern upon polymerization (Figure 8.7A). Unlike in the evaporating system, the mask is no longer

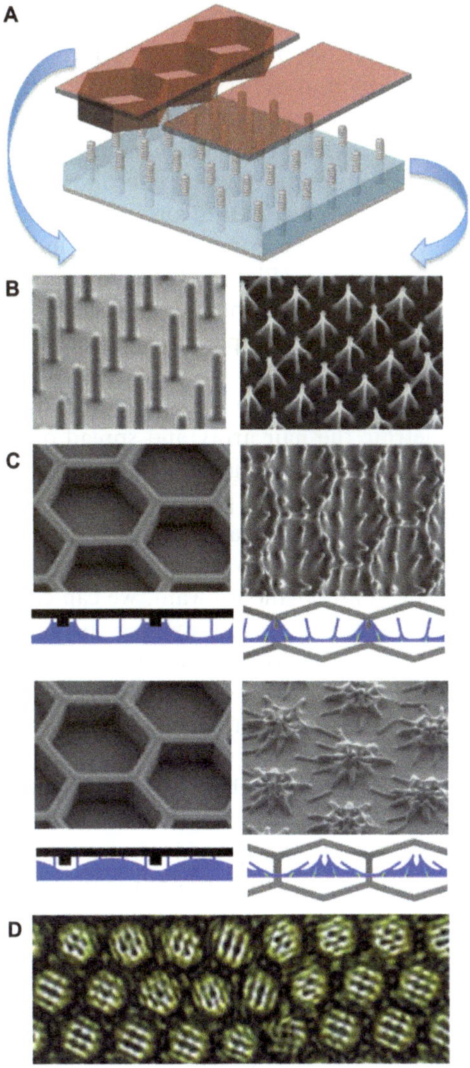

Figure 8.7 Patterning the gel surface with a textured mask during polymerization (A) creates variations in thickness that reversibly bend the hairs in programmed directions. A post-studded mask leads to clustering (B), while a honeycomb mask leads to opening or closing microflorets (C, upper and lower) depending on whether the mask is non-wetting or wetting (schematics below corresponding images). (D) A thickness defect causes florets to swirl chirally as they open or close.

needed after the initial construction step. Introducing different surface patterns makes it possible to translate a given stimulus into all sorts of different motions even if the array is the same in every case.[32,34] Using a mask with an array of nano posts to create thick spots at the center of every four hairs, we can program the hairs to repeatedly cluster and uncluster in groups of four as the gel contracts and swells (Figure 8.7B). A honeycomb pattern induces the hairs to bend away from each other into opening microflorets (Figure 8.7C, upper panels). However, the chemistry of the mask surface also plays a role in determining the final gel topography. If the mask has a non-wetting surface chemistry, the gel precursors contact only the raised parts of the pattern and become thickest in these areas; this mechanism is responsible for the honeycomb outlines that produce the opening microflorets. But if we make the mask wetting, the gel penetrates the texture and produces the inverse pattern. A gel patterned by a wetting honeycomb mask thus induces hairs to come together as closing microflorets (Figure 8.7C, lower panels). Combining regular surface patterns with a well-placed defect introduces yet another interesting effect: the microflorets open and close with a chiral twist (Figure 8.7D).

Yet we need not limit ourselves to surface topography. The liquid gel precursors initially penetrate the hair array and spread over the base, but we have developed a method for sculpting the gel in three dimensions by controlling exactly which parts of it polymerize. Using multiphoton lithography (MPL), we induce polymerization only where we focus the laser beam[38] (Figure 8.8A). This allows us to essentially write the gel onto the hair array. Since we can focus in any plane we want, we can control precisely how thick the gel is and can suspend the layer anywhere between the base and the tips of the hairs (Figure 8.8B). Controlling the thickness provides control over the rate of swelling, since thinner gels swell more quickly, while liberating the gel from its attachment to the base greatly enhances its ability to expand and bend the hairs. A suspended gel is free to swell laterally and thus bends the hairs outward, generating far more force than an equivalent gel restrained by the base.

This capacity for lateral expansion provides yet another way to program patterned hair motions. MPL allows us to sculpt the suspended gel into circles, hearts or any other arbitrary shape and size (Figure 8.8C). Since hairs bend in the direction of expansion, all hairs encompassed by the gel will bend outward from the center toward the boundaries of the shape as it swells. This allows us to create, for example, regions of hairs that collectively beat in the shape of a heart in response to changing temperature or other stimulus (Figure 8.8C). Since we can directly specify the shape, thickness, and vertical placement of the gel as well as the effective stiffness of the hairs, we can predict and control not only the direction but also the distance each hair bends as a function of its location in the gel. A force map for all the hairs embedded in a circular gel suspended at their tips (Figure 8.8D) shows that, in general, force and bending magnitudes are highest around the perimeter, where forces reach on the order of micronewtons.

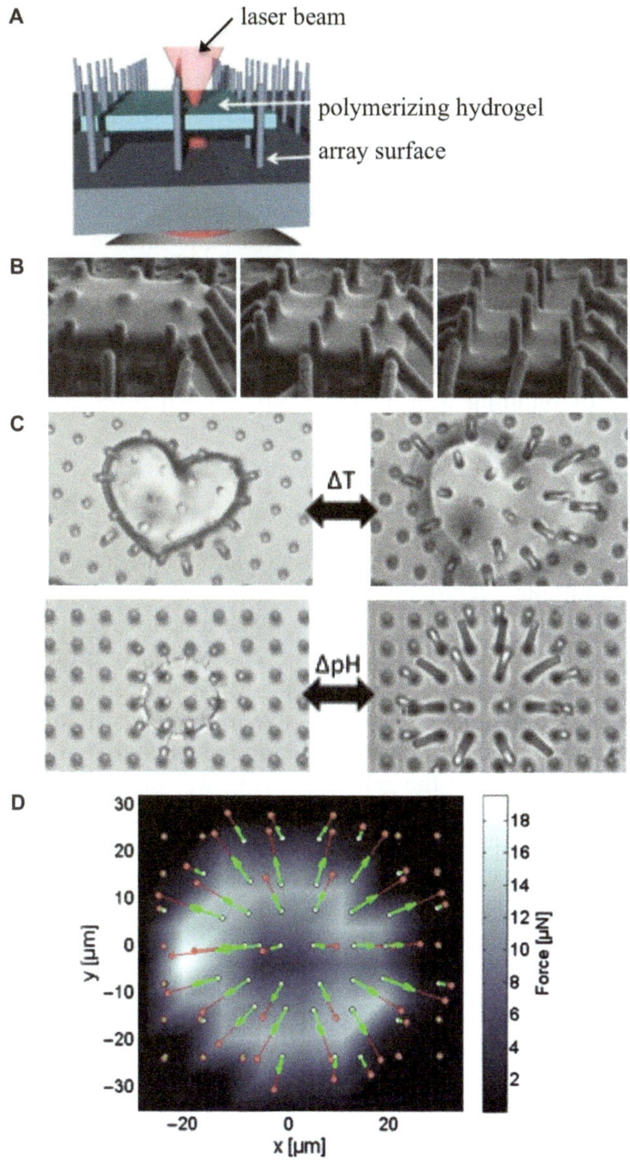

Figure 8.8 Multi-photon lithography (A) directly writes the gel onto the array and allows it to be sculpted in three dimensions. (B) A thin layer of gel can be suspended at the tips, middle or base of the hairs. (C) Suspended gels expand laterally and bend the hairs outward. (D) The force and extent of bending can be predicted for each hair and increase toward the edges of the gel.

8.4 HAIRS in the World

Once the nanohair–hydrogel duo is ready for action, we can put it to work just about anywhere. It can function while submerged in liquid,[37] using gels that sense pH, dissolved molecules such as glucose, or features of the physical environment such as temperature, light, or mechanical pressure, or we can design it to work in air in response to ambient conditions such as humidity.[32,33] At the same, the system is effectively self-powered and can be scaled up to any size without regard for the long-range distribution or cost of energy. Since all the energy required to drive the nanohair motions is derived from the changes in the surrounding environment, powering and sensing are both part of the same process: the surface locally scans its environment for changes, which provide both the energy and the instructions for driving local nanohair motions. As a self-scaling, self-powered, air- or liquid-compatible system, HAIRS can be adapted to work within the channels of a micro-fluidic chip,[37] in our bloodstream, on the side of a boat, on a window or building wall, inside a pipe, or wherever else it may come in handy.

8.4.1 Reading a Complex Environment

In even a tiny micro-fluidic channel, HAIRS will often be faced with a complex confusion of multiple overlapping signals each evolving over space and time. To enable the system to read and sort out as much of this information as possible, we can draw on several of our design strategies to program sensitivity to multiple stimuli. At the level of gel chemistry, we can modify the gel to include a series of functional groups sensitive to different stimuli, so that each stimulus evokes a characteristic response alone or in combination with others (Figure 8.9A). By sculpting the gel using MPL, we can go further and create interlocking zones composed of gels with distinct stimulus sensitivities. With just two interlocking puzzle pieces, one sensitive to pH and the other to temperature, we observe four distinct responses depending on whether the surroundings have a low pH and low temperature, low pH and high temperature, high pH and low temperature, or high pH and high temperature[38] (Figure 8.9B). Since expansion or contraction of one gel can locally compress or expand the other, the four different combinations can be distinguished at the level of single posts at the interface, as well as by the larger scale hair motions. On top of gel patterning, we can also design the hair tips to directly sense multiple stimuli by using micro-contact printing to pattern them with different surface chemistries[26] (Figure 8.9C). For example, if tips in adjacent regions were functionalized with different SAMs whose adhesive properties were sensitive to different chemical solutes, the system would be able to integrate these signals with those the gels sense.

HAIRS surfaces can also finely distinguish spatial and temporal patterns in the environmental stimuli. A pH-sensitive HAIRS surface incorporated into a micro-fluidic channel is able to translate adjacent streams of acid and base into adjacent regions of bent and upright nanohairs, with a sharp boundary reflecting the change from acid to base[37] (Figure 8.10). Stimulus-sensitive hair tips would potentially enable the system to distinguish stimulus differences between the base and tips of the hairs in addition to these lateral spatial

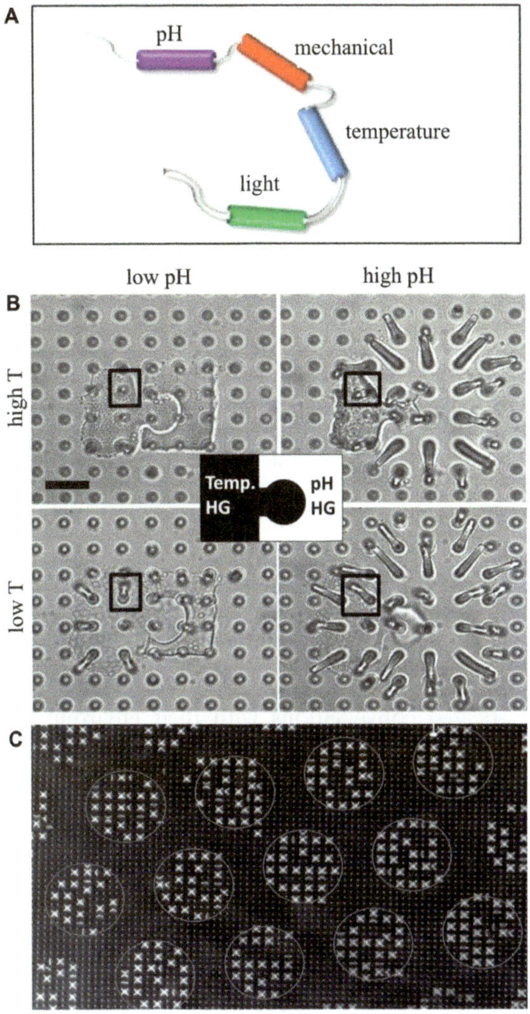

Figure 8.9 Reading a mixture of stimuli. (A) Multi-stimulus sensitivity can be introduced into a single gel by incorporating multiple functional groups. (B) Interlocking regions of gels sensitive to different stimuli lead to distinct hair reconfigurations for each possible combination of stimuli. (C) Analogously to (B), the hair array can be patterned with different surface chemistries that confer different stimulus sensitivities.

patterns. Since the gel is thin and the hairs are nanoscale, the response time is also fast enough to distinguish when and how quickly the stimulus changes. Switching from an acid to a base solution in a micro-fluidic channel induces an almost immediate nanohair response (Figure 8.10, right), while changing the pH slowly by electrochemically generated gradients produces a slower, gradual response reflecting the slower rate of change.

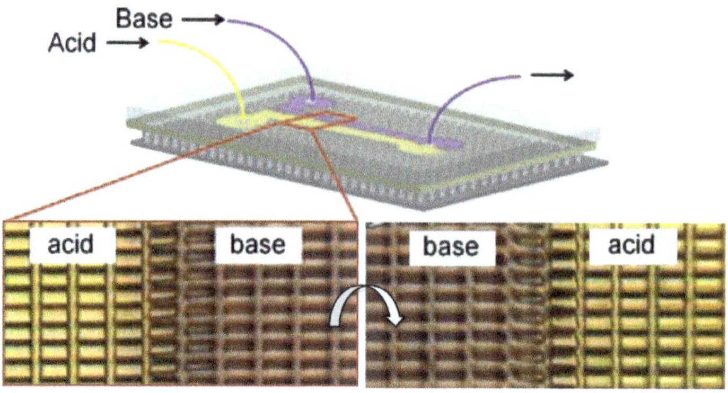

Figure 8.10 Discriminating spatial and temporal patterns. Hairs embedded in pH-responsive hydrogel and incorporated into a micro-fluidic channel contract under a stream of acid and straighten under a stream of base, with a sharp boundary between the two (left). Switching acid and base leads to fast switching between bent and upright regions (right).

8.4.2 Talking Back

With the ability to read a complex mix of changing signals, and the versatile nanohair mobility to respond in kind, HAIRS surfaces have all the elements to take part in sophisticated conversations with their environment (Figure 8.11). They can be programmed to detect the chemical chatter of settling bacteria, and intervene with motions that scramble the signals, scatter or trap the individuals, and turn the surface into a sea of dynamic patterns inhospitable for biofilm development.[82–84] Hair motions can potentially self-adapt to trap and foil a whole gamut of different-sized organisms by adjusting the scale of hierarchical swirling based on signature chemical cues. Living or non-living particles can be selectively trapped, released and propelled in different directions by chiral clusters that swirl together and apart with adjustable size and handedness.[23,85] Responsively patterned chiral motions could likewise enable a swimming particle to steer itself through a complex maze of signals and release its cargo upon reaching its destination. Moving hairs can also manipulate liquid flow by changing the wetting properties of the surface,[33] such as switching it from hydrophilic in a dry environment to super-hydrophobic in a humid environment or responsively changing droplets' rolling and sliding direction with dynamic anisotropic patterns.

At the nanoscale, moving hairs can capture and manipulate not only liquids and particles but also light. As our chirally swirling microflorets open and close, the reconfiguring hairs change the wavelengths of light absorbed and reflected and produce dramatic patterned color changes.[34] Responsively reconfiguring hairs can similarly be designed to transmit or reflect light in response to environmental conditions.[35] Clustering can also alter the spectral properties of trapped or tip-attached molecules,[86,87] changing the optical properties of the surface, or can start, stop, and pattern the flow of electrons if the hairs are conductive.[79] Chemically functionalized hair tips can directly catalyze chemical

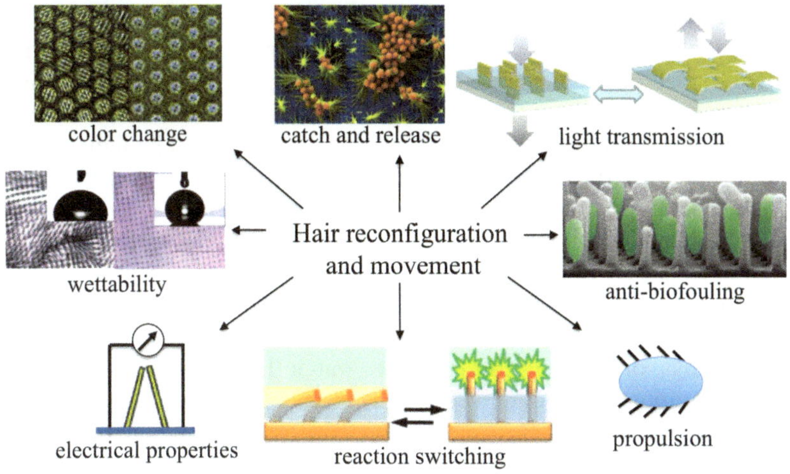

Figure 8.11 A sampling of surface outputs made possible by the reconfiguring hairs.

and biochemical reactions by moving into and out of a stream of reagents.[88] Since gels can be designed to sense any of these optical, electronic, and chemical outputs as stimuli, sensing and responding can be brought full circle to generate surfaces that dynamically adapt their hair motions as part of self-regulating feedback systems.[88]

8.5 Outlook: Small Hairs, Big Dreams

By shaping energy transactions at the nanoscale—harvesting the energy inherent in a dynamic environment and chemo-mechanically translating it into an intricate language of motion—muscular cilia provide a self-sustaining, adaptive platform that we hope will inspire imaginative approaches to current challenges across fields. The combination of multi-stimulus sensitivity and finely controlled cilia suggests a way to design ‘mechano-chemical drugs’ that not only navigate to their target but use nanohairs to perform cellular surgery and deliver mechanical signals as they reconfigure. Cell growth, cancer, neuronal activity, and most other processes depend on complex nanoscale pokes, prods and flows as well as biochemical cues,[89,90] but we have no mechanical equivalent to combinatorial libraries of small molecules; HAIRS design strategies make it possible to envision biochemically responsive ‘mechanical libraries’—or even mechano-chemical libraries with chemically functionalized hair tips—to study and manipulate such processes. With hierarchical patterned motions linking the subcellular through tissue scales, such libraries could likewise act as dynamic, responsive scaffolds for growing organs, for tackling biofouling in the complex environments of sinks, pipes and ships, or for selectively promoting biofilm in bioremediation reactors. As part of water transport and purification systems, they could be designed to regulate flow speed and direction and activate surface-based treatment or extraction mechanisms in response to specific impurities, or enable micro-irrigation systems to adapt delivery based on local moisture, salinity, and temperature, with continuous reports provided by accompanying color changes of the reconfiguring hairs. As bottom-up architectural materials, responsive hairs could create roofs that control run-off based on chemical composition, windows that let light in when it is cold and keep it out when it is warm, or walls that activate self-regulating thermo chemical reactions to maintain a constant temperature.

In the process, HAIRS can also teach us how the world works. Interconverting chemistry and motion is possibly the most fundamental business of life, not to mention the rest of nature, yet we are at a loss to explain exactly how the magic happens at the nanoscale. HAIRS gives us a chance to peer into the inner workings of a model system and watch how distinct chemical and structural components communicate across space and time as we vary their molecular, geometric, and mechanical properties and expose them to different environments. As we explore the multi-scale mechanisms of energy conversion, signal amplification, symmetry breaking, feedback, and other complex chemo-mechanical processes taking place within the hybrid systems, the possibilities for designing adaptive ciliated surfaces moving in synchrony with their environment will undoubtedly expand in new and unexpected directions.

Acknowledgement

This work was supported by the U.S. Department of Energy, Office of Basic Energy Sciences, Division of Materials Sciences and Engineering under Award #DE-SC0005247. This work is based on the research of Dr P. Kim, L. Zarzar, Dr B. Pokroy, Dr S.H. Kang, A. Epstein, Dr X. He, Dr M. Aizenberg, Dr M. Matsunaga, and Dr M. Khan.

References

1. M. Salathe, *Annu. Rev. Physiol.*, 2007, **69**, 401–422.
2. R. A. Lyons, E. Saridogan and O. Djahanbakhch, *Hum. Reprod. Update*, 2006, **12**, 363–372.
3. J. J. Essner, K. J. Vogan, M. K. Wagner, C. J. Tabin, H. J. Yost and M. Brueckner, *Nature*, 2002, **418**, 37–38.
4. Z. G. Zhou and Z. W. Liu, *J. Bionic Eng.*, 2008, **5**, 358–365.
5. A. R. Shields, B. L. Fiser, B. A. Evans, M. R. Falvo, S. Washburn and R. Superfine, *Proc. Natl. Acad. Sci. U.S.A.*, 2010, **107**, 15670–15675.
6. M. Vilfan, A. Potocnik, B. Kavcic, N. Osterman, I. Poberaj, A. Vilfan and D. Babic, *Proc. Natl. Acad. Sci. U.S.A.*, 2010, **107**, 1844–1847.
7. S. N. Khaderi, C. B. Craus, J. Hussong, N. Schorr, J. Belardi, J. Westerweel, O. Prucker, J. Ruhe, J. M. J. den Toonder and P. R. Onck, *Lab Chip*, 2011, **11**, 2002–2010.
8. A. Ogura and K. Takahashi, *Nature*, 1976, **264**, 171–172.
9. R. A. Bloodgood, *J. Cell Sci.*, 2010, **123**, 505–509.

10. A. S. Shah, Y. Ben-Shahar, T. O. Moninger, J. N. Kline and M. J. Welsh, *Science*, 2009, **325**, 1131–1134.
11. R. J. Shao, M. Nutu, L. Karlsson-Lindahl, A. Benrick, B. Weijdegard, S. Lager, E. Egecioglu, J. Fernandez-Rodriguez, K. Gemzell-Danielsson, C. Ohlsson, J. O. Jansson and H. Billig, *Am. J. Physiol. – Cell Physiol.*, 2009, **297**, C140–C151.
12. S. M. King, *Cytoskeleton*, 2010, **67**, 207–213.
13. P. Rompolas, R. S. Patel-King and S. M. King, *Mol. Biol. Cell*, 2010, **21**, 3669–3679.
14. D. M. Woolley, *Biol. Rev.*, 2010, **85**, 453–470.
15. A. Hilfinger and F. Julicher, *Phys. Biol.*, 2008, **5**, 016003-15.
16. C. J. Brokaw, *Cell Motil. Cytoskel.*, 2009, **66**, 425–436.
17. T. J. Mitchison and H. M. Mitchison, *Nature*, 2010, **463**, 308–309.
18. T. Sanchez, D. Welch, D. Nicastro and Z. Dogic, *Science*, 2011, **333**, 456–459.
19. J. Howard, *Annu. Rev. Biophys.*, 2009, **38**, 217–234.
20. H. Hess, *Annu. Rev. Biomed. Eng.*, 2011, **13**, 429–450.
21. J. Howard, S. W. Grill and J. S. Bois, *Nat. Rev. Mol. Cell Biol.*, 2011, **12**, 392–398.
22. B. Pokroy, A. K. Epstein, M. C. M. Persson-Gulda and J. Aizenberg, *Adv. Mater.*, 2009, **21**, 463.
23. B. Pokroy, S. H. Kang, L. Mahadevan and J. Aizenberg, *Science*, 2009, **323**, 237–240.
24. A. K. Epstein and J. Aizenberg, *Mater. Res. Soc. Symp. Proc.*, 2010, **1236E**, 1236–SS1209–1207.
25. S. H. Kang, B. Pokroy, L. Mahadevan and J. Aizenberg, *ACS Nano*, 2010, **4**, 6323–6331.
26. M. Matsunaga, M. Aizenberg and J. Aizenberg, *J. Am. Chem. Soc.*, 2011, **133**, 5545–5553.
27. A. Grinthal, S. H. Kang, A. K. Epstein, M. Aizenberg, M. Khan and J. Aizenberg, *Nano Today*, 2012, **7**, 35–52.
28. P. Kim, W. E. Adorno-Martinez, M. Khan and J. Aizenberg, *Nat. Protoc.*, 2012, **7**, 311–327.
29. P. Kim, A. K. Epstein, M. Khan, L. D. Zarzar, D. J. Lipomi, G. M. Whitesides and J. Aizenberg, *Nano Lett.*, 2012, **12**, 527–533.
30. S. H. Kang, N. Wu, A. Grinthal and J. Aizenberg, *Phys. Rev. Lett.*, 2011, **107**, 177802.
31. J. Paulose, D. R. Nelson and J. Aizenberg, *Soft Matter*, 2010, **6**, 2421–2434.
32. A. Sidorenko, T. Krupenkin, A. Taylor, P. Fratzl and J. Aizenberg, *Science*, 2007, **315**, 487–490.
33. A. Sidorenko, T. Krupenkin and J. Aizenberg, *J. Mater. Chem.*, 2008, **18**, 3841–3846.
34. P. Kim, L. D. Zarzar, X. H. Zhao, A. Sidorenko and J. Aizenberg, *Soft Matter*, 2010, **6**, 750–755.
35. P. Kim, L. D. Zarzar, M. Khan, M. Aizenberg and J. Aizenberg, *Proc. SPIE*, 2011, **7927**, 792705.

36. P. Kim, L. D. Zarzar, X. M. He, A. Grinthal and J. Aizenberg, *Curr. Opin. Solid State Mat.*, 2011, **15**, 236–245.
37. L. D. Zarzar, P. Kim and J. Aizenberg, *Adv. Mater.*, 2011, **23**, 1442–1446.
38. L. D. Zarzar, P. Kim, M. Kolle, J. Brinker, J. Aizenberg and B. Kaehr, *Angew. Chem. Int. Ed.*, 2011, **50**, 9356–9360.
39. K. S. Liu and L. Jiang, *Nano Today*, 2011, **6**, 155–175.
40. C. Y. Hui, A. Jagota, Y. Y. Lin and E. J. Kramer, *Langmuir*, 2002, **18**, 1394–1407.
41. D. Chandra and S. Yang, *Acc. Chem. Res.*, 2010, **43**, 1080–1091.
42. A. S. Kozlov, J. Baumgart, T. Risler, C. P. C. Versteegh and A. J. Hudspeth, *Nature*, 2011, **474**, 376–379.
43. A. C. LeBoeuf, D. O. Maoileidigh and A. J. Hudspeth, *Biophys. J.*, 2011, **101**, 1316–1325.
44. A. S. Kozlov, T. Risler, A. J. Hinterwirth and A. J. Hudspeth, *J. Physiol. (London)*, 2012, **590**, 301–308.
45. S. Nonaka, S. Yoshiba, D. Watanabe, S. Ikeuchi, T. Goto, W. F. Marshall and H. Hamada, *PLoS Biol.*, 2005, **3**, 1467–1472.
46. N. J. Glassmaker, A. Jagota, C. Y. Hui and J. Kim, *J. R. Soc. Interface*, 2004, **1**, 23–33.
47. Y. P. Zhao and J. G. Fan, *Appl. Phys. Lett.*, 2006, **88**, 103123–103123-3.
48. J. L. Liu, X. Q. Feng, R. Xia and H. P. Zhao, *J. Phys. D Appl. Phys.*, 2007, **40**, 5564–5570.
49. D. Chandra, S. Yang, A. A. Soshinsky and R. J. Gambogi, *ACS Appl. Mater. Inter.*, 2009, **1**, 1698–1704.
50. M. De Volder, S. H. Tawfick, S. J. Park, D. Copic, Z. Z. Zhao, W. Lu and A. J. Hart, *Adv. Mater.*, 2010, **22**, 4384.
51. H. G. Duan, J. K. W. Yang and K. K. Berggren, *Small*, 2011, **7**, 2661–2668.
52. D. Kuckling, *Colloid Polym. Sci.*, 2009, **287**, 881–891.
53. I. Tokarev, M. Motornov and S. Minko, *J. Mater. Chem.*, 2009, **19**, 6932–6948.
54. L. Ionov, *J. Mater. Chem.*, 2010, **20**, 3382–3390.
55. M. A. C. Stuart, W. T. S. Huck, J. Genzer, M. Muller, C. Ober, M. Stamm, G. B. Sukhorukov, I. Szleifer, V. V. Tsukruk, M. Urban, F. Winnik, S. Zauscher, I. Luzinov and S. Minko, *Nat. Mater.*, 2010, **9**, 101–113.
56. P. Calvert, *Adv. Mater.*, 2009, **21**, 743–756.
57. R. Messing and A. M. Schmidt, *Polym. Chem. UK*, 2011, **2**, 18–32.
58. S. K. De, N. R. Aluru, B. Johnson, W. C. Crone, D. J. Beebe and J. Moore, *J. Microelectromech. S*, 2002, **11**, 544–555.
59. K. Kataoka, H. Miyazaki, M. Bunya, T. Okano and Y. Sakurai, *J. Am. Chem. Soc.*, 1998, **120**, 12694–12695.
60. H. G. Schild, *Prog. Polym. Sci.*, 1992, **17**, 163–249.
61. S. Sugiura, A. Szilagyi, K. Sumaru, K. Hattori, T. Takagi, G. Filipcsei, M. Zrinyi and T. Kanamori, *Lab Chip*, 2009, **9**, 196–198.
62. T. Satoh, K. Sumaru, T. Takagi and T. Kanamori, *Soft Matter*, 2011, **7**, 8030–8034.

63. R. M. Luo, H. Li, E. Birgersson and K. Y. Lam, *J. Biomed. Mater. Res. A*, 2008, **85A**, 248–257.
64. C. M. Kingsbury, P. A. May, D. A. Davis, S. R. White, J. S. Moore and N. R. Sottos, *J. Mater. Chem.*, 2011, **21**, 8381–8388.
65. T. J. Kucharski and R. Boulatov, *J. Mater. Chem.*, 2011, **21**, 19413–19415.
66. O. Okay and S. Durmaz, *Polymer*, 2002, **43**, 1215–1221.
67. O. Kuksenok, V. V. Yashin, M. Kinoshita, T. Sakai, R. Yoshida and A. C. Balazs, *J. Mater. Chem.*, 2011, **21**, 8360–8371.
68. V. V. Yashin and A. C. Balazs, *Science*, 2006, **314**, 798–801.
69. J. Boissonade, *Eur. Phys. J. E*, 2009, **28**, 337–346.
70. J. Horvath, I. Szalai, J. Boissonade and P. De Kepper, *Soft Matter*, 2011, **7**, 8462–8472.
71. R. Yoshida, *Colloid Polym. Sci.*, 2011, **289**, 475–487.
72. S. J. DuPont, R. S. Cates, P. G. Stroot and R. Toomey, *Soft Matter*, 2010, **6**, 3876–3882.
73. M. Guvendiren, J. A. Burdick and S. Yang, *Soft Matter*, 2010, **6**, 2044–2049.
74. J. Kim, J. Yoon and R. C. Hayward, *Nat. Mater.*, 2010, **9**, 159–164.
75. R. Marcombe, S. Q. Cai, W. Hong, X. H. Zhao, Y. Lapusta and Z. G. Suo, *Soft Matter*, 2010, **6**, 784–793.
76. O. Tabata, H. Hirasawa, S. Aoki, R. Yoshida and E. Kokufuta, *Sens. Actuators A – Phys.*, 2002, **95**, 234–238.
77. P. Dayal, O. Kuksenok, A. Bhattacharya and A. C. Balazs, *J. Mater. Chem.*, 2012, **22**, 241–250.
78. S. Saha, D. Copic, S. Bhaskar, N. Clay, A. Donini, A. J. Hart and J. Lahann, *Angew. Chem. Int. Ed.*, 2012, **51**, 660–665.
79. M. De Volder, S. H. Tawfick, D. Copic and A. J. Hart, *Soft Matter*, 2011, **7**, 9844–9847.
80. W. Hong, X. H. Zhao and Z. G. Suo, *J. Appl. Phys.*, 2008, **104**, 084905-9.
81. W. H. Wong, T. F. Guo, Y. W. Zhang and L. Cheng, *Int. J. Solids Struct.*, 2010, **47**, 2034–2042.
82. J. Branscomb and A. Alexeev, *Soft Matter*, 2010, **6**, 4066–4069.
83. A. I. Hochbaum and J. Aizenberg, *Nano Lett.*, 2010, **10**, 3717–3721.
84. A. K. Epstein, A. I. Hochbaum, P. Kim and J. Aizenberg, *Nanotechnology*, 2011, **22**, 494007.
85. A. Bhattacharya, G. A. Buxton, O. B. Usta and A. C. Balazs, *Langmuir*, 2012, **28**, 3217–3226.
86. M. Hu, F. S. Ou, W. Wu, I. Naumov, X. M. Li, A. M. Bratkovsky, R. S. Williams and Z. Y. Li, *J. Am. Chem. Soc.*, 2010, **132**, 12820–12822.
87. F. S. Ou, M. Hu, I. Naumov, A. Kim, W. Wu, A. M. Bratkovsky, X. M. Li, R. S. Williams and Z. Y. Li, *Nano Lett.*, 2011, **11**, 2538–2542.
88. X. He, M. Aizenberg, O. Kuksenok, L. D. Zarzar, A. Shastri, A. Balazs and J. Aizenberg, *Nature*, 2012, **487**, 214–218.
89. D. E. Ingber, *FASEB J.*, 2006, **20**, 811–827.
90. V. Vogel and M. P. Sheetz, *Curr. Opin. Cell Biol.*, 2009, **21**, 38–46.

CHAPTER 9

Microwalkers

CHARLES E. SING[a,b] AND ALFREDO ALEXANDER-KATZ*[b]

[a] Department of Materials Science and Engineering, Northwestern University, 4021 Cook Hall, 2220 Campus Dr., Evanston IL 60208, USA, Email: charles.sing@northwestern.edu; [b] Department of Materials Science, Massachusetts Institute of Technology, Cambridge, MA 2139, USA
*Email: aalexand@mit.edu

9.1 Introduction: Walking and Friction

The process of walking is intimately related to the concept of surface friction, and in fact, walking on flat surfaces is only possible if friction exists. The reason for this lies at the heart of physical conservation laws. In the particular case of human walking, we know by experience that if we try to walk on ice we tend to slip much more easily than if we walk on cement. The reason is because when we walk we are converting a torque (angular momentum) into translational motion, and the efficiency of such conversion relies on the amount of friction our feet experience with the surface. This friction has a tangential character (parallel to the surface) so it allows the torque acting on the leg due to the muscle contractions to be converted into a net translational and directional motion (see Figure 9.1 for an schematic). The magnitude of the friction between our feet and the surface is also important and can be modulated in several ways. For example, a slippery surface will considerably reduce the friction force experienced, and thus it will lead to slipping behavior that ultimately reduces the net speed of walking. Nevertheless, the degree of friction can also be controlled by the weight being applied on the surface and the area on which it is applied. As we know, walking under low-gravity conditions is almost

RSC Nanoscience & Nanotechnology No. 30
Artificial Cilia
Edited by Jaap M J den Toonder and Patrick R Onck

Published by the Royal Society of Chemistry, www.rsc.org

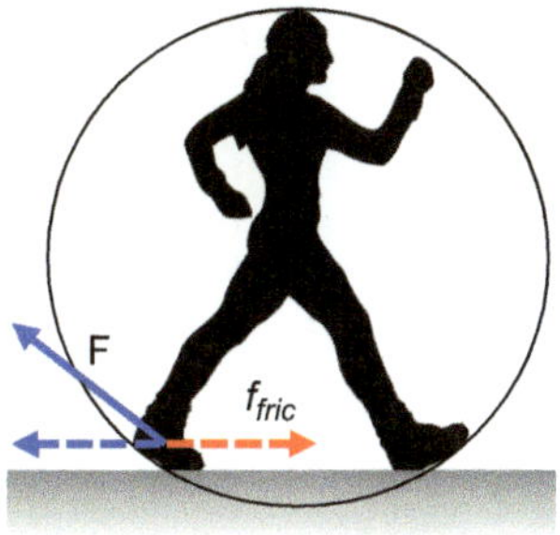

Figure 9.1 Schematic representation of walking. A force **F** tangential to the circle is the result of muscle contractions and forces. The translational motion of the person is due to the projection of this force on the surface (dashed blue curve). In response to the applied force, a friction force acts which ultimately leads to translation. The size of the friction force is dependent on the surface friction coefficient and the load, typically. In ice, for example, the friction force is smaller than compared to asphalt for the same load conditions.

impossible, and looks much more like jumping instead of walking. These same principles govern the motion of cars, where clearly one imposes a torque on the wheel and the interaction with the surfaces rectifies it and translates it into motion. However, in slippery conditions such as mud, one can get stuck!

The coupling between walking and friction occurs for almost all known kinds of walking organisms, ranging from humans all the way to insects that walk on water. In all cases there is the appearance of an effective friction source. Interestingly, cellular and molecular motility also rely on microscopic sources of friction due to chemical binding interactions. Contrary to the macroscopic case where friction is a 'static' quantity for given conditions, as for example a given contact area, in cellular motility and molecular motor movements the friction term becomes dynamic and stochastic. Nevertheless, this dynamic friction allows these biological constructs to move. Of particular interest are molecular motors which literally perform a walking dance along their tracks.[1–12] Two common examples of such motor proteins are dynein and kinesin which transport nutrients and cargo inside the cell and are critical for the active remodeling of the cell itself (see Figure 9.2).[1,3] Molecular motors are crucial in many biological functions such as the formation of the spindle during mitosis, the protrusions of cilia for sensing and cell motility, and the contractile forces appearing in muscle to name a few. The effective friction between these motors and the substrate is dictated by the strength of the binding potential between their 'feet' and the track, consisting typically of cytoskeleton filaments. The 'efficiency' of walking for molecular motors, or how good walkers they are, something referred in the biological jargon as processivity, is in fact directly related to the 'friction' force arising from their specific binding propensity. Such molecular motors do not walk in the absence of a track, even if they are combusting chemical energy in the form of ATP. Such a scenario is equivalent to a person trying to walk under water if there is no surface. The motion of molecular motors is also dependent on the hydrodynamic conditions in the cell,

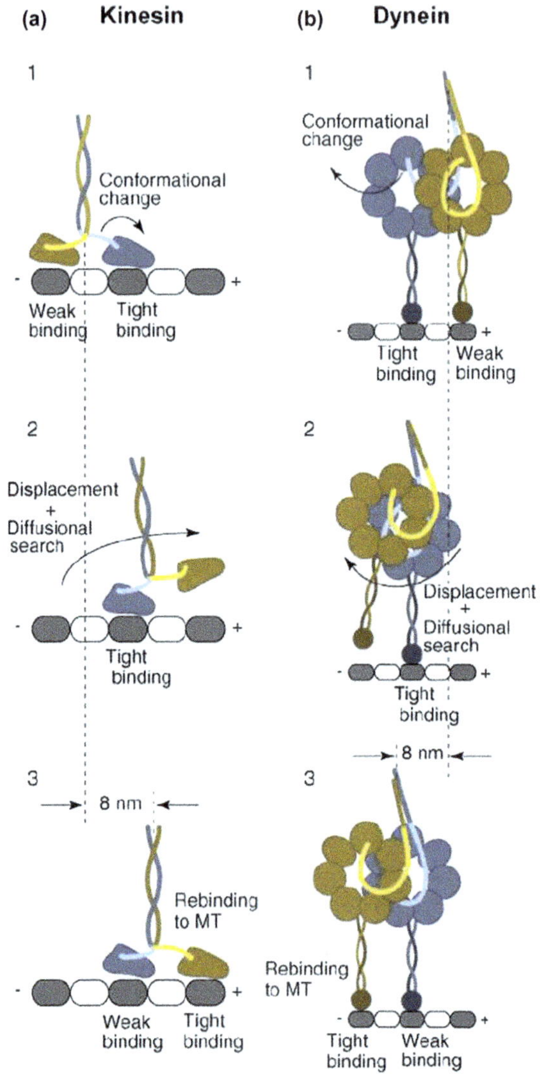

Figure 9.2 Molecular motors walk on tracks due to the effective surface friction caused by the periodic and well-coordinated tight binding of the leading 'foot'. This mechanism is at the origin of molecular recognition, molecular motors, and active soft matter in general.
(Figure reproduced, with permission, from Block.[1])

as for example the effective viscosity, the size of the cargo that will increase the drag force on the motor, or cooperative schemes of motion which are necessary in some instances.[13–24]

The striking resemblance of walking motions across systems that differ by nine orders of magnitude in size is remarkable, and understanding from a unified point of view such walking motions is thus of interest. This chapter explores the motion of a class of microscopic walkers (microwalkers) with similarities to those described above using a highly simplified model system that contains all the essential ingredients needed to understand walking in the microscopic realm. The model system is a chain of colloids self-assembled by magnetic forces upon which a torque acts. Symmetry in the absence of boundaries only allows such a construct to rotate, but near a wall they start to walk due to friction forces. An interesting aspect of such model systems is that they can be tailored and controlled in very particular ways such that one can study the behavior of a single microwalker or the collective dynamics of a 'swarm' of these walkers. Such simple model microwalkers have already been developed and have received wide attention in recent years. They can be used for a multitude of tasks, such as fully controlled two-dimensional walking[25–29] creating controlled flows at the micron scale,[25] or for the manipulation of nanoscopic objects.[28] The latter two parts share similarities with the transport of fluids performed by cilia; however, we refer the reader interested in this topic to the other chapters in this book as they are directly related to this problem. For pedagogical reasons, in the first part of this chapter we will start by analyzing a single microwalker, while in the second part we deal with collective motion. In the final section we discuss potential new avenues of research in this area.

9.2 Single Microwalkers

9.2.1 Assembly and Motion of Microwalkers

Microwalkers can easily be created by applying a homogeneous magnetic field to a liquid suspension of (super)paramagnetic colloids that can range in size from nanometers to micrometers. Upon the application of the field, the colloids acquire a magnetic moment in the direction of the field. Some of the larger colloids are not precisely super-paramagnetic, so they actually retain a residual magnetic moment and the chains do not disperse once the field is turned of. However, they can be considered super-paramagnetic in the sense that they are free to rotate and point in the direction of the field. The colloidal chains form due to dipole–dipole interactions. A schematic of the process of formation is shown in Figure 9.3a and b. In order to induce the chains to walk, there are a few different ways all based on the same underlying approach: rotation of the magnetic field. In the simplest possible scenario, a homogeneous magnetic field in space will be rotated at a fixed frequency ν. The colloidal chains rotate at the same frequency because they prefer to be aligned with the field, but given their slightly heavier nature, they rotate near the bottom surface and walk (see Figure 3c). In most of this chapter we will be dealing with glass surfaces since these experiments are conducted on slides that are hydrophilic, but the phenomenon is general. In this respect, it is well established that friction forces exist in the vicinity of surfaces due to the no-slip condition. Even for a very slippery surfaces, the slip length tends to be of the order of a a few nanometers to a few tens of nanometers, which for the large colloids is small compared to their size.[30,31] Thus, without losing generality we can consider that there will

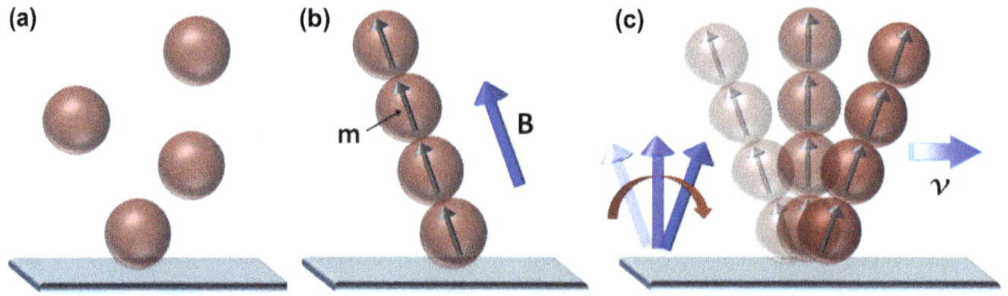

Figure 9.3 Schematic of microwalkers. Super-paramagnetic beads undergo Brownian motion in solution in the absence of a magnetic field (a). When the field **B** is turned on, the beads acquire a magnetic moment **m** and form chains due to magnetic dipole–dipole interactions (b). Finally, when the field rotates, the chains follow and rotate as well. Due to the enhanced friction near the surface the chains start 'walking' (c). This happens only when rotational and translational motions are coupled, as in the case near surfaces.

always be a source of friction stemming from an effective no-slip boundary condition. This friction can be modulated by how close the beads actually rotate near the surface or if they are chemically functionalized. For smaller beads, and in the absence of any external force that pulls them close to the surface, there is always a small liquid gap in between because the gravitational force is not strong enough to push the liquid out of the gap. This effectively yields a friction that can be described in hydrodynamic terms only, since it does not depend directly on the nature of the interaction between the beads and the surface, *i.e.* these systems have not reached the molecular limit. However, if the beads are larger or there is a force pulling them towards the surface, such as an inhomgeneous magnetic field, then one starts to find that the friction force is modulated strongly.[28,32]

Given the enhancement of friction near the surface, the rotating chains start to walk. A schematic of such process is shown in Figure 9.3c. An important measure of the walking ability of such chains is the velocity v they can attain as a function of the frequency of rotation ν and number of beads per chain N. In Figure 9.4 one can see that the number of beads dictates in a strong fashion the velocity of the microwalker. This behavior is not difficult to understand unless one introduces the existence of long range hydrodynamic forces at play, particularly for the case of high frequencies where the chain does not have time to sediment. As will be derived theoretically in the next section the increase in velocity is linear in the number of beads, and experiments seem to confirm this trend.

The microwalkers that we have described here are completely self-assembled from magnetic forces, and thus they can break. This property becomes important at high frequencies where the fluid friction becomes large. At this point the chain fragments into a hierarchical object with multiple smaller chains rotating with the field and around each other at a different frequency much smaller that the field rotation, which leads to decaying velocities. To understand this, simply think that the smaller fragments rotate with the

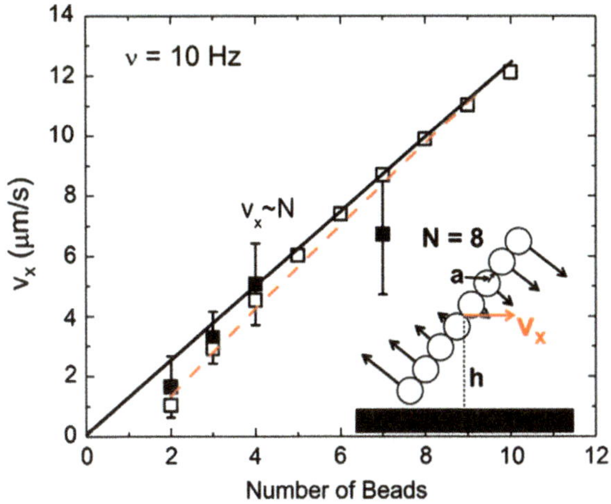

Figure 9.4 Walking velocity as a function of the length of the chain. The length of the chain is proportional to the number of beads. The experimental results are filled symbols, while the open symbols are simulations. The velocity is approximately linear on the number of beads, which is the asymptotic scaling result (solid black curve). Deviations from this curve are well captured by a more precise calculation (dashed red curve) incorporating the finite size of the beads and of the chain. Derivation of such a curve is given in section 9.3. (Figure from Sing *et al.*[25])

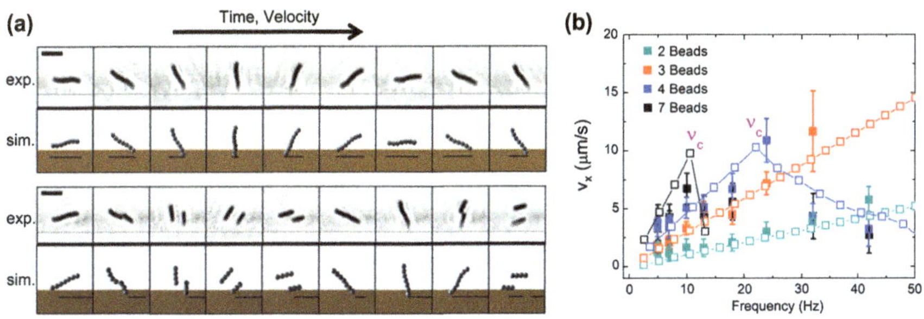

Figure 9.5 Conformation of the chains below and above the critical frequency of breakup. In (a) we show snapshots of a single chain of seven beads undergoing rotation. In the upper traces the chain is rotating below the break-up transition. The upper trace shows experimental images, while the lower curve shows simulation images. In the bottom traces, the frequency is above the transition and the chain splits into two sections and reforms within a cycle. The velocity drastically drops above the transition frequency, as shown in (b). Solid symbols are experiments and open symbols are simulation results.
(Figure taken, with permission, from Sing *et al.*[25])

field, but that the rotational frequency of the whole is much lower than that from the magnetic field. Microwalkers in both the single chain and the fragmented chain configuration are shown in Figure 9.5, along a graph that shows the velocity of a few different chains. Notice that above a critical frequency ν^*

the velocity suddenly decreases. There are methods to keep the microwalkers intact, such as binding the beads with complementary sDNA strands[33] or using stronger magnetic fields, which will lead to a larger ν^*. In the theory section we show that the important parameter that controls the break-up of the aggregates is the Mason number, which compares the applied torque to the viscous torque on the chain.

Other designs of microwalkers have included a large and a small bead, and the actuation protocol has not been on a single plane, but rather the magnetic field motion describes a conical surface.[27,29] In such systems it is also possible to have full control of the direction of motion, and these walkers have been taken around arbitrary tracks (see Figure 9.6).

Enhancing friction can be accomplished in different ways. Morimoto and co-workers have used a paramagnetic substrate and have been able to tailor the friction force from weak to strong.[28] A schematic diagram of the different regimes is shown in Figure 9.7. In the weak case one expects slip, so the trajectory of one of the end beads should display a shape as shown in Figure 9.7a. On the other hand, if the friction force is strong, the motion of the walker proceeds mostly by a hinge-type mechanism, and shown in Figure 9.7b. If the magnetic field is inhomogeneous, larger 'below' the substrate, the chains are pulled towards the substrate and friction can be strongly enhanced. Such as scheme has been used by Stetten and co-workers to move microwalkers across a bend in a counter-flow assay, as shown in Figure 9.8. The idea here is that the chain would have to slip in order to be able to align with the strongest field lines. Instead, and because of strong friction, the chain aligns with the lateral field lines that are coming out at an angle from the permanent magnet. Large

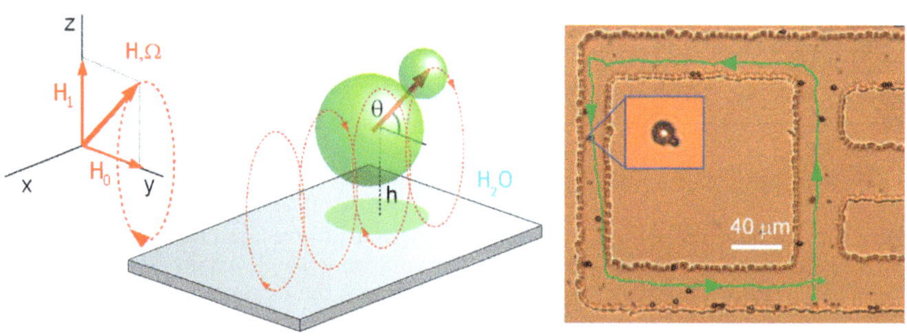

Figure 9.6 Different microwalker architecture and actuation protocol. Tierno *et al.* have used a doublet consisting of a large and a small bead. The actuation protocol is different compared to the previous case. Here, the surface that the magnetic field vector depicts is a cone instead of a circle. One can go from the lattr to the former by simply applying a constant magnetic field in the horizontal direction perpendicular to the circle plane. Using this protocol the authors controlled the motion of such walkers and made them walk labyrinth-like situations. Microwalkers can be controlled in general to perform walking in any direction in two or three dimensions as long as there is a surface present for walking.
(Figures used, with permission, from Tierno *et al.*[26,27])

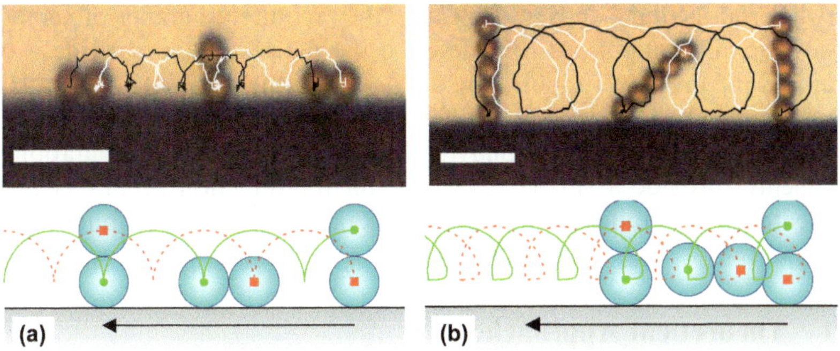

Figure 9.7 Controlling the velocity of a single microwalker. In (a), the friction is large and the walkers can only hinge on one of its beads. This is the fastest per cycle that the microwalkers can move. In (b) the friction is not enough and the walker slips noticeably, leading to lower velocities per cycle.
(Figures reprinted, with permission, from Morimoto *et al.*[28])

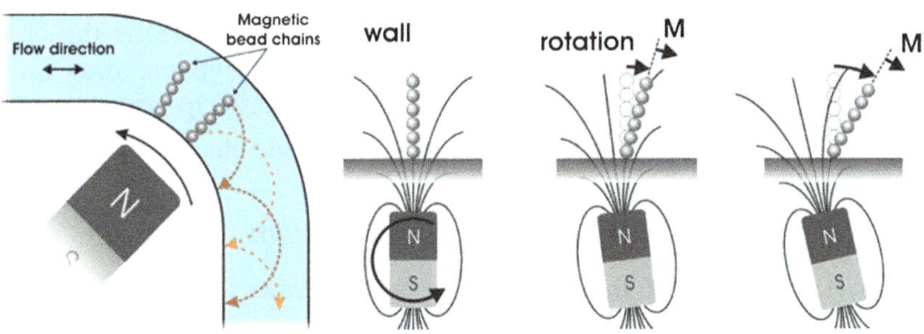

Figure 9.8 Counter-clockwise motion of microwalkers. When the magnetic field is not homogeneous, a strong force pulls the chains near to the surface. Upon rotation of the magnet, the chain is unable to align with the maximum strength field line and instead aligns with the line that passes by the 'anchoring point'. This leads to counterclockwise motion. High walking velocities reaching millimeters per second have been achieved using this method.
(Figure reprinted, with permission, from Karle *et al.*[32])

fields are difficult to attain with air coils, but in permanent magnets they can be as high as 0.5 T. At the moment, this is the state of the art in single microscopic walkers, but as will be discussed in future avenues, this field is ripe for new advances.

9.3 Computational and Theoretical Approaches to Study Single Microwalkers

While the study of microwalkers is a field that is still in its infancy, most of its principles are based on well-established physical concepts that allow for a quantitative description of experimental observations of these systems *via* either analytical or numerical solution of theoretical models. These models are somewhat rudimentary, in that as of now they require a number of geometric and situational assumptions, but describe the data well and provide a clear conceptual picture that supports the discussion in the previous section. In the following, we take a look at two different (but similar) approaches to understanding microwalker systems: analytic theory and numerical computation. We subsequently provide a small perspective on the direction of this field, and where enterprising students could look to make their own contribution!

9.3.1 Theoretical Approaches

Theoretical treatments of this microwalker platform rely on the interplay of two different realms of physical ideas: magnetic materials and hydrodynamics. We will largely focus on the hydrodynamic behaviors in this following section, and provide more information on the magnetic properties when discussing numerical computation. What is a prerequisite to know about the magnetic forces for the purposes of our theoretical calculations boils down to two considerations: the field is what applies a rotational torque to the chain; however these forces are exactly balanced by the hydrodynamic drag forces, and the total magnetic torque scales as $\tau_M \sim \vec{m} \times \vec{B} \sim NB^2$ where B is the magnetic field and N is the number of beads in a single chain.

9.3.1.1 Fundamental Hydrodynamic Equations

The novel properties of the microwalker system arise from the interaction between the forces placed on the walker (through the magnetic field, though in principle any type of field could be used) and the surrounding fluid. Understanding this system, then, involves understanding the principles of hydrodynamics that result from applying a local force on a fluid. These ideas stem from a set of 'entities' outlined and compiled by Bachelor (though developed by others) known collectively as *hydrodynamic singularities*. The idea is to describe a number of perturbation types to an infinite fluid, and then use collections of these perturbations (things such as point forces, point rotations, *etc.*) to describe more complicated systems with multiple bodies, surfaces, and externally applied flow and pressure fields.

The most basic of these hydrodynamic singularities is the flow perturbation (change in fluid velocity at any point $\mathbf{r}_i$) due to a point force of strength F at $\mathbf{r}_j$. This can be described using the ubiquitous Navier–Stoke's equation for an incompressible, Newtonian fluid:

$$\rho\left(\frac{\partial \vec{v}}{\partial t} + \vec{v} \cdot \nabla \vec{v}\right) = -\nabla P + \eta \nabla^2 \vec{v} + \rho \delta(\mathbf{r}_j) F \tag{9.1}$$

where η is the viscosity and $\delta(x)$ is the Dirac delta function. This is a rather difficult equation to solve, but fortunately we have physical approximation on our side. There is a dimensionless number, known as the Reynold's number (Re), that compares the magnitude of the first term (associated with 'inertial' forces of the fluid) to the remainder of the terms (the 'viscous' forces on the fluid):

$$Re = \rho \mid \vec{v} \mid L/\eta \tag{9.2}$$

where L is some characteristic length scale. This number essentially compares the term on the left with the second term on the right of the above Navier–Stokes equation. For our purposes, this number is small! So small, in fact, that we will take the first term (the one on the left) to be zero. This approximation is widely used in the field of small-length scale hydrodynamics, and simplifies the situation greatly. Let us now write our updated equation:

$$0 = -\nabla P + \eta \nabla^2 \vec{v} + \rho \delta(\mathbf{r}_j) F \tag{9.3}$$

While this is still a nonlinear differential equation that is rather difficult to solve, it is solvable using Fourier transform methods. The final result is a tensor known as the 'Oseen tensor' $\mathbf{\Omega}_{ij}$, which describes the velocity field surrounding an applied point force in a fluid:

$$\mathbf{v}(\mathbf{r}_i) = \mathbf{\Omega}_{ij}\mathbf{F}(\mathbf{r}_j) = \frac{1}{8\pi\eta \mid \boldsymbol{r} \mid}\left[\mathbf{I} + \frac{\mathbf{rr}}{\mathbf{r}\cdot\mathbf{r}}\right]\mathbf{F}(\mathbf{r}_j) \tag{9.4}$$

where $\mathbf{r} = \mathbf{r}_j - \mathbf{r}_i$. This coupling will be widely used, along with the well-known Stokes law for the drag force $\boldsymbol{F}_D$ on a bead of radius a:

$$6\pi\eta a\mathbf{v} = -\mathbf{F}_D \tag{9.5}$$

Theoretical descriptions of microwalkers rely on the behavior of this tensor upon considering its treatment in the presence of a planar surface. A surface in hydrodynamics represents what is known as a 'no-slip boundary condition.' This means that the surface must tend towards the velocity of the surface's motion (usually zero velocity). This is due to molecular-level interactions at the interface of the fluid and the solid surface that prevent the free translation of the molecules directly adjacent to the surface. This condition must be met by any description of fluid flow in the vicinity of a surface, and therefore the above Oseen tensor must be modified to accommodate this requirement.

The appropriate modification for the Oseen 'hydrodyamic singularity' was derived by Blake in the 1970s by using an image system that accomplishes the correct boundary conditions upon the appropriate inclusion of a series of other flow singularities. The Oseen tensor, representing what is also known as a 'stokeslet' is given in most notations as:

$$\mathbf{S}_{ij}(r) = \mathbf{\Omega}_{ij}(\mathbf{r}) \tag{9.6}$$

where the left hand side $\mathbf{S}_{ij}$ stands for 'stokeslet'. Two of these singularities are used in the case of a point force near the surface, which we demonstrate in Figure 9.9. The primary one is located at the point force itself ($\mathbf{S}_{ij}(\mathbf{r})$), while another stokeslet is located at the same position except on the opposite side of the surface at $z = 0$ ($\mathbf{S}_{ij}(\mathbf{r} - 2z_0)$), where z_0 is the z-component of the point force position). This does exactly as one would imagine, the $x - y$ translational motion due to one stokeslet at the surface plane will be exactly counter-balanced

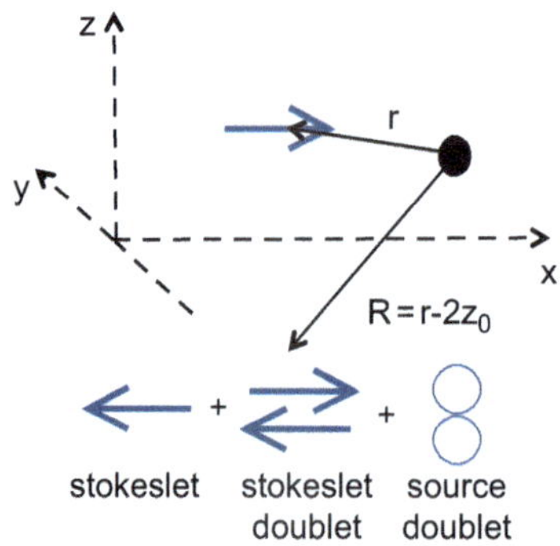

Figure 9.9 Blake image system for a stokeslet near a surface.

by the opposite, or 'image', stokeslet. This is not complete, however. If one looks at the form of the equations, it will become clear that the velocity is *not* zero at the surface due to a rotation around the midpoint between the stokeslet and its image. This rotation must be counteracted, so it is important to include a term that directly cancels out the gradient of the original stokeslet velocity field. This is given by what is known as a 'stokeslet doublet' $\mathbf{S}_{ij}^{D}(\mathbf{r})$, which is:

$$\mathbf{S}_{ij}^{D}(\mathbf{r}) = (1 - 2\delta_{i3})\nabla\mathbf{S}_{ij} = r_3\mathbf{P}_{ij}^{D}(r) + \frac{1 - 2\delta_{i3}}{4\pi\eta}\frac{\delta_{j3}r_i - \delta_{i3}r_j}{|\mathbf{r}|^3} \tag{9.7}$$

This term introduces a similar contribution that must also be included, which is the 'potential dipole' $\mathbf{P}_{ij}^{D}(\mathbf{r})$, which represents the gradient of the pressure field around the stokeslet:

$$\mathbf{P}_{ij}^{D}(\mathbf{r}) = (1 - 2\delta_{i3})\nabla\frac{\mathbf{r}}{4\pi\eta\ |\mathbf{r}|^3} = \frac{1 - 2\delta_{i3}}{4\pi\eta\ |\mathbf{r}|^3}\left[\mathbf{I} - 3\frac{\mathbf{rr}}{\mathbf{r}\cdot\mathbf{r}}\right] \tag{9.8}$$

These terms can be summed to obtain the overall mobility tensor $\Omega_{B,ij}(\mathbf{r})$ (dubbed the 'Blake' tensor, after its creator) for a point force near a surface:

$$\Omega_{B,i0}(\mathbf{r}, z_0) = \mathbf{S}_{ij}(\mathbf{r}) - \mathbf{S}_{ij}(\mathbf{r} - 2z_0) - z_0\mathbf{S}_{ij}^{D}(\mathbf{r} - 2z_0) + z_0^2\mathbf{P}_{ij}^{D}(\mathbf{r} - 2z_0) \tag{9.9}$$

which, once again, links the force at a point $F(\mathbf{r}_0)$ to the velocity at a point $v(\mathbf{r}+\mathbf{r}_0)$ through the equation $v(\mathbf{r}+\mathbf{r}_0)=\Omega_{B,i0}F(\mathbf{r}_0)$. While this is a rather complicated tensor, it is nonetheless analytically tractable for simple systems and is appropriate for use under situations where the entities are both not too close to the surface or to each other. For situations where the system is dense and this is possible and, in the computational section of this chapter, we will show how modifications are made.

9.3.1.2 Dimer-curial

These principles drive the motion of walkers on the surface, and we start by deriving the motion of a single dimer as it rotates in the vicinity of the surface.

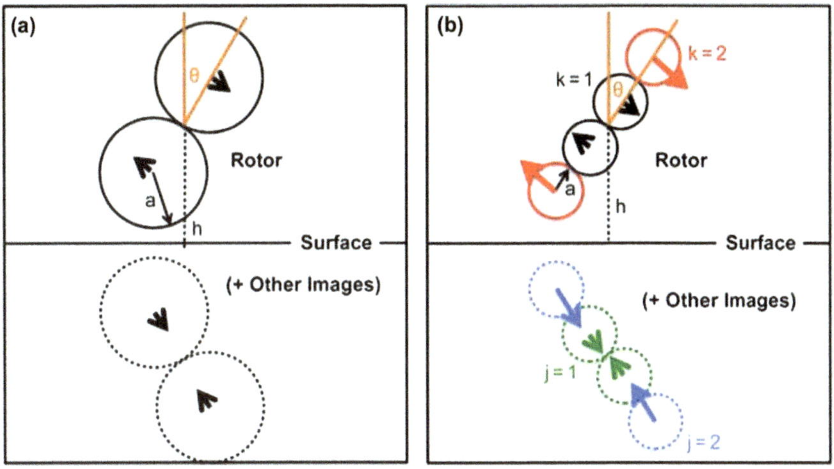

Figure 9.10 Schematic of the forces being applied to the beads and the image forces appearing in the fluid due to the proximity with the surface. In (a) we show a doublet, and in (b) we show a longer chain with four beads. (Figure reprinted, with permission, from Sing *et al.*[25])

This case is demonstrated in Figure 9.10a, which schematically illustrates our geometrical parameters: two beads of radius a are located at an average height h from a no-slip surface, and they are at an angle θ from the direction normal to the surface. The instantaneous translational velocity v_x of the rotor relates to the force F on each bead in the indicated directions perpendicular to the dumbbell axis via the equation:

$$v_x(\theta) = \hat{\mathbf{e}}_\mathbf{x} \cdot (\Omega_{B,12}F_2 + \Omega_{B,21}F_1) \tag{9.10}$$

where $\hat{\mathbf{e}}_\mathbf{x}$ is the unit vector in the x-direction. The first thing to notice upon expanding the Blake tensor to all its components via eqn (9.9) is that the non-image term (the stokeslet) is there even in the case of an infinite fluid. Let us subsequently notice that, for the case of a dimer, the vector $\mathbf{r}$ that describes the position of bead 1 to bead 2 is necessarily opposite the vector that describes the position of bead 2 to bead 1 ($\mathbf{r}_{12} = -\mathbf{r}_{21}$). This allows us to consider the effect of this symmetry operation on the Oseen tensor, which is clearly that $\Omega_{12} = \Omega_{21}$. Furthermore, via the picture it is clear that the forces are both equal and opposite, so $F_2 = -F_1$. Therefore, in the bulk, $v_x = 0$ due to the geometric symmetry of the system. We can subsequently ignore this term in the our calculations and consider only the image terms in the Blake tensor throughout the remainder of this chapter.

The velocity of the walker is calculated using the above tensors to be:

$$v_x(\theta) = \frac{F}{8\pi\eta}\left[\frac{3}{4}\cos\theta\left(\frac{1}{\lambda_-} - \frac{1}{\lambda_+}\right) - \frac{3}{2}\sin^2\theta\left(\frac{ah\lambda_-\lambda_+}{\Gamma^{5/2}}\right)\right] \tag{9.11}$$

which introduces the convenient functions $\lambda_{-}=h-a\cos\theta$, $\lambda_{+}=h+a\cos\theta$, and $\Gamma=h^2+a^2\sin^2\theta$. Eqn (9.11) can be integrated over an entire rotation to yield the averaged velocity of the dimer walker $\left[\langle v_x\rangle=(1/2\pi)\int_0^{2\pi}v_x(\theta)\mathrm{d}\theta\right]$. This form exhibits the essential ingredients to reveal a convenient scaling relationship that applies for large values of N. If we assume that the height h is proportional to the length of the chain N (a condition that is observed in both experiment and simulation), we can separate this equation into:

$$v_x=\frac{F}{8\pi\eta h}f(\theta) \tag{9.12}$$

The force on a single dumb-bell bead $F=6\pi\eta a\times(va)$ is the drag force times the velocity of the bead along the circumference of the rotational path with an angular frequency of v. This is substituted into eqn (9.12) along with the substitution $a\rightarrow Na$ to obtain the scaling relationship:

$$v_x\sim\frac{vN^2a^2}{h} \tag{9.13}$$

9.3.1.3 N-*mer Walking*

This expression for a dimer can be generalized to the case of a chain of N beads, and the above scaling relationship can be corroborated. We limit ourselves to even values of N, though a similar expression could be written for odd values. To build up to this case, we consider a chain that is a sequence of dipole rotors separated by increasing integer distances from the center dimer. These distances are generally given by the values $d_k=a(2k+1)$, where the integer value k describes a specific dipole rotor. Every dipole k will have an effect on a dipole l that likewise exists with a radius of $d_l=a(2l+1)$. This geometry is indicated in Figure 9.10b. To determine the overall translational velocity of the rotor, we sum over all the possible combinations of k and l since the image of each rotor k has an effect on all rotors l. The force likewise also depends on k, since $F=6\pi\eta vad_k$ due to the drag force on the beads at an applied frequency v. The final equation is rather complicated, so we define the following auxilliary functions:

$$\Delta_i=\begin{cases}d_k-d_l & i=\text{odd}\\ d_k+d_l & i=\text{even}\end{cases} \tag{9.14}$$

$$m_i=\begin{cases}+1 & i\leq 2\\ -1 & i>2\end{cases} \tag{9.15}$$

$$\alpha_i=2h+m_i\Delta_{i+1}\cos\theta \tag{9.16}$$

$$\beta_i=\alpha_i^2+\Delta_i^2\sin^2\theta \tag{9.17}$$

$$\lambda_i=h+m_id_k\cos\theta \tag{9.18}$$

$$\gamma_i = \begin{cases} (h + d_l \cos\theta)(h + d_k \cos\theta) & i = 1 \\ (h - d_l \cos\theta)(h + d_k \cos\theta) & i = 2 \\ (h - d_l \cos\theta)(h - d_k \cos\theta) & i = 3 \\ (h + d_l \cos\theta)(h - d_k \cos\theta) & i = 4 \end{cases} \tag{9.19}$$

Given these equations, it is possible to calculate the force f_x on dipole l due to the images of dipole k:

$$\begin{aligned} f_{x,kl} \sim \eta v a^2 d_k \Bigg[&\cos\theta \sum_i^4 \left(2m_i\gamma_i \left(\frac{3\Delta_i^2 \sin^2\theta - \beta_i}{\beta_i^{5/2}} - \frac{\Delta_i^2 \sin^2\theta + \beta_i}{\beta_i^{3/2}} \right) \right) \\ &- \sin^2\theta \sum_i^4 \Delta_i \left(\frac{\alpha_i - 2\lambda_i}{\beta_i^{3/2}} + \frac{6\gamma_i\alpha_i}{\beta_i^{5/2}} \right) \Bigg] \end{aligned} \tag{9.20}$$

The translational velocity of this rotor is related to the force $f_{x,kl}$ by summating both indices k and l, and then dividing by the mobility of the overall assembly:

$$v_x \sim \frac{1}{\eta N a} \sum_{1=0}^{N/2} \sum_{k=0}^{N/2} f_{x,kl} \tag{9.21}$$

The resulting v_x is once again a function of the geometric parameters h and θ, the latter of which allows the analytical description of the instantaneous velocity as a function of time for a given v. This is shown in Figure 9.4, which also plots the traces from simulation data (which we will discuss in the next section). As is apparent, the analytical theory matches well to the simulation results. The comparison between this result and simulation results will be made more thoroughly in the discussion of the simulation results. Like the dimer case, we can average the velocity over a single rotation:

$$\langle v_x \rangle = \frac{1}{2\pi} \int_0^{2\pi} v_x \mathrm{d}\theta \tag{9.22}$$

9.3.1.4 Chain Fragmentation Scaling

Since these rotors are held together entirely by magnetic forces, there is a possibility for the drag forces during rotation to pull apart the aggregate. This break-up can be theoretically described by considering a comparison between the magnetic and frictional torques. The torque τ_B applied by the magnetic field $\mathbf{B}$ is given by the relationship:

$$\tau_B = \mathbf{m} \times \mathbf{B} \sim |\mathbf{m}||\mathbf{B}| \tag{9.23}$$

where $|\mathbf{B}|$ is the magnitude of the magnetic field and $\mathbf{m}$ is the magnetic moment vector of length $|\mathbf{B}|$. The magnitude of the magnetic moment is given by the susceptibility of the material to the applied field strength $|\mathbf{B}|$ and scales with the number of beads M, so $|\mathbf{m}| \sim N|\mathbf{B}|$. This leads to a scaling expression for the magnetic torque:

$$\tau_B \sim NB^2 \tag{9.24}$$

where $B = |\mathbf{B}|$. If the torque due to the frictional forces overcomes the magnetic torque, the system will become unstable. The scaling for the frictional torque τ_F can be also calculated. For a given bead i:

$$\tau_{F,i} \sim d_i v_i \eta \sim v d_i^2 \eta \tag{9.25}$$

where v_i is the bead velocity, v is the rotor frequency, d_i is the distance of bead i from the center of mass, and η is the viscosity. This expression can be integrated over the length of the chain, yielding the overall frictional torque τ_F:

$$\tau_F \sim \int_0^N v d_i^2 \eta \mathrm{d}(d_i) \sim v N^3 \eta. \tag{9.26}$$

Comparison of τ_B and τ_F yield a characteristic frequency v_c above which break-up will occur:

$$v_c \sim \frac{\tau_F}{\tau_B} \sim \frac{B^2}{N^2 \eta}. \tag{9.27}$$

In the computational portion, we will see that this scaling behavior is indeed observed as shown in Figure 9.11 where we plot the average number of fragments as a function of the rescaled frequency v/v_c. Notice that all the data collapses into a universal curve.

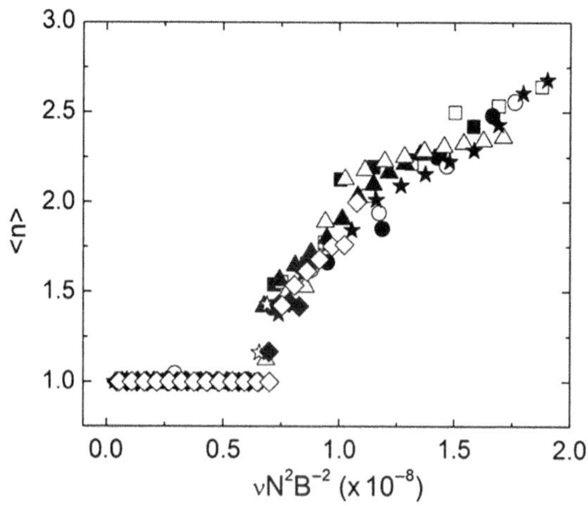

Figure 9.11 Simulations the time-average number of chain fragments <n> for chains of $N = 7$–10 at $B = 3$–7 mT as a function of the rescaled frequency (see eqn (9.27)).
(Figure reprinted, with permission, from Sing *et al.*[25])

9.3.2 Computational Approaches

This system, due to its geometric simplicity and analytical tractability, is likewise amenable to incorporation into traditional computational methods. While a number of simulation methods could be used, the disparity between the size of the super-paramagnetic beads and the surrounding medium suggests that an implicit solvent method is possible (we can represent the liquid as a continuum without molecular-level detail). We also desire a simulation that faithfully represents the dynamics of the system, since we are interested in the evolution of the system over time rather than finding an equilibrium state. We therefore use what are known as 'Brownian dynamics' simulations, which track the motion of beads using the Langevin equation:

$$\frac{\partial \mathbf{r}_i}{\partial t} = \sum_j^N \left(\mu_{\mathbf{ij}} \cdot \nabla_{\mathbf{r}_j} U(t) + \nabla_{\mathbf{r}_j} \cdot \mathbf{D}_{ij}\right) + \xi_i(t) \tag{9.28}$$

where $\mu_{\mathbf{ij}}$ and $\mathbf{D}_{ij} = k_B T \mu_{ij}$ are the mobility and diffusion tensors respectively, and $\xi_i(t)$ is a random force that satisfies the Fluctuation–Dissipation theorem $(\langle \xi_i(t)\xi_j(t')\rangle = 2k_B T \mu_{ij}\delta(t-t'))$. For these situations, we can ignore both the noise term and subsequently the diffusion divergence term due to the disparity of length scales between the solvent molecules and the beads, and we are only left with the potential gradient term. While this renders such simulations deterministic, we will show that this behavior well reproduces both theory and experiment.

The mobility matrix $\mu_{\mathbf{ij}}$ accounts for the hydrodynamic interactions, and has the same meaning as the Oseen and Blake tensors described above. Since calculation of these tensors is tedious in our analytic theory, the computer specializes in these sorts of long and repetitive processes! Therefore, we actually introduce an even more complicated description of the Blake tensor for our walkers near the surface. This represents a relaxation of the assumptions of the original, theoretical simulation; while those calculations introduced complexities due to the presence of the surface, we still utilize point forces due to the long-ranged interactions of the images with the rotor. We do not need to make this assumption, however, and introduce the Rotne—Prager–Yamakawa tensor that represents the first-order correction to account for beads of finite size. To account for beads of finite size, the point force is expanded in a multipole expansion so that it is distributed on a sphere of radius a and acts upon a sphere also of radius a:

$$\langle \mu_{ij}(\mathbf{r}+2\mathbf{a})\rangle_{\mathbf{a}} \approx \mu_{ij}(\mathbf{r}) + \langle \mathbf{a}\rangle \cdot \nabla \mu_{\mathbf{ij}}(\mathbf{r}) + \frac{1}{2}\langle \mathbf{aa}\rangle : \nabla\nabla \mu_{ij} \dots \tag{9.29}$$

The derivation is tedious, but the final result for the Rotne–Prager–Yamakawa expansion of the Oseen tensor is:

$$\frac{\mu_{ij}}{\mu_0} = \left(\begin{array}{ll} \dfrac{3a}{4r_{ij}}\left(\left(1+\dfrac{2a^2}{3r_{ij}^2}\right)\mathbf{I} + \left(1-\dfrac{2a^2}{r_{ij}^2}\right)\dfrac{\mathbf{r}_{ij}\mathbf{r}_{ij}}{r_{ij}^2}\right) & r_{ij} \geq 2a \\ \left(1-\dfrac{9r_{ij}^2}{32a^2}\right)\mathbf{I} + \dfrac{3}{32}\dfrac{\mathbf{r}_{ij}\mathbf{r}_{ij}}{ar_{ij}} & r_{ij} < 2a \end{array} \right. \tag{9.30}$$

where $\mu_0 = 1/(6\pi\eta_S a)$ is the Stokes mobility of a single bead. The Rotne–Prager–Yamakawa–Blake tensor follows from this same analysis, and it has been described in detail in the literature. The are alternative methods to reproducing these hydrodynamics, for example the RPYB tensor is an analytical method that works well in the presence of a solid surface; however it becomes much more difficult to maintain an analytical method when more complicated boundary conditions must be observed to the point of being entirely intractable. Recent advances in parallel processing has led to the rise in use of the 'Lattice–Boltzmann' simulations that represent the fluid as a lattice of populations of fluid particles spite the artificial nature of this picture, first-order hydrodynamics can be retained in a facile manner in a way that allows for the incorporation of complicated boundary conditions. This is, for small systems such as this one, computationally intensive. If the system includes an increasing number of particles; however, the linear scaling of such a system means that it quickly becomes advantageous with the use of parallel processing.

The force on the beads, $F = -\nabla_{\mathbf{r}j} U(t)$, is the time-dependent field that describes the driving force for the motion of the beads (the direction and desire of the beads to move; the ability is given by the mobility). This specific model contains four contributions; however, it is in principle to add in more contributions as necessary. The relevant contributions for magnetically actuated microwalkers include the excluded volume between two beads F_b, the wall repulsion F_w, the gravitational force that confines the bead aggregates to the surface F_g, and the magnetic dipole–dipole force F_B that is induced upon application of the field B. The excluded volume potential is:

$$U_b(r) = \begin{cases} \infty & r \leq 2a \\ 0 & r > 2a \end{cases} \tag{9.31}$$

where a is the radius of a particle and r is the bead–bead distance. While this potential is convenient in most theoretical treatments, an iterative process such as the one developed in this section would not be stable due to the possibility of a discrete bead 'jump' into the region of infinite potential. Rather, it is necessary to use a potential that is smooth but still sufficiently steep that beads do not overlap. One possibility is a modified Lennard-Jones potential:

$$F_b = \frac{\varepsilon}{r - 2a}\left[\left(\frac{\sigma}{r - 2a}\right)^{12} - \left(\frac{\sigma}{r - 2a}\right)^{6}\right] \tag{9.32}$$

where $\sigma = 0.1a$ sets a length scale beyond the bead radii where that represents the position of the potential. The depth ε of the potential, since the beads are not attractive through any force other than the magnetic force (which we explicitly include later), can be kept very small. The simulation results presented in this chapter utilize $\varepsilon = 0.02k_BT$ such that this is essentially negligible. The wall force, since it is also based on an excluded volume potential, is similar to the bead–bead potential except it acts solely in the z-direction:

$$F_w(z) = \frac{\varepsilon}{r_z - a}\left[\left(\frac{\sigma}{r_z - a}\right)^{12} - \left(\frac{\sigma}{r_z - a}\right)^{6}\right] \tag{9.33}$$

where r_z is the distance between the wall and the center of mass of the bead and the rest of the parameters are the same as for F_b.

The applied magnetic field B induces magnetic dipoles on each bead. Since we are dealing with relatively small fields, it is standard to assume that the resulting magnetic moment $\mathbf{m}$ is both in the direction of and proportional to the applied field. This is not rigorously true, since the induced magnetic moments of the surrounding beads influence the magnetic field at the point of a given bead; however, we consider this contribution to be negligible. It is also assumed, though recent work on large beads has demonstrated that it may be relevant in certain situations, that there is no 'lag time' between the time-dependency of the magnetic field and the re-orientation time of the moment within the magnetic bead. Upon incorporating these assumptions, the moment is simply:

$$\mathbf{m} = \frac{V_c \Delta_\chi}{\mu_0}\mathbf{B} \tag{9.34}$$

where μ_0 is the magnetic permittivity, V_c is the effective volume of the bead that is paramagnetic, and $\Delta\chi$ is the magnetic susceptibility difference between the bead and the medium (in the case corresponding to the experimental results in this chapter, the difference between the bead and water is $\Delta\chi = 0.7$). The force $F_{B,ij}$ between two induced moments $\mathbf{m}_i$ and $\mathbf{m}_j$ is given by:

$$F_{B,ij} = \frac{3\mu_0}{4\pi r^4}\left(\left(\frac{\mathbf{m}_i \cdot \mathbf{r}_{ij}}{r}\right)\mathbf{m}_j + \left(\frac{\mathbf{m}_j \cdot \mathbf{r}_{ij}}{r}\right)\mathbf{m}_i - \left(\frac{5(\mathbf{m}_j \cdot \mathbf{r}_{ij})(\mathbf{m}_j \cdot \mathbf{r}_{ij})}{r^2} - (\mathbf{m}_i \cdot \mathbf{m}_j)\right)\frac{\mathbf{r}_{ij}}{r}\right) \tag{9.35}$$

where $\mathbf{r}_{ij}$ is the position vector between beads i and j, with magnitude r. Due to the order of magnitude of these values in SI becoming drastically distant from 1 (r^4 for a bead radius of 1 μm is 1×10^{-24}, for example) it is customary to render the values dimensionless by normalizing them by characteristic values. We typically represent these values using tildes; for distances, we define $\tilde{\mathbf{r}} = \mathbf{r}/a$ and $\tilde{\mathbf{m}} = \mathbf{m}/m_0$, where $m_0 = V_c\Delta\chi B/\mu_0$. If we also define a force parameter $F_0 = 4\pi(af\Delta\chi B)^2/(3\mu_0)$ (where f is the fraction of the bead volume that is paramagnetic), we can rewrite the force in dimensionless terms:

$$\tilde{F}_{B,ij} = \frac{F_0}{\tilde{r}^4}\left(\left(\frac{\tilde{\mathbf{m}}_i \cdot \tilde{\mathbf{r}}_{ij}}{\tilde{r}}\right)\tilde{\mathbf{m}}_j + \left(\frac{\tilde{\mathbf{m}}_j \cdot \tilde{\mathbf{r}}_{ij}}{\tilde{r}}\right)\tilde{\mathbf{m}}_i - \left(\frac{5(\tilde{\mathbf{m}}_j \cdot \tilde{\mathbf{r}}_{ij})(\tilde{\mathbf{m}}_i \cdot \tilde{\mathbf{r}}_{ij})}{\tilde{r}^2} - (\tilde{\mathbf{m}}_i \cdot \tilde{\mathbf{m}}_j)\right)\frac{\tilde{\mathbf{r}}_{ij}}{\tilde{r}^2}\right) \tag{9.36}$$

The gravitational force (though other force fields such as electric fields could be considered as well) that confines the aggregates to the surface is simply $F_g(z) = -\Delta\rho V g$, where the gravitational acceleration is $g = 9.8$ m s^{-2} and the differential density is $\Delta\rho = 0.8$ g cm^{-3}.

The Langevin equation can be iteratively solved in its discretized, dimensionless form:

$$\tilde{\mathbf{r}}_\mathbf{i}(\tilde{t} + \Delta\tilde{t}) = \tilde{\mathbf{r}}_\mathbf{i}(\tilde{t}) + \Delta\tilde{t}(\tilde{\mu}_\mathbf{ij} \cdot (\tilde{\mathbf{F}}_{j,b} + \tilde{\mathbf{F}}_{j,w} + \tilde{\mathbf{F}}_{j,B} + \tilde{\mathbf{F}}_{j,g})) \tag{9.37}$$

where the dimensionless value of time $\tilde{t} = t/\tau_D$ is normalized by the diffusion time τ_D of a single bead $\tau_D = 6\pi\eta a^3/(k_B T)$. Due to the lack of a noise term, however, this choice of a dimensionless time is completely arbitrary.

9.3.2.1 Computational Results

The literature on this system has shown excellent agreement between theory, experiment, and simulation for the calculated values. For example, Figure 9.5b shows very good agreement between simulation (open symbols) and experiment (filled symbols) for the aggregate translational velocity $\langle v \rangle$ as a function of rotational frequency of the magnetic field. Extended outwards in both simulation and theory for larger aggregates, the linear scaling of the velocity predicted by the scaling law is demonstrated in the detailed theory, simulation, and experiment (Figure 9.4). The success of this theory allows for predictive modeling of a single microwalker as it moves across the surface. Other parameters follow predictions, as well. For example, break-up of aggregates can be characterized by the average number of chains $\langle n \rangle$ that the overall aggregate breaks into as a function of frequency. For an intact aggregate, $\langle n \rangle = 1.0$; however, this number quickly rises upon increasing frequency v or decreasing field strength B. For a wide number of situations, rescaling the frequency v by the v_c (as per our prediction) reveals a universal curve that is only dependent on the rescaled frequency vN^2/B^2 (see Figure 9.11). Finally, the translational velocity as a function of aggregate orientation θ matches the theoretical prediction (see Figure 9.12) for a number of values in simulation. For these microwalkers, we clearly have an excellent understanding of the single-chain behavior.

9.4 Microwalker 'Armies'

While the current state of the theory can deal with single magnetic microwalkers, the situation becomes drastically more complicated upon the consideration of an ensemble of these aggregates. The flow induced by a single rotor now has the opportunity to affect other rotors, which may not be ordered or even dispersed evenly within a solution. Armies of microwalkers effectively produce a surface current and can translate objects much larger than the walkers themselves, as shown in Figure 9.13a where a large unilamellar vesicle

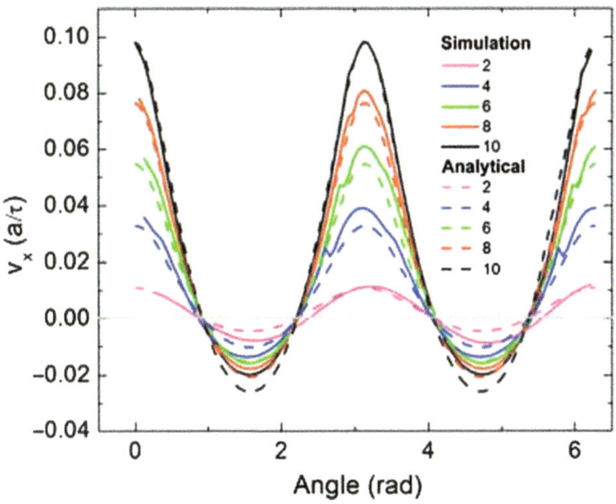

Figure 9.12 The analytical and simulation results describing the angular dependence on the rotor's translational velocity v_x as a function of the number of beads in the chain, N. Notice that both results are in good agreement. (Figure reprinted with permission from Sing *et al.*[25])

is moved at 25 μm s^{-1} by the underlying rotors. While this complicated problem is likely intractable, we can still calculate the effect of multiple rotors in the homogenous limit by spatial averaging of the hydrodynamic tensors. With this approximation, we find a velocity profile that can be simply computed as:

$$v(z) = \rho_S \beta(z) \tag{9.38}$$

where ρ_S is the surface density of rotors, and the function $\beta(z)$ is an integral of the velocity field v_x along a horizontal plane spanning the x- and y-directions:

$$\beta(z) = \int v_x(x, y, z)\mathrm{d}x\mathrm{d}y \tag{9.39}$$

This approach is a mean-field method that effectively considers the velocity of the fluid at a height z away from the surface on average assuming that we can simply superimpose the effects of individual rotors without considering their interactions. This can be calculated with simulation by simply summing up over a discrete grid around the rotor; however, the long-range nature of the hydrodynamic interactions causes this summation to converge incredibly slowly. Alternatively, it is possible to derive the area-averaged Blake mobility tensor analytically. This has been performed in the literature,[34] and effectively applies a force to the solid plane of liquid at height z whenever there is a force that acts at a height z. The form of the velocity profile for a given force is:

$$v_x(z_i) = \rho_S \int dz_i \mu_{ij}(z_i, z_j) F_x(z_j)/a \tag{9.40}$$

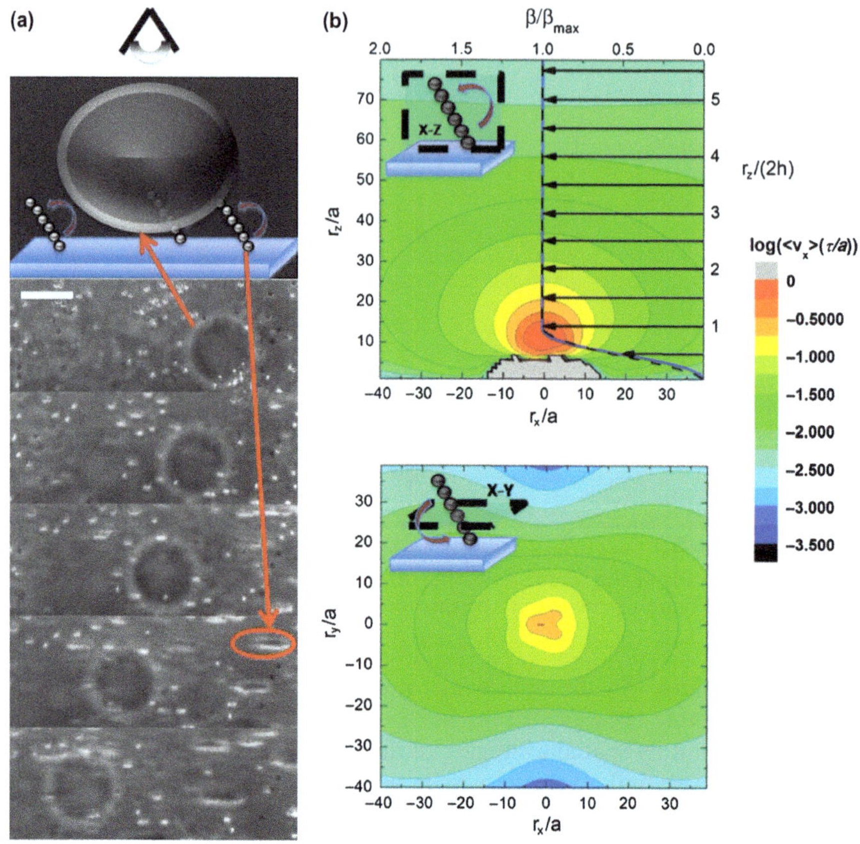

Figure 9.13 (a) The presence of a velocity field created by the rotors enables the transport of the large, dispersed objects such as the vesicles in this figure (large dark circle). The flow is created by a carpet of surface walkers positioned (bright short chain-like regions) actuated at 32 Hz. The snapshots are taken from the top, as indicated in the cartoon. The scale bar represents 20 μm, and the time between frames is 0.5 s. (b) The logarithmic intensity of the x-direction of the velocity field immediately surrounding a single seven-bead rotating chain (measured from the chain center of mass), in both the x–z (b) and x–y (c) planes (geometry shown in the cartoon in the upper right corners). We also plot on the x–z plane the function β (normalized by $\beta_{\max}$) as a function of z, which is proportional to the mean velocity profile of a collection of rotors as represented in eqn (9.46). Both the simulation (black dashed) and analytical (blue solid) results for β are shown with the arrows representing the direction of the flow. Both profiles are indistinguishable above $z=h$. We note that β tends towards zero near the surface, due to the non-slip condition, and resembles a plug flow profile. The gray area in (b) represents flow that opposes the predominating flow direction and cannot be plotted in the logarithmic scale.
(Figure reprinted, with permission, from Sing *et al.*[25])

where $\rho_S F_x(z)/a$ is the laterally averaged force density applied at height z. The Blake–Oseen tensor can be averaged analytically to yield $\mu_{ij}(z_i,z_j)$:

$$\mu_{ij}(z_i, z_j) = \begin{pmatrix} z_j/\eta & z_i > z_j \\ z_i/\eta & z_i < z_j \end{pmatrix} \tag{9.41}$$

This result is much more accurate since it does not require the prohibitively large spatial averaging necessary to calculate from eqn (9.39). Figure 9.13b plots the calculation of β as a function of z for $N=7$ from simulation data. At low Reynolds number, the effect of a rotor is straightforward on the surrounding fluid. So long as the averaged velocity is considered, and the measurement occurs beyond the layer of microwalkers, the boundary condition at the surface goes from no-slip to a constant velocity dictated by the rotor motion. This mobile surface is what drives the vesicle movement seen in the first part of this chapter.

It is possible to derive an analytical form of β using this same approach. We limit ourselves to the case that $h=Na$, which is the observed geometry, and assume that the rotor behaves as a rigid rod of length $2h$. We first calculate the force density $F(z)/a$ to be independent of the rotor angle, since even though there is more force when the rotor is vertical there is more area per segment dz as the rotor moves away from this angle. Thus:

$$F(z)/a = 6\pi\eta v(z-h) \tag{9.42}$$

While this is constant with θ, the actual range of z values that feel a force is very much dependent on θ. For example, if the rotor is fully vertical, a force is on the fluid everywhere between $z=0$ and $z=2h$. If the rotor is horizontal, the force is only applied at an infinitely small region at $z=h$. We incorporate this into the expression for the force density by multiplying $F(z)/a$ by the fraction of time $f(z)$ that a force is being applied to a given height z. The form for $f(z)$ is:

$$f(z) = \begin{cases} 0 & z > 2h \\ \dfrac{2}{\pi}\arccos\left(\dfrac{z-h}{h}\right) & 2h > z > h \\ \dfrac{2}{\pi}\arccos\left(\dfrac{h-z}{h}\right) & h > z > 0 \end{cases} \tag{9.43}$$

It is not easy to analytically integrate the arccosine function, so to simplify the integration and final form for $\beta(z)$, we approximate $f(z)$ as:

$$f(z) = \begin{cases} 0 & z > 2h \\ \left(\dfrac{2h-z}{h}\right) & 2h > z > h \\ \left(\dfrac{z}{h}\right) & h > z > 0 \end{cases} \tag{9.44}$$

This underestimates the force of the rotor by a small amount; however, it still retains the limits of $f=0$ at $z=0$, $2h$, and $f=1$ at $z=h$. These functions can then be integrated to obtain a form for $\beta(z)$:

$$\beta(z) = \int_0^\infty \mathrm{d}z_j \mu_{ij}(z, z_j) F_x(z_j) f(z_j)/a \tag{9.45}$$

$$\beta(z) = \begin{cases} \pi v z^3 \left(1 - \frac{z}{2h}\right) & z < h \\ \pi v \left(h^3 - 4h^2 z + 6hz^2 - 3z^3 + \frac{z^4}{2h}\right) & h < z < 2h \\ \pi v h^3 & z > 2h \end{cases} \tag{9.46}$$

This approximate equation matches well the form of $\beta(z)$ determined by simulation data, and we have plotted both in Figure 9.13b. These results also match the dependence of the maximum value of the flow profile β_{max} as the rotor size N is changed. There is a small amount of deviation; however, such differences are quite small considering the approximations made (the approximate form for $f(z)$ and the replacement of the flexible bead chain by a rigid rod). This finalizes our treatment of an assembly of walkers. More work is necessary in this respect to produce uniform samples, but this might not be technologically relevant as it is time intensive.

9.5 Summary and Outlook

Microwalkers are a new class of mobile synthetic microscopic 'robots' that walk on surfaces due to the excess friction near the surface. They were discovered originally by three groups (one in Japan, another in Spain, and another in the US/Germany) working independently between 2007 and 2008.[25,27,28] These so-called microwalkers are a new paradigm in artificial mobility on surfaces because they bear similitudes with molecular motors but can be controlled exactly. Also, they can function as pumps in a similar way as artificial cilia works. Since its inception, this area has seen strong growth, although it is still in its beginnings, with many possible interesting avenues starting to appear. There are several parts of this system that can be modified and are eagerly awaiting to be studied. Here, we do not do a comprehensive list of different systems that could be studied, but rather suggest interesting avenues for future research. The basis of the microscopic walkers described before are a magnetic field to assembly and actuate the walkers, friction with a surface, and cooperative effects between multiple walkers. In regards to the former ingredient, the magnetic field, there has not been enough research on the different actuation protocols possible to manipulate such walkers. For example, there are interesting questions when it comes to using periodic oscillating fields with different forward and backward frequencies. In theory, such schemes should not lead to motion because of the time invariance of the Stokes equation; however, coupling with sedimentation and velocity dependent friction coefficients may break the symmetry and lead to interesting insights into the fluid mediated surface–surface interactions in shear. Another interesting aspect that has not been studied is an optimal actuation protocol as experiments have shown that allowing the rotor sediment might not be the best solution. As a matter of fact, the linearity observed as a function of frequency and chain length seems to suggest that other hydrodynamic mechanisms are as important as the direct contact friction, suggesting the it might be possible to exploit them using optimized actuation protocols. In this respect, choosing a substrate that can reduce the backward motion (due to hydrodynamic forces) by raising the chains further away from the surface while they become parallel to the surface is a must. In this respect, Morimoto *et al.*[28] used a paramagnetic substrate that served this purpose; however, it also enhanced the surface-on-surface friction due to the attractive nature of the substrate during the parts of the cycle where the chain is oriented perpendicular to the substrate. More research in this direction is needed to understand the limits of the actuation protocols.

The substrate can play a predominant role in the motion of microwalkers since it dictates the effective friction experienced by the beads. Modifying such friction by means of applying external fields such as electromagnetic[35] or 'gravitional'[36] to get the beads closer to the substrate is a possible avenue for tailoring the friction experienced by the walkers. Another one is to change the material composition of the substrate and the bead surface. This can be done easily using commercially available beads used in biological separations protocols. For example, a polymer brush can be incorporated using a biotinylated motif at one of the polymer ends. Such products are widely used in biology and thus can be easily obtained. Research in this area has shown that brushes can change the direct friction coefficient by orders of magnitude, in particular, charged brushes seem to excel in their lubrication properties.[37–41] This, in fact, brings us to another avenue which is that of cooperative motion. Given the fact that microwalkers are magnetically assembled, or at least respond to magnetic forces, they behave differently dependent on the environmental magnetic field and the intrinsic local properties of the substrate. This means that if one considers multiple microwalkers, the local environment plays an important role in their walking properties. For example, if one had a disordered substrate with regions of high friction and regions of low friction, the flows within and around such regions would be different compared to cases where the substrate is uniform. Such heterogeneities can bring in some interesting physics with regards to the fluid dynamics of random active soft-matter systems. We have just highlighted some of the topics that we consider are of much interest. Readers are encouraged to postulate their own questions and find their own interest. The field of microwalking is just beginning, so this is an exciting time to work on such systems, be creative, and discover new phenomena.

Acknowledgements

We are graeful to Jaap den Toonder and Patrick Onck for organizing an amazing session on artificial cilia in Toulouse 2010, and later inviting us to contribute this chapter. We would also like to thank Roland Netz and Matthias Schneider with whom we have had enlighting conversations during the past years. Of course, most of this work would have not been possible without the funds provided by an NDSEG fellowship (C.E.S.) and NSF CAREER award DMR-1054671 (A.A.-K.).

References

1. S. M. Block, *Biophys. J.*, 2007, **92**, 2986–2995.
2. R. Lipowsky and S. Klumpp, *Phys. A – Stat. Mech. Appl.*, 2005, **352**, 53–112.
3. A. Gennerich and R. D. Vale, *Curr. Opin. Cell Biol.*, 2009, **21**, 59–67.
4. C. L. Asbury, *Curr. Opin. Cell Biol.*, 2005, **17**, 89–97.
5. S. A. Burgess, M. L. Walker, H. Sakakibara, P. J. Knight and K. Oiwa, *Nature*, 2003, **421**, 715–718.
6. R. D. Vale, *Cell*, 2003, **112**, 467–480.
7. Z. Okten and M. Schliwa, *Nature*, 2007, **450**, 625–626.
8. M. Schliwa and G. Woehlke, *Nature*, 2003, **422**, 759–765.
9. N. Mizuno, S. Toba, M. Edamatsu, J. Watai-Nishii, N. Hirokawa, Y.Y. Toyoshima and M. Kikkawa, *EMBO J.*, 2004, **23**, 2459–2467.
10. N. Hirokawa, *Science*, 1998, **279**, 519–526.
11. N. Hirokawa, Y. Noda and Y. Okada, *Curr. Opin. Cell Biol.*, 1998, **10**, 60–73.
12. U. Henningsen and M. Schliwa, *Nature*, 1997, **389**, 93–96.
13. M. Schliwa, *Nature*, 1999, **397**, 204–205.
14. H. Hagiwara, H. Yorifuji, R. Satoyoshitake and N. Hirokawa, *J. Biol. Chem.*, 1994, **269**, 3581–3589.
15. R. Lipowsky, J. Beeg, R. Dimova, S. Klumpp and M. J. I. Muller, *Phys. E – Low-Dimensional Syst. Nanostruct.*, 2010, **42**, 649–661.
16. M. J. I. Muller, S. Klumpp and R. Lipowsky, *Biophys. J.*, 2010, **98**, 2610–2618.
17. S. Klumpp, Y. Chai and R. Lipowsky, *Phys. Rev. E*, 2008, **78**.
18. S. Klumpp and R. Lipowsky, *Proc. Natl. Acad. Sci. U.S.A.*, 2005, **102**, 17284–17289.
19. S. Klumpp, M. J. I. Muller and R. Lipowsky, *Traffic and Granular Flow '05*, A. Schadschneider, T. Pöschel, R. Kühne, M. Schreckenberg and D. E. Wolf, (eds.), Springer, 2007, 251–261.
20. M. J. I. Muller, S. Klumpp and R. Lipowsky, *Proc. Natl. Acad. Sci. U.S.A.*, 2008, **105**, 4609–4614.
21. S. P. Gross, M. Vershinin and G. T. Shubeita, *Curr. Biol.*, 2007, **17**, R478–R486.
22. V. Soppina, A. K. Rai, A. J. Ramaiya, P. Barak and R. Mallik, *Proc. Natl. Acad. Sci. U.S.A.*, 2009, **106**, 19381–19386.
23. J. Gagliano, M. Walb, B. Blaker, J. C. Macosko and G. Holzwarth, *Eur. Biophys. J. Biophys. Lett.*, 2010, **39**, 801–813.

24. D. B. Hill, V. Swaminathan, A. Estes, J. Cribb, E. T. O'Brien, C. W. Davis and R. Superfine, *Biophys. J.*, 2010, **98**, 57–66.
25. C. E. Sing, L. Schmid, M. F. Schneider, T. Franke and A. Alexander-Katz, *Proc. Natl. Acad. Sci. U.S.A.*, 2010, **107**, 535–540.
26. P. Tierno, R. Golestanian, I. Pagonabarraga and F. Sagues, *J. Phys. Chem. B*, 2008, **112**, 16525–16528.
27. P. Tierno, R. Golestanian, I. Pagonabarraga and F. Sagues, *Phys. Rev. Lett.*, 2008, **101**, 218304.
28. H. Morimoto, T. Ukai, Y. Nagaoka, N. Grobert and T. Maekawa, *Phys. Rev. E*, 2008, **78**, 021403.
29. P. Tierno, O. Guell, F. Sagues, R. Golestanian and I. Pagonabarraga, *Phys. Rev. E*, 2010, **81**, 011402.
30. D. M. Huang, C. Sendner, D. Horinek, R. R. Netz and L. Bocquet, *Phys. Rev. Lett.*, 2008, **101**, 226101.
31. C. Sendner, D. Horinek, L. Bocquet and R. R. Netz, *Langmuir*, 2009, **25**, 10768–10781.
32. M. Karle, J. Wohrle, J. Miwa, N. Paust, G. Roth, R. Zengerle and F. von Stetten, *Microfluid. Nanofluid.*, 2011, **10**, 935–939.
33. R. Dreyfus, J. Baudry, M. L. Roper, M. Fermigier, H. A. Stone and J. Bibette, *Nature*, 2005, **437**, 862–865.
34. Y. W. Kim and R. R. Netz, *J. Chem. Phys.*, 2006, **124**, 114709.
35. T. Courcier, H. Joisten, P. Sabon, S. Leulmi, T. Dietsch, J. Faure-Vincent, S. Auffret and B. Dieny, *Appl. Phys. Lett.*, 2011, **99**, 093107.
36. K. Halvorsen and W. P. Wong, *Biophys. J.*, 2010, **98**, L53–L55.
37. J. Klein, *Annu. Rev. Mater. Sci.*, 1996, **26**, 581–612.
38. B. Bhushan, J. N. Israelachvili and U. Landman, *Nature*, 1995, **374**, 607–616.
39. U. Raviv, S. Giasson, N. Kampf, J. F. Gohy, R. Jerome and J. Klein, *Nature*, 2003, **425**, 163–165.
40. M. Chen, W. H. Briscoe, S. P. Armes, H. Cohen and J. Klein, *Eur. Polym. J.*, 2011, **47**, 511–523.
41. M. Chen, W. H. Briscoe, S. P. Armes and J. Klein, *Science*, 2009, **323**, 1698–1701.

CHAPTER 10

Artificial Flagellar Micro-swimmers

RÉMI DREYFUS,*[a] JEAN BAUDRY[b] AND JÉRÔME BIBETTE[b]

[a] Compass Lab, UMI 3251, CNRS/Rhodia/UPenn, Bristol, Pennsylvania, USA; [b] Laboratoire Colloides et Materiaux Divises, UMR 7610, CNRS/UPMC, Paris, France
*Email: remi.dreyfus@gmail.com

10.1 Introduction

The prominent rise of micro- and nano-technologies based on the predictions that there is 'plenty of room at the bottom'[1] has recently led scientists to envision and develop new strategies for building tools or device on the micro-/nanoscale that can transform energy into motion,[2–6] This sudden interest has reached several scientific communities. In fluid mechanics, scientists have been involved in understanding the physical principles that govern propulsion on a micro-scale level.[7–14] Biologists have been able to describe and analyze the common modes of propulsion that are encountered in nature.[15–18] Physicists have devised new types of motion that allow for optimized propulsion[2–4,19,20] while chemists have been designing molecules or DNA supramolecular assemblies that allow for motion while experiencing changes in conformation.[21–23] The reasons for this increase in interest lie in the wide area of applications of nano- or micromachines. Especially in the field of biomedical engineering where we could imagine that future drugs will be delivered not only by passive diffusion but by active machines able to swim or swarm at a fast

RSC Nanoscience & Nanotechnology No. 30
Artificial Cilia
Edited by Jaap M J den Toonder and Patrick R Onck

Published by the Royal Society of Chemistry, www.rsc.org

velocity. We could also imagine adding such engines to surfaces that will become 'smart' and able to displace fluids, cells, or to mix viscous fluids at small scales. Nanomotors could also be part of future nanofactories that will synthesize and process materials.

Here in this chapter we will show how, by taking inspiration from natural swimming organisms, we were able to leverage our knowledge in materials science in order to create artificial micro-swimmers that have some of the features of flagella. These engines were also used to confirm previous predictions made by scientists in the field of fluid mechanics.

10.2 Swimming at Low Reynolds Number

Three main types of swimming motions are encountered at the micro-scale in nature.[16] Two of them are shortly represented in Figure 10.1a and b. A very common type of propulsion relies on the rotational motion of a corkscrew-like flagellum (Figure 10.1a). Bacteria flagella are considered as being stiff. Their rotational motion is induced by the presence of small motors at the surface of the bacteria cells. Another very common encountered type of propulsion is the propulsion induced by flagella. A flagellum can be seen as a bundle of stiff microtubules all bound together by proteins. Between these microtubules, molecular motors are inducing small local forces that can bring the overall flagellar motion.[24] For flagella, the motion is best represented by a wave that propagates from one end of the flagella, usually the head, to the free end (Figure 10.1b). The third type of motion is a ciliary motion;[10] however, this motion does not correspond to a propagating wave, but rather to a stroke-like motion. This type of motion is encountered at the surface of some micro-organisms. Ciliae are also present in lungs and ensure the circulation of mucus. The physical reason why these three kinds of deformation can induce motion at the micro-scale, where viscous effects are dominating, can be explained in simple terms by the slender-body theory.[25–29] The slender body theory has been developed in order to derive the expression for the viscous drag $\boldsymbol{f}$ that experiences a slender body moving at a velocity $\boldsymbol{v}$ in a viscous fluid at small Reynolds number. If you consider a slender body of length $2L$ that is large compared to its width a, moving at a velocity $\mathbf{v} = v_s\mathbf{s} + v_n\mathbf{n}$ projected along vectors $\mathbf{s}$ and $\mathbf{n}$ where $\mathbf{s}$ and $\mathbf{n}$ are, respectively, the tangential and normal vectors in the Frenet geometry, the viscous drag per unit length is written as:

$$\begin{cases} f_s = -\zeta_= v_s \\ f_n = -\zeta_\perp v_n \end{cases} \tag{10.1}$$

For a slender body, the coefficients $\zeta_=$ and $\zeta_\perp$ are not equal. As a first approximation, there is approximately a factor of 2 between the two coefficients: $\zeta_\perp \approx 2\zeta_=$. The anisotropy of the tensor that relates the forces to the velocities can explain qualitatively the reason why the deformations described previously allow for the motion of micro-organisms. Indeed let us consider the

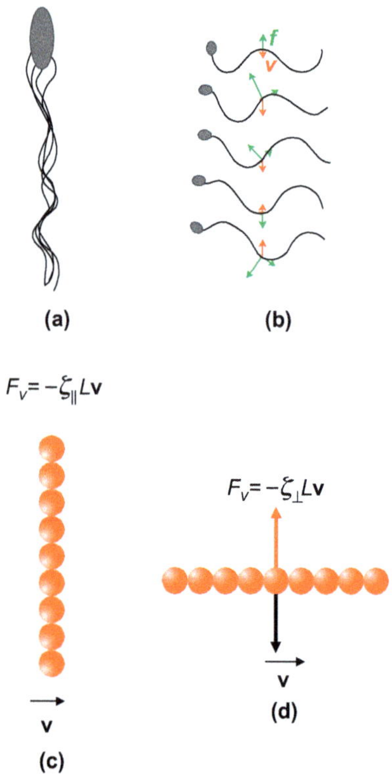

Figure 10.1 (a) Scheme of bacteria with bundles of helical-shaped rigid flagella. (b) Flagellar motions of spermatozoa. A wave of deformation propagating from the head to the end can be clearly seen. The white arrows indicate the velocity of a small portion (in red) of the flagellum that can be considered as a slender body. The green arrows indicate the parallel and perpendicular components of the viscous drag. Over a cycle, the parallel component vanishes but the perpendicular does not vanish and induces a propulsive force in the direction opposite to the wave propagation. (c) Vertical motion of a rod and its corresponding viscous drag. (d) Horizontal motion of a rod and its corresponding viscous drag.

example given in Figure 10.1b. For deformations of flagella, arrows indicate velocities. The viscous forces corresponding to the deformations are also represented in the same figure. It appears clearly that, because of the anisotropy between $\zeta_{=}$ and $\zeta_{\perp}$, over a cycle, the resultant of the viscous force does not vanish. Therefore an overall propulsive force arises and allows for the swimming motion of the micro-organism.

A natural approach to create man-made artificial machines that have the ability to propel themselves or to swim can consist in mimicking some of the motions encountered in nature. Here we decided to try to recreate mobility by utilizing the strategies used by flexible flagella, such as those used by cilia or spermatozoa.

10.3 Fabrication and Modeling of Magnetic Flexible Filaments

The technical challenge consists in creating on the micro-scale level filament structures that are (1) flexible with a controlled elastic modulus, and (2) actuated externally by an external field such as a magnetic or an electric field. Since the pioneering work by Furst *et al.*,[30] many groups have been able to create flexible filament shaped metamaterials.[31–36] In the work presented here, the protocol for filament creation and characterization are directly inspired by this work. Indeed, we have decided to self-assemble these filament structures using magnetic colloids. Magnetic colloids are particles that have diameters comprised between hundreds of nanometers to a micrometer. They are easily and readily observable under regular microscopy. The colloids are usually made of a porous matrix made of polystyrene. Within the pores of the matrix, small nano-domains of ferromagnetic iron oxides, Fe_2O_3, are grown. These magnetic domains confer specific magnetic properties to the colloids. The colloids are said to be super-paramagnetic. This means that they exhibit a paramagnetic behavior: under a magnetic field, they become magnetized. As soon as the magnetic field is turned off, the magnetization vanishes and the materials have lost their magnetic properties. They are considered *super*-paramagnetic because the magnetic susceptibility χ of the colloids is of the order of unity, which is quite high compared to the regular values of susceptibility for paramagnetic materials (usually of the order of 10^{-3}). When magnetic colloidal particles that are suspended in solution are subjected to a magnetic field, they become magnetized and attract, in the same way as small magnets would do. Under this induced attraction, colloids align and form long chains that grow in the same direction as that of the external magnetic field. This process is schematically shown in Figure 10.2a. As the particles are paramagnetic, the chains that are formed are not permanent, meaning that as soon as the magnetic field is turned off, particles dissociate from the chain and redistribute themselves homogeneously in solution after a certain time.[37–39] A strategy for creating flexible filaments can then consist in first self-assembling these non permanent magnetic filament by submitting magnetic colloids to an external magnetic field and then somehow find a way to irreversibly link the particles in the chains together. Here we have used two different strategies. The first one consists in adding a small amount of polyacrylic acid (PAA) in solution. It has been shown that the polymer PAA adsorbs on the particles.[33] If the particles are kept close to each other in a chain under a magnetic field for a sufficient time, the PAA can bridge two adjacent particles together and glue them together (Figure 10.2b). Another strategy consists in using particles that have streptavidin on their surface and DNA with biotin at each end as binding agents.[40] As the coupling between biotin and streptavidin is strong, long chains of particles can be formed with each particle glued to their neighbors (Figure 10.2c). This process allows for the creation of a new magnetic metamaterial, which consists of long magnetic filaments made of magnetic colloids that can be seen under the microscope (Figure 10.2d).

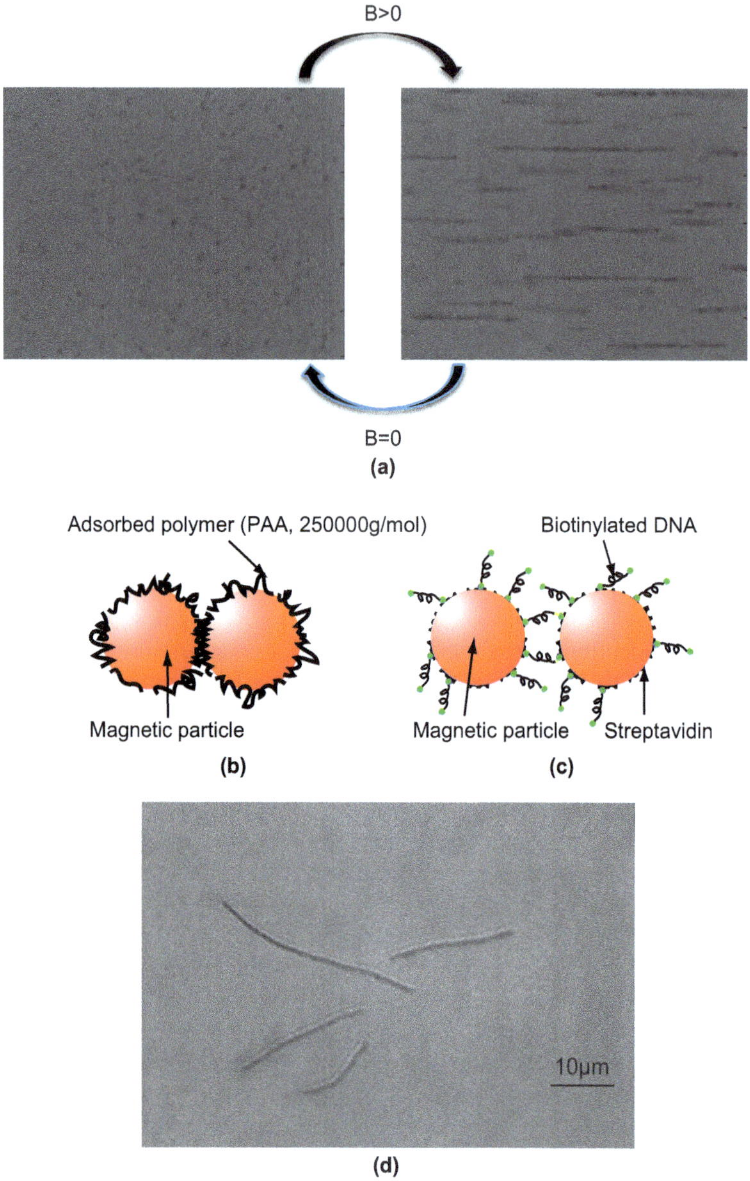

Figure 10.2 Creation of magnetic permanent filaments. (a) Magnetic colloids are suspended in solution. As soon as a magnetic field is applied, long chain of colloids are formed. As soon as the magnetic field is turned off, the particles diffuse and are suspended homogeneously in solution. (b) Schematic bridging of two adjacent particles by adsorbed PAA. (c) Schematic bridging of two adjacent particles by biotinylated DNA. (d) Microscopy pictures of permanent magnetic chains. The chains are not aligned in a single direction, showing that no magnetic field is applied.

The filaments that are formed this way seem to be good candidates as starting materials for creating artificial flagella, as they are microscopic entities that are magnetic, and therefore their motion can be controlled and actuated externally. As mentioned in the introduction, flexibility seems also to be an important feature in order to insure propulsion. The filament that were formed here were actually also flexible. This could be readily observed from the experiments that are shown in Figure 10.3a. In Figure 10.3a, magnetic filaments are initially aligned in a certain direction using the magnetic field. The field is then turned off for a very short time, which barely allows the filament to relax due to Brownian motion. Suddenly, a magnetic field is applied in a direction, which is orthogonal to the original direction in which the filaments are aligned. Interestingly, we can observe that filaments do not relax the same way as a magnetic rigid rod would do: for long filaments, the two ends independently rotate in order to align in the direction of the magnetic field (Figure 10.3a and b). This observation is mostly true for long chains; short chains have a tendency to rotate as a block. This experiment and the calculations relative to this instability, are described in detail by Roper *et al.*[41] Once the instability is fully developed, filaments end in a state where hairpins are formed (Figure 10.3a–c). Depending on the length of the filament, the filaments end in a state of a single hairpin with one loop or multiple hairpins with more that one loop. Long filaments have the ability to form many loops. This final hairpin structures are clearly metastable states as loops are clearly energetically unfavorable compared to a single chain aligned in the direction of the filed. However, one the instability is fully developed, the relaxation of a chain to its equilibrium state of lowest energy has never been observed, at least on the time scale of a week. The presence of loops and the existence of the hairpin instability clearly show that the filaments are not rigid but effectively flexible, which is one of the features that was looked for, in order to mimic artificial swimmers. Single hairpin structures were actually further used to measure the flexibility of the chains.[32,41] Indeed, it appeared immediately that some information could be extracted from the shape of the hairpin. Once the instability is fully developed and the hairpin created, the shape of the hairpin can be changed by increasing the external magnetic field. As can be seen in Figure 10.3b, when the magnetic field is increased, due to the attraction between the magnetic moment in the particles, they deform to align as many magnetic moments as possible, leading to an increase of the curvature of the hairpin loop. Therefore, when the magnetic field is applied, the magnetic energy tends to decrease the curvature, inducing higher elastic energy stored in the loop. The final shape of the hairpin is the shape obtained when the magnetic energy $\boldsymbol{E}_{\mathrm{m}}$ balances the elastic energy $\boldsymbol{E}_{\mathrm{e}}$: $\boldsymbol{E}_{\mathrm{m}} \sim \boldsymbol{E}_{\mathrm{e}}$. A simple scaling shows that the magnetic energy scales like $\boldsymbol{B}^2$, where $\boldsymbol{B}$ is the external applied magnetic field, while the elastic energy scaled like $\boldsymbol{C}^2$, where $\boldsymbol{C}$ is the curvature of the loop, showing that $\boldsymbol{C} \sim \boldsymbol{B}$. The coefficient relating these last two entities should be related to the elastic modulus of the chain. In order to fully determine the prefactor in the scaling, more extensive calculations are needed. The details of the calculations can be found in the literature.[32,41] Here we will very shortly summarize the main results. Let us start

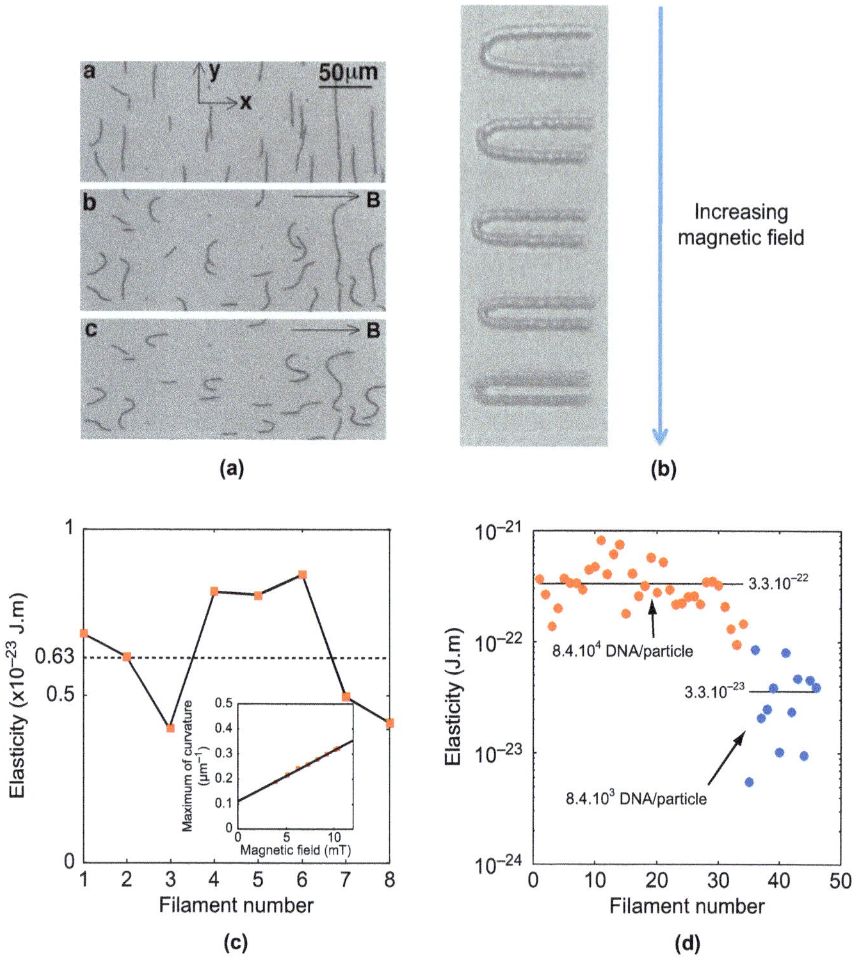

Figure 10.3 Flexibility of magnetic filament. (a) Snapshots of pictures showing the behavior of filaments as soon as the magnetic field is turned on in the direction perpendicular to the filaments. An instability develops to form hairpin shapes. (Reprinted, with permission, from Goubault *et al.*[32] Copyright 2003 by the American Physical Society, http://prl.aps.org/abstract/PRL/v91/i26/e260802 . (b) Evolution of a single hairpin as the magnetic field in increased from 2 mT to 10 mT. The magnetic field is parallel to the arms of the hairpin. The radius of curvature of the hairpin loop is decreasing as the strength of the field is increasing. (c) Measurement of the elasticity of a chain made of particles bound by PAA for eight different hairpins. The inset shows the evolution of the curvature of the loop as a function of the applied magnetic field. (d) Measurement of the elasticity of a chain made of particles bound by DNA for different initial concentrations of DNA in solution.

in the case of static hairpin by intuitively derive the non-dimensional number, which governs the physics of the system. As all the experiments are performed under homogeneous fields, the magnetic energy involved in the system consists of the interaction between two induced magnetic dipolar moments. Therefore this energy can be approximately written as:

$$E_{\mathrm{m}} \approx \mu_0 \frac{m^2}{a^3} \approx \frac{a^3 B^2}{\mu_0} \tag{10.2}$$

where μ_0 is the vacuum magnetic permeability, a is the bead radius, and m is dipolar moment. The elastic energy stored in the system $\boldsymbol{E}_{\mathbf{e}}$ can be estimated by:

$$E_l \approx a \frac{K_b}{L^2} \tag{10.3}$$

where E_{l} is Elastic energy, L is the chain length, and K_{b} is the elastic modulus. This leads us to define the so-called magneto elastique number M_{n}, which compares the magnetic contribution to the elastic contribution. The exact expression for M_{n} is given by:

$$M_{\mathrm{n}} = \frac{\pi (a\chi B)^2 L^2}{6\mu_0 K_{\mathrm{b}} \left(1 - \frac{\chi}{12} - \frac{\chi^2}{72}\right)} \tag{10.4}$$

where χ is the magnetic susceptibility. By writing the mechanical equilibrium between the elastic torques and the magnetic torques that are applied along the magnetic chain, Roper *et al.*[41] were able to derive the analytical expression for the shape of the hairpin structures and show that all along the chain, the angle θ between the tangent to the chain and the direction of the magnetic field follows the following equations:

$$\theta = 2\tan^{-1}\left(e^{-\sqrt{M_{\mathrm{n}}}s}\right) \tag{10.5}$$

where s is the curvilinear coordinate. From this last equation, we can show that the maximum of curvature of the loop $C_{\max}$ is:

$$C_{\max} = 2\sqrt{M_{\mathrm{n}}} \tag{10.6}$$

This last equation shows that the bending rigidity of the chain can be measured by analyzing the evolution of the curvature as a function of the magnetic field. Using this technique we are able to measure the bending rigidity of PAA chains (Figure 10.3c) and DNA chains (Figure 3d). The linear relationship between the curvature and the applied magnetic field is shown in the inset of Figure 3c for one chain of PAA. By changing the initial amount of DNA in solution, we are able to tune the rigidity of the chain. We can also show that the rigidity depends on the nature of the link between the particles. It is also interesting to note that from chain to chain, there is a significant variability in bending rigidity, which is due to some defects in the chain, that are very hard to avoid.

10.4 Artificial Micro-swimmers

Flexible magnetic filament that can be actuated externally in a controlled way can thus be fabricated. Every swimming motion encountered in nature is periodical, meaning that the deformation of the propelled body is repeated in a similar way every cycle. In order to induce such a cyclic deformation, flexible filaments are first aligned in the direction of a constant magnetic field $\boldsymbol{B}_c$. In the direction perpendicular to the constant field $\boldsymbol{B}_c$, an alternative field $\boldsymbol{B}_a$ is applied. Deliberately, the amplitude of $\boldsymbol{B}_a$ is kept smaller than the amplitude of $\boldsymbol{B}_c$ to avoid the chain to align in the direction parallel to $\boldsymbol{B}_a$. The frequency of the alternative magnetic field usually ranges from 0 to 50 Hz. Figure 10.4a shows the right half of a chain subjected to such a field at a frequency equal to 20 Hz. It clearly shows that the periodic magnetic field induces a periodic deformation of the chain. Interestingly, for such a long chain as the one presented in Figure 10.4a, all the chain does not deform or move in order to follow and align in the direction of the magnetic field. Only the end actually has the ability to experience the deformation. The reason for that simply comes from the fact that at high frequency, the chain experiences itself a viscous drag, which acts against the displacement of the chain. The more rigid the chain is, the longer is the end that has to be deformed to align in the direction of the magnetic field, the higher the drag is and the smaller is the deformation. This way, an additional non-dimensional parameter appears which, this time, compares the magnitude of the viscous forces to the magnitude of the elastic forces. This number has been initially introduced by Wiggins *et al.*[42,43] and is the so-called 'sperm' number. The 'sperm number' is defined as:

$$S_\mathrm{p} = L\left(\frac{2\zeta_{=}\omega}{K_\mathrm{b}}\right)^{1/4} \qquad (10.7)$$

where ω is the pulsation of the alternate magnetic field. Note that by combining the sperm number and the magneto-elastic number, another number can be defined that will compare the viscous forces to the magnetic forces.

By further analyzing the deformations that are induced by the magnetic field on the magnetic chain, it can be seen in Figure 10.4a that the deformation actually corresponds to a wave of deformation that propagates from the end of the filament towards the middle of the filament. The red arrow of Figure 10.4a indicates the propagation of the wave. In contrast to the wave of deformation that spermatozoa experience, the deformation wave is quickly damped along the chain due to viscous friction. However, as a wave, this deformation is strong enough to induce a propulsive force, which is opposed to the direction of the wave propagation. From symmetry argument, we can see that a perfectly uniform chain is submitted on its end to two propulsive force, acting in opposite direction. Therefore a homogeneous chain cannot swim as an alternate magnetic field is applied. However, once the symmetry is broken, either by forming a hairpin (Figure 10.4b),[44] by adding a red blood cell on its

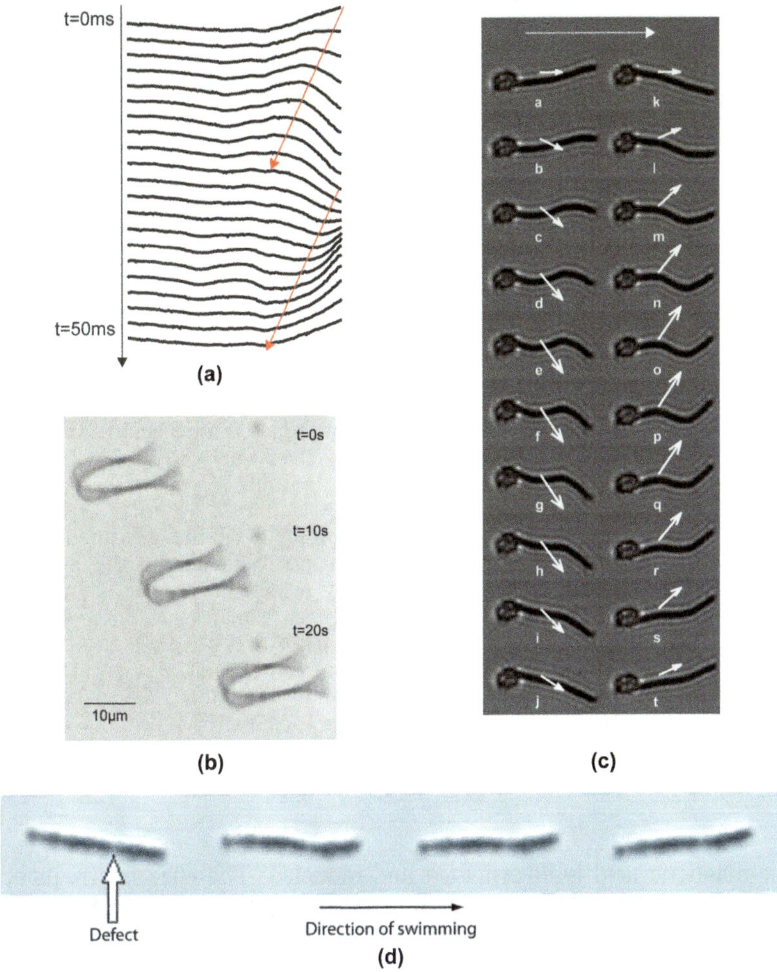

Figure 10.4 (a) Deformation of the right end of a flexible filament. The magnetic field oscillates at a frequency of 20 Hz. (Reproduced, with permission, from Dreyfus *et al.*[45]) (b) Motion of a hairpin swimmer (Reproduced, with permission, from Roper *et al.*[44]) (c) Deformation of a swimmer made of a filament attached to a red blood cell. (Reproduced, with permission, from Dreyfus *et al.*[45]) (d) Filament that swims due to a defect in its structure.
(Reproduced, with permission, from Roper *et al.*[41,44])

end (Figure 10.4c)[45] or by having a defect present in the chain (Figure 10.4d),[44] the filament can be used as artificial flexible micro-swimmers.[45]

More attention was given to the swimmer of Figure 10.4c, which actually shows a proof of concept that magnetic filaments can be used to transport a cell from one point to another in a sample.

This swimmer could also be used to check early theoretical and numerical predictions. In order to check these predictions, the influence of the frequency

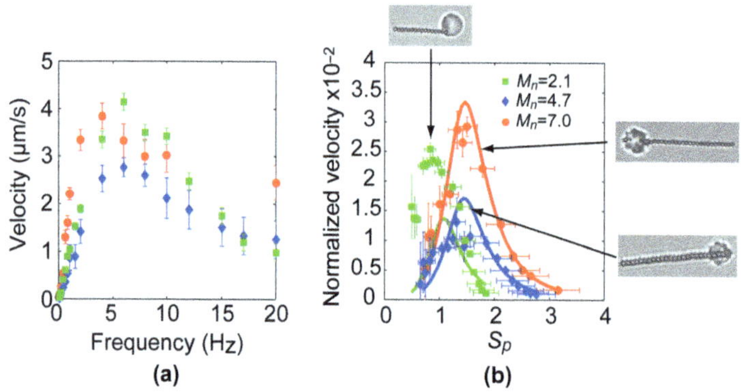

Figure 10.5 (a) Effect of the frequency on the velocity of the swimmer. We note b_0 the ratio between the amplitude of the alternative field to the constant field. Green squares: $L = 6.5\,\mu m$, $B_c = 8.7\,mT$, $B_a = 9.3\,mT$, $b_0 = 1.07$, $M_n = 2.1$; blue diamonds: $L = 10.5\,\mu m$, $B_c = 8.7\,mT$, $B_a = 9.3\,mT$, $b_0 = 1.07$, $M_n = 4.7$; red dots: $L = 12\,\mu m$, $B_c = 8.9\,mT$, $B_a = 10.3\,mT$, $b_0 = 1.16$, $M_n = 7$. (b) Comparison between the measured velocities normalized by $L\omega$ and the velocities obtained by the model. The theoretical curves are adjusted by fitting the distance h between the wall and the red blood cell. Green squares: $K_b = 3.3 \times 10^{-22}$ J m, $L = 6.5\,\mu m$, $B_c = 8.7\,mT$, $B_a = 9.3\,mT$, $b_0 = 1.07$, $M_n = 2.1$, $h = 300\,nm$; blue diamonds: $K_b = 3.3 \times 10^{-22}$ J m, $L = 10.5\,\mu m$, $B_c = 8.7\,mT$, $B_a = 9.3$ mT, $b_0 = 1.07$, $M_n = 4.7$, $h = 80\,nm$; red dots: $K_b = 3.3 \times 10^{-22}$ J m, $L = 12\,\mu m$, $B_c = 8.9\,mT$, $B_a = 10.3\,mT$, $b_0 = 1.16$, $M_n = 7$, $h = 120\,nm$. (Reproduced, with permission, from Dreyfus *et al.*[45])

of the magnetic field is investigated and modeled. The effect of frequency is shown in Figure 10.5a. It shows that, at low frequency and at high frequency, the swimming velocity vanishes. The reason for that can be explained qualitatively. At low frequency, the viscous drag on the chain is small enough so that the chain has the time to follow the direction of the magnetic field: the filaments under these conditions move as a rigid rod. Their deformation shows no propagating waves, their motion is time reversible.[46] Therefore there is no propulsive force acting on the filaments and their velocity is zero. At high frequency, the filament has basically no time to follow the magnetic field and the deformations are so small that they also cannot induce any net motion. In between, we can see that there is an optimum of swimming velocity around 5 Hz. When velocity is scaled by the typical velocity $L\omega$ and plotted as a function of the sperm number as in Figure 10.5b, we find a bell-like curve that confirms some earlier predictions performed by Wiggins *et al.*,[42,43] Lagomarsino *et al.*[47] and Lowe.[48] As for natural spermatozoa, our swimmers exhibit some optimum of scaled velocity as a function of sperm number. In our case it occurs at relatively low S_p, between 1 and 2. Roper *et al.*[44] also performed the modeling of the swimmers and found reasonable agreement (solid lines in Figure 10.5b) with the experimental data.

10.5 Conclusion

As a conclusion, and to summarize, this work has allows us to extensively study the behavior of flexible magnetic filaments under a magnetic field. It has showed how it possible to exploit the existence of the hairpin instability to measure the bending rigidity of a chain. When an oscillating magnetic field is applied to magnetic chains only the ends experience deformations, which have the effect of distorting the chain and creating a wave that propagates from the ends to the center of the chain. Finally, three different ways are proposed to break the symmetry of a chain and to promote the emergence of a propulsive force. Although known and studied since the 1950s, swimming at low Reynolds number has resulted in only very few experimental studies. Based on our experience and expertise in fluid mechanics and materials science, we were able to relate these two separate fields to create the first example of microscopic artificial swimmers. It allowed us to study some features of the propulsion system, to verify some known theoretical predictions, and also to propose an effective rationale to create artificial flagella, which are going to have many applications in mechanical and biomedical engineering in the future. It is indeed clear that there will be an increasing need in the future to develop effective ways to propel objects at every scale in viscous media. As such, this work will open many opportunities in that direction.

References

1. Richard Feynman, APS meeting lecture, *Caltech Engineering and Science*, 1960, **23**(5), 22–36.
2. J. E. Avron, O. Gat and O. Kenneth, *Phys. Rev. Lett.*, 2004, **93**, 1860011–1860014.
3. J. E. Avron, O. Kenneth and D. H. Oaknin, *New Journal of Physics*, 2005, **234**, 1–8.
4. L. E. Becker, S. A. Koehler and H. A. Stone, *J. Fluid Mech.*, 2003, **490**, 15–35.
5. R. Dreyfus, J. Baudry and H. A. Stone, *Eur. Phys. J. B*, 2005, **47**, 161–164.
6. A. Najafi and R. Golestanian, *Phys. Rev. E*, 2004, **69**, p0629011–0629014.
7. H. C. Berg, *Phys. Today*, 2000, **53**, 24–29.
8. H. C. Berg and R. A., *Nature*, 1973, **245**, 380–382.
9. C. Brennen and H. Winet, *Annu. Rev. Fluid Mech.*, 1977, **9**, 339–398.
10. J. Gray and G. J. Hancock, *J. Exp. Biol.*, 1955, **32**, 802–814.
11. G. J. Hancock, *Proc. R. Soc. London Ser. A, Math. Phys. Sci.*, 1953, **217**, 96–101.
12. D. F. Katz, *J. Fluid Mech.*, 1974, **64**, 33–49.
13. G. Taylor, *Proc. R. Soc. London Ser. A, Math. Phys. Sci.*, 1951, **209**, 447–461.
14. G. Taylor, *Proc. R. Soc. London Ser. A, Math. Phys. Sci.*, 1952, **211**, 225–239.
15. D. L. Ringo, *J. Cell Biol.*, 1967, **33**, 543–571.
16. D. Bray, *Cell Movements: From Molecules to Motility*. Garland Publishing, New York, 1992.

17. P. Satir, *J. Cell Biol.*, 1965, **26**, 805–834.
18. P. Satir, *J. Cell Biol.*, 1968, **39**, 77–94.
19. D. Tam and A. E. Hosoi, *Phys. Rev. E*, 2011, **83**, 0453031–0453034.
20. D. Tam and A. E. Hosoi, *Phys. Rev. Lett.*, 2007, **98**, 0681051–0681054.
21. S. P. Fletcher, F. Dumur, M. M. Pollard and B. L. Feringa, *Science*, 2005, **310**, 80–82.
22. G. Haberhauer, *Angew. Chem. Int. Ed.*, 2011, **50**, 7529–7532.
23. T. Omabegho, R. M. Sha and N. C. Seeman, *Science*, 2009, **324**, 67–71.
24. C. J. Brokaw, *Proc. Natl. Acad. Sci. U.S.A.*, 1975, **72**, 3102–3106.
25. G. K. Batchelor, *J. Fluid Mech.*, 1970, **44**, 419–440.
26. J. R. Blake, *J. Eng. Math.*, 1974, **8**, 113–104.
27. D. F. Katz, J. R. Blake and S. L. Paverifontana, *J. Fluid Mech.*, 1975, **72**, 529–540.
28. S. Broersma, *J. Chem. Phys.*, 1960, **32**, 1632–1635.
29. R. G. Cox, *J. Fluid Mech.*, 1971, **45**, 625–657.
30. E. M. Furst, C. Suzuki, M. Fermigier and A. P. Gast, *Langmuir*, 1998, **14**, 7334–7336.
31. L. Cohen-Tannoudji, E. Bertrand, L. Bressy, C. Goubault, J. Baudry, J. Klein, J. F. Joanny and J. Bibette, *Phys. Rev. Lett.*, 2005, **94**, 0383011–0383014.
32. C. Goubault, P. Jop, M. Fermigier, J. Baudry, E. Bertrand and J. Bibette, *Phys. Rev. Lett.*, 2003, **91**, 2608021–2608024.
33. C. Goubault, F. Leal-Calderon, J. L. Viovy and J. Bibette, *Langmuir*, 2005, **21**, 3725–3729.
34. S. L. Biswal and A. P. Gast, *Phys. Rev. E*, 2003, **68**, 0214021–0214029.
35. S. L. Biswal and A. P. Gast, *Phys. Rev. E*, 2004, **69**, 0414061–0414069.
36. A. Cebers and I. Javaitis, *Phys. Rev. E*, 2004, **70**, 0214041–0214046.
37. J. Philip, O. Mondain-Monval, F. Leal-Calderon and J. Bibette, *J. Phys. D, Appl. Phys.*, 1997, **30**, 2798–2803.
38. J. H. E. Promislow and A. P. Gast, *Langmuir*, 1996, **10**, 4095–4102.
39. R. Dreyfus, D. Lacoste, J. Bibette and J. Baudry, *Eur. Phys. J. E*, 2009, **28**, 113–103.
40. A. Koenig, P. Hebraud, C. Gosse, R. Dreyfus, J. Baudry, E. Bertrand and J. Bibette, *Phys. Rev. Lett.*, 2005, **95**, 1083011–1083014.
41. M. Roper, R. Dreyfus, J. Baudry, M. Fermigier, J. Bibette and H. A. Stone, *J. Fluid Mech.*, 2006, **554**, 167–190.
42. C. H. Wiggins and R. E. Goldstein, *Phys. Rev. Lett.*, 1998, **80**, 3879–3882.
43. C. H. Wiggins, D. Riveline, A. Ott and R. E., *Biophys. J.*, 1998, **74**, 1043–1060.
44. M. Roper, R. Dreyfus, J. Baudry, M. Fermigier, J. Bibette and H. A. Stone, *Proc. R. Soc. A*, 2008, **464**, 877–904.
45. R. Dreyfus, J. Baudry, M. L. Roper, M. Fermigier, H. A. Stone and J. Bibette, *Nature*, 2005, **437**, 862–865.
46. E. M. Purcell, *Am. J. Phys.*, 1977, **45**, 3–11.
47. M. C. Lagomarsino, F. Capuani and C. P. Lowe, *J. Theor. Biol.*, 2003, **224**, 215–225.
48. C. P. Lowe, *Philos. Trans. R. Soc. London Ser. B, Bio. Sci.*, 2003, **358**, 1543–1550.

Section 3
The Potential of Artificial Cilia: Experimental Evidence

CHAPTER 11

Fluid Manipulation by Artificial Cilia

J. HUSSONG,*[a] W.-P. BREUGEM[b] AND J. WESTERWEEL[b]

[a] Ruhr-University Bochum, Universitätsstraße 150, 44801 Bochum, Germany; [b] Delft University of Technology, Leeghwaterstraat 21, 2626 CA Delft, The Netherlands
*Email: jeanette.hussong@rub.de

11.1 Introduction

Cilia are periodically beating slender cell extensions growing on microorganisms or organ surfaces. To induce a net flow at low Reynolds numbers a single cilium has to exhibit a configurationally asymmetric beat cycle, *i.e.* the cilium stretches during the forward stroke and bends or twists during the recovery stroke.[1] In nature, adjacent cilia beat slightly out of phase which results in travelling waves that move over the cilia tip surface, called a metachronal wave. In real cilia systems metachronal coordination is the outcome of hydrodynamic interaction between densely packed cilia.[2,3] However, it was recently shown that metachronal coordination can act as an independent fluid transport mechanism for artificial cilia when they perform a slightly phase-shifted beat cycle.[4] This is because the collective motion creates propagating waves of symmetry-breaking spatial porosity and solid phase velocity distributions inside the cilia layer resulting in an overall net fluid drag.

To date, artificial cilia systems are built to perform a synchronous but configurationally asymmetric cilia motion.[5–9] In such systems the generation of a significant asymmetry during the beat cycle is the key mechanism to create net flow. To date, magnetic actuation is the most common method to actuate

RSC Nanoscience & Nanotechnology No. 30
Artificial Cilia
Edited by Jaap M J den Toonder and Patrick R Onck

Published by the Royal Society of Chemistry, www.rsc.org

artificial cilia. An exception is the work of den Toonder *et al.* who produced rectangular-shaped artificial cilia that were actuated electrostatically to mix non-conducting fluids in micro-devices.[10] While both magnetic and electrostatic fields are non-intrusive actuation principles, magnetic actuation is favourable for biomedical applications since it avoids heat generation and works in both conducting and non-conducting fluids. Magnetically actuated artificial cilia are typically seeded with or assembled from super-paramagnetic particles, which cause the artificial cilia to respond to an external magnetic field. Numerical studies helped to gain a good understanding of the interplay between magnetic, cilia and fluid forces during the actuation.[11–13] Naturally, results are strongly dependent on the specific characteristics of a cilia system, such as material properties, shape and cilia distribution.

Up to date the experimental verification of artificial cilia induced unidirectional flow is challenging. One of those very few experimentally successful studies is that of Shields *et al.*[7] who actuated cylindrically shaped hair-like artificial cilia. The structures are made of a rubbery matrix seeded with super-paramagnetic particles. A rotating permanent magnet that enforces a three-dimensional paddling motion of the flexible pillar-shaped cilia that creates a net flow with maximum fluid velocities of up to approximately 9 μm s^{-1}.[7] Vilfan *et al.*[8] reported on magnetically actuated micro-bead rods that achieve fluid velocities of almost 5 μm s^{-1} at an actuation frequency of 0.5 Hz by performing a three-dimensional paddling motion similar to that reported by Shields *et al.*

In this chapter we report on the fluid manipulation induced by rectangular-shaped, artificial cilia performing a plane but asymmetric beat cycle. Similar to the structures reported on by Shields *et al.* and Vilfan *et al.* the present cilia are composed of a super-paramagnetic micro-bead polymer dispersion and are actuated magnetically.[14] The phase-averaged and time-resolved flow is quantified by means of phase-locked micro-Particle Image Velocimetry (μPIV) measurements at increasing actuation frequencies and the maximum pressure head and maximum volume flow rate are calculated.

11.2 Materials and Methods

11.2.1 Measurement Set-up

Figure 11.1a shows a schematic drawing of the experimental set-up, consisting of the channel device, the actuation unit and the measurement set-up. The measurement set-up includes an upright fluorescence microscope (LEICA MZ 16 FA FLUO COMBI III), a dual-cavity laser (New Wave Pegasus-PIV-30W), two 12-bit double-frame CCD cameras of 1376 × 1040 pixels resolution (Imager Intense, LaVision) and a built-in PC trigger and timing unit (LaVision PTU 8). An external trigger signal serves as reference time for the trigger and timing unit which synchronizes the laser, the cameras and the data acquisition. The laser, the cameras and the microscope are operated with commercial PIV software (Davis 7.2, LaVision GmbH). After the software is initiated a recording is taken with a prefixed time delay to the incoming trigger signal of the

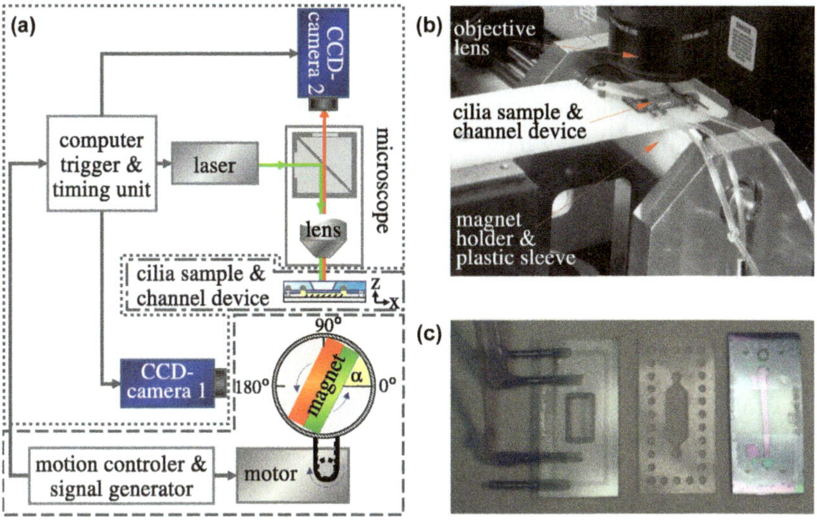

Figure 11.1 (a) Schematic drawing of the experimental set-up with cilia sample integrated into a micro-channel, actuation system and measurement device of two cameras, an upright microscope, laser and a trigger and timing unit.[15] (b) Photo of cilia device and actuation system. (c) Three-component micro-fluidic channel device. Right: Substrate with sacrificial layer embedded artificial cilia. Middle: A poly(*N*,*N*-dimethylacrylamide) (PDMAA) cartridge with voids for the fluid and vacuum channel. Left: Glass cover with integrated channel inlet and outlet as well as a vacuum channel.

positioning controller. Rhodamine-B, a fluorescent dye, and polyethylene glycol (PEG)-coated polystyrene particles of a mean bead diameter of 1.28 μm (Microparticles GmbH) are used as fluid tracers during the μPIV measurements. A PEG coating makes the tracer surface hydrophilic so that water molecules form a shielding layer around the particle, which reduces the sticking of particles to surfaces. Approximately 10 mL distilled water (BRADYSAFE TM) is seeded with 20 μL tracer solution (2.5% w/v aqueous dispersion).

11.2.2 Channel Device

The channel device includes the substrate with cilia, the channel that is assembled around the substrate with cilia, connecting tubing, and a syringe pump. The channel device is clamped onto a plastic table that is mounted to a *xyz*-traversing mechanism (see Figure 11.1b). The micro-channel used for the experiments is 500 μm high, 5 mm wide and 2 cm long. The cilia array at the bottom of the micro-fluidic channel covers an area of $5 \times 10\,\mathrm{mm}^2$. The channel device is assembled from three components shown in Figure 11.1c: The substrate with cilia (right), a cartridge (middle) and a glass cover (left). The cartridge is made of poly(*N*,*N*-dimethylacrylamide), hence on referred to as PDMAA. It is 500 μm thick with a rectangular void of 10 mm length and 5 mm width in its middle. The cartridge is placed on top of the cilia substrate such that the edges of the void form the side walls of the micro-fluidic channel. The channel is closed with the 3 mm thick glass cover that is equipped with an integrated viewing window of 1 mm thickness. This glass thickness provides a sufficient robustness during handling while a reduced thickness of the viewing window enables an optical access for lenses of shorter working distance. Inlet and outlet channels are integrated into the glass cover. A connection to a vacuum channel allows sealing the micro-fluidic channel by under pressure during usage.

11.2.2.1 Cilia Production

In the following we summarize the main production steps that are given in detail by Belardi *et al.*[14] The artificial cilia structures are fabricated in a two-colour lithographic process. The three main production steps are shown schematically in Figure 11.2a–f. First, a silicium wafer is modified with monolayer of silane to create a hydrophilic surface Figure 11.2a. Second, a layer of PDMAA, with photo-reactive groups with an activation wavelength of $\lambda \leq 250$ nm is spin-coated on the substrate Figure 11.2b. Afterwards, the sample is aligned under a mask and exposed to UV light (250 nm). The PDMAA layer is cross-linked in the light-exposed regions, which will function as anchor regions for the cilia structures (Figure 11.2c). Third, a poly *n*-butylacrylate composite is spin-coated on the substrate (Figure 11.2d). Paramagnetic nanoparticles (Fe_3O_4) with a nominal diameter of 10–20 nm are dispersed in the composite, which also contains photo-reactive side groups that can be activated when exposed by light with a wavelength $\lambda \leq 365$ nm. To achieve homogeneous nanoparticle–rubber dispersions the nanoparticles are made compatible with the rubber by a surrounding surfactant layer. The photo-reactive groups are activated by light with a wavelength of 365 nm under a second mask, such that the rubber composite layer is patterned. The anchor layer beneath the cilia layer

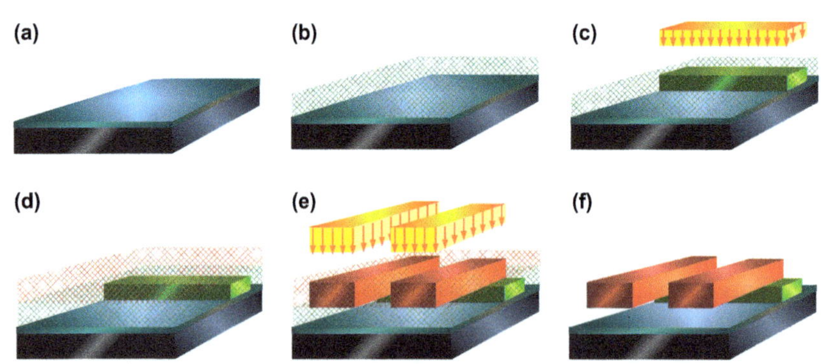

Figure 11.2 Two-colour lithography steps for the production of the artificial cilia.[15] (a)–(c) Fabrication of the anchor strips. (d)–(f) Fabrication of the cilia structures.

is not affected by the second illumination process, since the PDMAA-layer side groups do only respond to light with wavelengths less than 250 nm (Figure 11.2e). The cross-linked rubber composite regions of $10 \times 80\,mm^2$ form the cilia structures including the anchor region. The produced cilia are embedded in a layer of unlinked PDMAA and rubber material hence on referred to as the sacrificial layer.

11.2.2.2 Cilia Release

The sacrificial layer is dissolved as soon as it gets in contact with the working fluid. This occurs during the filling of the micro-fluidic channel which is hence on called the cilia release. The micro-channel is filled with the working fluid (distilled water seeded with tracer particles) at a syringe pump flow rate of approximately 2–4 mL s^{-1}. After the release, the artificial cilia are flap-shaped 70 μm long, 20 μm wide and approximately 0.9 μm thick, elastic polymer structures that are attached at one end to the surface of the substrate.[14] The cilia are located 20 μm apart in span-wise direction while the stream-wise distance from cilia tip to cilia tip equals 100 μm (Figure 11.2f). In the absence of external forces a water-submerged cilium curls in the *xz*-plane due to an increased swelling ratio of the bottom layer of the cilium[14] such that only approximately half of the cilium's length is visible from a top view.

11.2.2.3 Magnetic Actuation

A cuboid-shaped magnet of $10 \times 25 \times 50\,mm^3$ is used for the magnetic actuation of the cilia filaments. It rotates counter-clockwise around its *y*-axis (Figure 11.3b) creating a clockwise-rotating magnetic field of 105–185 mT. Henceforth we denote the rotation angle of the magnet as α and the rotation angle of the magnetic field vector at the location of the cilia as β. When α increases over time, β will decrease, and *vice versa*.

The permanent magnet is positioned in the middle of a cylindrical-shaped aluminium holder (see Figure 11.1a) below the micro-channel device. The magnet holder is driven by an electro-motor through a belt (maxon DC motor RE 25, Maxon Motor Benelux BV). The torque delivered by this motor allows revolution frequencies between 1 and 20 Hz. The motor is connected to a positioning controller that generates a trigger signal after each full revolution of the motor. Since the number of revolutions of the motor and the magnet holder are identical, the angular velocities of the motor and of the holder can be assumed to be identical.

In an external magnetic field the cilium aligns with the magnetic field vector due to dipole–dipole interaction of the super-paramagnetic nanoparticles embedded in the cilia polymer matrix.[14] In a static field the cilia adopt an equilibrium shape where magnetic and elastic bending forces counterbalance. A top view of released cilia under a static magnetic field at 60° magnet orientation are shown in Figure 11.3a.

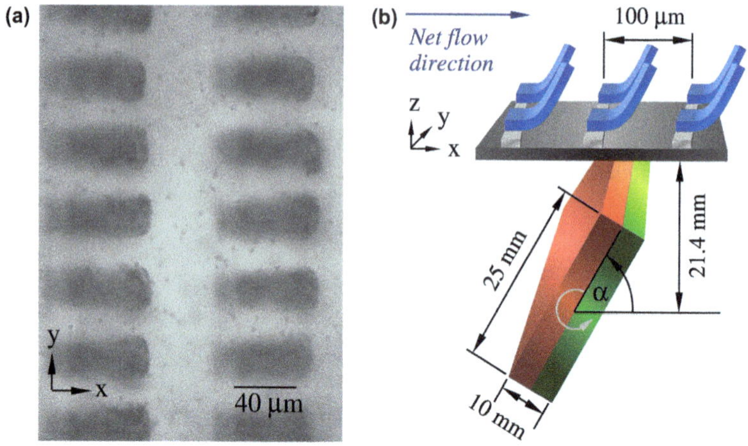

Figure 11.3 (a) Brightfield image of a micro-channel detail with two rows of artificial cilia.[15] The cilia tips point in the positive x-direction; (b) Schematic drawing of the cuboid-shaped magnet below the cilia array. The magnetic field vector at the location of the cilia rotates clockwise with a decreasing orientation angle β while the magnet rotates in counter-clockwise direction with an increasing orientation angle α.

11.3 Brightfield and μPIV Measurements

The micro-channel is closed at the inlet and outlet before actuation. During the measurements two synchronized cameras are used. Camera 1 records the magnet orientation, while camera 2 simultaneously takes brightfield images of the cilia or epifluorescent particle images of water dispersed fluid tracers. This allows to link both the in-plane cilia beat cycle and the cilia-induced flow to the magnetic field orientation. Measurements are performed in the middle of the channel. The field of view spans an area on the channel's centre-line of $400 \times 300\,\mu m^2$. To record the in-plane cilia beat cycle single frame brightfield images are taken by camera 2 at different phases of the actuation cycle. For the μPIV measurements camera 2 acquires one double frame during one magnet revolution. For actuation frequencies beyond 5 Hz double frames are acquired every second, third or fourth magnet revolution, respectively. The phase-averaged fluid velocities are derived by cross-correlating the first frames of consecutive double frames. The time delay between those image pairs is thus equal to the rotation frequency of the magnet. A set of 100 images are recorded per measurement plane since multiple recordings allow ensemble averaging for the vector field which increases the reliability of the measurement results. The principles and details of the μPIV measurement method are explained elsewhere.[16–18]

To measure the dependency of the induced phase-averaged flow rate on the actuation frequency f, μPIV measurements are carried out up to $f = 20$ Hz at one fixed measurement plane above the channel floor. The phase-averaged velocity profile over the channel height is reconstructed by measuring the flow in different parallel planes with 20 μm spacing.[19]

By step-wise increases of the time delay between the trigger signal of the electro-motor's positioning controller and the image acquisition time, double frames are acquired at 20 phases during the actuation cycle. By cross-correlating consecutive images of each double frame we can infer the instantaneous fluid velocities at each of the 20 phase positions during the actuation cycle (note that one actuation cycle induces two cilia strokes[14]). The time delay Δt between the first frame and second frame is adjusted to the flow velocities induced by the cilia to achieve particle displacements of 10–30 pixels for a 64×64 pixels interrogation window size.[16,20]

Images are pre-processed before computation of the flow field. The signal of fluorescent tracer particles sticking to the channel walls and the cilia is removed and the effect of fluorescent light reflected off the cilia as well as fluorescent light of out-of-plane particles is minimized. This reduces the occurrence of any spurious velocity data in the final measurement result.

11.4 Results

11.4.1 Time-resolved Cilia and Fluid Motion

Two representative vector fields of the cilia-induced flow during the effective stroke and the recovery stroke are shown in Figure 11.4a and b. The figure shows the results of ensemble-averaged PIV data for an actuation frequency of 10 Hz in a measurement plane 60 μm above the channel floor. The vector fields are superimposed on averaged image data of the fluorescence signal of the cilia. Figure 11.4a shows the velocity field during the effective cilia stroke at a magnet orientation angle of 74°. A typical velocity field during the cilia recovery stroke

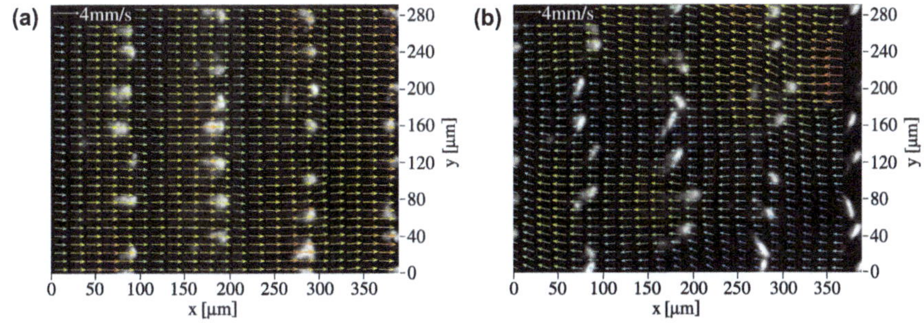

Figure 11.4 Result of ensemble-averaged μPIV data in a plane 60 μm above the cilia-covered channel floor and at an actuation frequency of 10 Hz are superimposed on averaged image data (light spots indicate average images of cilia at two phases of the actuation cycle).[15] (a) Velocity field during the effective cilia stroke at a magnet orientation angle of 74°; (b) Velocity field during the cilia recovery stroke for a magnet orientation angle of 138°.

is shown in Figure 11.4b for a magnet orientation angle of 138°. The flow direction is reversed and the velocity magnitudes are of the same order of magnitude as during the effective stroke. A comparison of Figure 11.4a and b shows that a more homogeneous and unidirectional flow is induced during the active stroke as compared to the recovery stroke. This is because the cilia perform a synchronized forward motion while following the magnetic field during the active stroke. In contrast, the recovery stroke is dominated by the bending stiffness and initial shape of each cilium and slight variations in one or the other tend to alter the onset and evolution of a cilium's recovery stroke.

In total, 20 velocity fields were determined during different phases of the actuation cycle in a plane 60 μm above the cilia-covered channel floor. The median velocity of those velocity fields is plotted as a function of the magnetic field orientation in Figure 11.5a for $f = 10$ Hz actuation frequency. Note that all frequencies are given in terms of actuation frequencies f which equals half the cilia beat frequency. Since a cilium performs two identical beat cycles

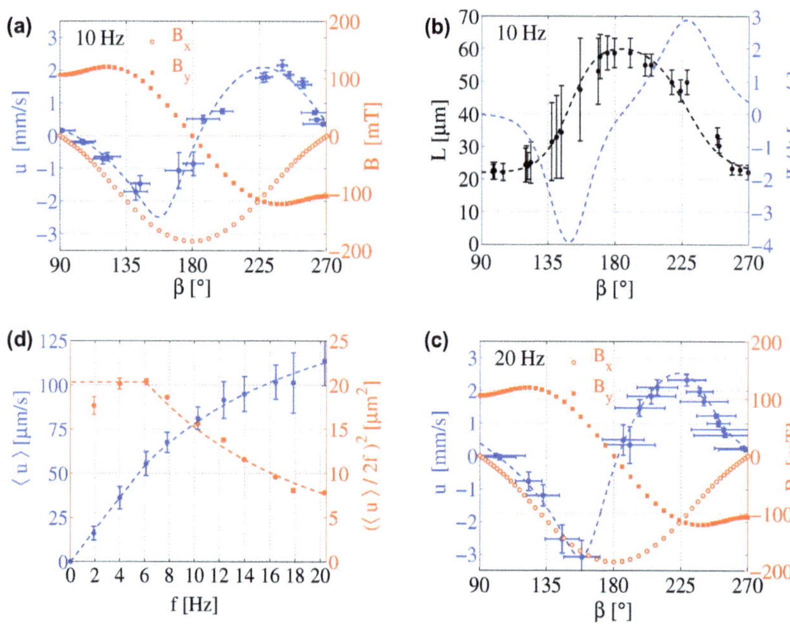

Figure 11.5 (a) Artificial cilia-induced fluid velocities at 10 Hz actuation frequency are plotted as a function of the field orientation angle. The chosen field of view lies in the middle of the channel 60 μm above the cilia-covered channel floor. (b) Projected cilia length at 10 Hz actuation frequency. In the absence of external forces the 70 μm long cilia are curved with a projected length of 34 ± 5 μm. (c) Cilia-induced fluid velocities at 20 Hz actuation frequency are plotted as a function of the field orientation angle 60 μm above the cilia-covered channel floor. (d) Median phase-averaged fluid velocities and resulting power spectrum are plotted as a function of the actuation frequency. The chosen field of view lies in the middle of the channel 90 μm above the cilia-covered channel floor.

during each magnet revolution the 20 velocity results are plotted as a function of half an actuation cycle with the magnetic field altering its orientation angle in time from $270° > \beta > 90°$. Horizontal error bars indicate the time delay between two consecutive particle images. Vertical error bars show the median absolute deviation of the measured fluid velocities. The wall normal and wall parallel magnetic flux density components are indicated in red.

The fluid velocities can be directly related to the cilia motion. Cilia are recorded from a top view throughout the actuation cycle as shown in Figure 11.3a. The apparent distance between cilia root to cilia tip is hence on referred to as the projected cilia lengths. Figure 11.5b shows projected cilia lengths as function of the magnetic field angle β at 10 Hz actuation frequency. Note that the channel floor fixated anchor region of 10 μm length is excluded from the data. Thus, the maximum projected cilia length of a fully horizontally stretched cilium is 70 μm. Error bars indicate the median absolute deviation of the measured values. The blue curve shows the in-plane cilia tip velocity throughout a beat cycle derived by differentiating the projected cilia length fitting function. Comparing Figure 11.5a and b indicates that the measured velocity field at 60 μm distance from the cilia-covered channel floor stays in-phase with the actual cilia tip velocities.

It indicates that transient inertial forces of the fluid are small compared to shear forces of the fluid. The ratio between both forces is generally known as the Womersley number *Wo* and it can be estimated with the cilia length and the angular cilia tip velocity to:

$$Wo^2 = \omega L^2/\nu \approx \pi \times 2f \times L^2/\nu \approx 0.6 \quad \text{for } f_{\max} = 20\,\text{Hz} \tag{11.1}$$

Thus for the whole range of investigated actuation frequencies viscous fluid forces dominate over inertial fluid effects. For the estimate of the angular cilia tip velocity we consider that a cilium rotates at maximum approximately 90° around its anchor point during a beat cycle. Therefore, the angular velocity is estimated to $\omega \sim 2\pi f$. A qualitative support for this choice can be given when we compare the estimated angular cilia tip velocity, $v = \omega \mathrm{L} \approx 2\pi f L$ with the measured maximum in plane cilia tip velocity $|v|_\mathrm{m}$ as given in Figure 11.5b for $f = 10$ Hz actuation frequency which results in $v \approx 4.4\,\mathrm{mm\,s^{-1}}$ and $|\mathrm{v}|_\mathrm{m} \approx$ $4\,\mathrm{mm\ s^{-1}}$, respectively.

When the magnet rotates from a horizontal to a vertical orientation ($\alpha = 0$–$90°$), the reverse rotating magnetic field tilts from the vertical orientated *yz*-plane into the horizontally orientated *xy*-plane ($270° > \beta > 180°$), followed by the cilia which perform their effective stroke (Figure 11.3b). In this phase the induced fluid velocities are positive coinciding with the forward motion of the cilia during the effective stroke. While the magnetic field continues rotating in a clockwise direction $180° > \beta > 90°$ (with a counter-clockwise rotating magnet between $\alpha = 90$–$180°$), the cilium performs its recovery stroke. It curls up, driven by its bending stiffness and the magnetic torque that gradually reverses its sign along the cilia filament with altering field and cilia orientation inducing negative fluid velocities. The beat cycle is completed after the cilium is aligned again with the magnetic field that has now rotated over 180°. Thus, the cilia perform two beat cycles during a full revolution of the magnetic field.

Even though the cilia recovery stroke is more rapid than the forward motion, maximum fluid velocities of up to 2.1 ± 0.1 mm s^{-1} were measured during the effective stroke while maximum velocities of 1.8 ± 0.3 mm s^{-1} have been measured during the recovery stroke. We anticipate that this originates from an insufficient resolution in acquired measurement data during the most rapid phase of the recovery stroke between approximately $150° < \beta < 170°$ field orientation angle.

Figure 11.5c shows cilia-induced fluid velocities at 20 Hz actuation frequency which is twice as fast as compared to the actuation frequency used during the measurements shown in Figure 11.5a. As in Figure 11.5a the chosen measurement plane is located 60 μm above the cilia-covered channel floor and fluid velocities are plotted as a function of the field orientation angle β. Even though the actuation frequency is doubled in the measurements shown in Figure 11.5c, compared to Figure 11.5a only weakly increased maximum fluid velocities are measured. Since the cilia beat frequency is directly proportional to the actuation frequency a moderate increase in instantaneous velocities is likely to result from a change in cilia beat cycle such as decreased cilia beat amplitude that prevents an increase in angular cilia velocity. Such a damping effect of the cilia beat cycle can by caused by drag effects of the surrounding fluid. Note that the observed effect is partially over-estimated due to a large time delay between particle images pairs leading to an under-estimation of the actual fluid velocities close to the velocity extremes.

Phase-averaged velocities $\langle u \rangle$ have been measured 90 μm above the cilia-covered channel floor. The results are plotted in blue in Figure 11.5d as function of the actuation frequency. The resulting power spectrum of the effective kinetic fluid energy per cilia beat cycle $(\langle u \rangle / 2f)^2$ is plotted in red. The factor 2 accounts for the fact that the cilia beat frequency equals twice the actuation frequency. The power spectrum indicates two basic flow regimes: at low actuation frequencies up to approximately 6 Hz the kinetic energy of the flow appears to be constant while it decreases exponentially for actuation frequencies above 6 Hz. The energy input by the magnet per revolution stays constant at all times and so must the total work done by the cilia after one beat cycle. Due to their size and stiffness inertial effects of the cilia are negligible throughout the range of experimentally verified actuation frequencies. Furthermore, viscous energy dissipation of the fluid is negligible at low actuation frequencies such that the cilia perform a quasi-static motion that is determined by the interplay of magnetic and elastic bending forces. In this state the cilia beat cycle is independent of the actuation frequency and the flow scales linearly with the actuation frequency as indicated by the blue curve in Figure 11.5d.

Since we showed in eqn (11.1) that viscous fluid forces dominate over inertial fluid effects throughout the investigated actuation range, we anticipate that when the actuation frequency is increased beyond 6 Hz viscous fluid forces increasingly damp the cilia motion leading to a change in the cilia beat cycle. Thus, an increasing part of the kinetic cilia energy is dissipated by the fluid and hence the net flow per actuation cycle decreases. This is evident from the exponential decrease of the effective kinetic energy of the flow per cilia beat cycle shown in Figure 11.5d which in turn explains the saturation reached in phase-averaged velocities (blue curve). The velocity and kinetic energy of the phase averaged flow in the quasi static and damped regime can be approximated to:

$$\langle u \rangle \approx U_0 f \qquad \text{for } f < 6\,\text{Hz} \tag{11.2a}$$

$$\langle u \rangle \approx U_0 f\left(1 - e^{-f/f_0}\right) \qquad \text{for } f > 6\,\text{Hz} \tag{11.2b}$$

$$\frac{1}{2}\left(\frac{\mathrm{d}\langle u \rangle}{\mathrm{d}f}\right)^2 \approx \left(\frac{U_0}{2}\right)^2 \qquad \text{for } f < 6\,\text{Hz} \tag{11.2c}$$

$$\frac{1}{2}\left(\frac{\mathrm{d}\langle u \rangle}{\mathrm{d}f}\right)^2 \approx \left(\frac{U_0}{2f_0}\right)^2 e^{-2f/f_0} \qquad \text{for } f > 6\,\text{Hz} \tag{11.2d}$$

where f is the actuation frequency of the magnetic field, U_0 and f_0 are constants that depend on the geometrical properties of the system as well as the elastic and magnetic properties of the cilia. Note that U_0 varies with changing distance to the cilia bottom surface, while f_0 is expected to stay constant over the channel height. Increasing viscous fluid forces will gradually alter the cilia beat cycle and keep decreasing the cilia beat amplitude until the cilia-induced flow reaches a global maximum at a finite actuation frequency.[11,21] Therefore, eqns (11.1a–d) approximate only the investigated frequency response of the cilia system.

11.4.2 The Cilia-induced Maximum Volume Flow Rate

The performance of a micro-pump can be characterized from two basic operational conditions. The first is a zero back-pressure condition under which the pump achieves its maximum volume flow rate. In the second case a back-pressure-induced flow is imposed that counter-balances the pump flow. In the present study the latter condition was chosen to obtain the generated pressure head of the artificial cilia pump from which in turn the cilia-induced maximum volume flow rate was calculated. For this the micro-channel was closed at the inlet and outlet during the experiments forming a micro-fluidic chamber of $20 \times 5 \times 0.5\,\text{mm}^3$. Velocity profiles were measured over the channel height along the channel's symmetry plane at 26 equidistantly spaced measurement planes with $20\,\mu\text{m}$ spacing and a field of view spanning an area of $400 \times 300\,\mu\text{m}^2$. The resulting velocity profiles at 10 Hz actuation frequency are depicted in Figure 11.6a and b for oppositely rotating magnetic actuation fields.

The rotation sense and actuation frequency of the magnetic field that leads to the instantaneous cilia and fluid motion shown in Figure 11.5a and b also creates the phase-averaged velocity profile shown in Figure 11.6a. Since the micro-channel is closed at the inlet and outlet during the experiments the cilia-induced flow inside the micro-fluidic chamber is compensated by a pressure driven backflow in the upper channel half. This leads to a velocity

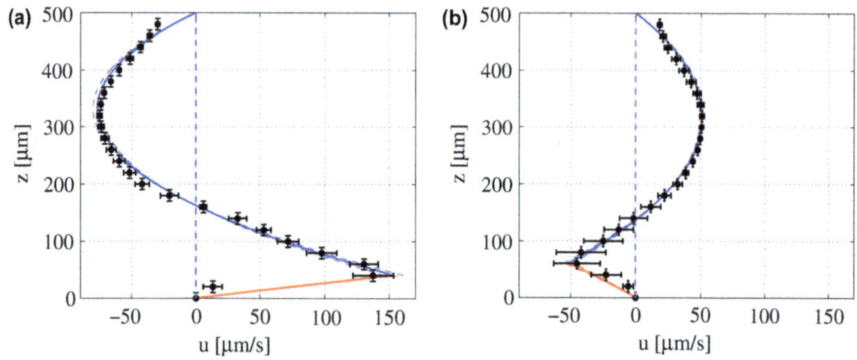

Figure 11.6 Velocity profile measured in a channel of 500 µm height in a closed channel configuration. The cilia-induced flow is counter-balanced by a pressure driven back-flow. (a) Velocity profile measured at 10 Hz actuation frequency.[15] (b) Velocity profile measured for a reversed sense of rotation of the actuation field as compared to Figure 11.6a.

maximum approximately 40–60 µm above the channel bottom and a reversed flow above approximately 130–170 µm spacing to the channel floor. We anticipate that the flattening of the velocity profile at the top of the channel originates from a large depth of correlation of the microscope objective that exceeds the theoretical estimate of 20 µm such that a systematic error of the velocity results is induced.[22] The flow is viscosity dominated and the channel width-to-height aspect ratio is approximately 10 : 1. Therefore, the velocity profiles as shown in Figure 11.6a and b can be approximated as the sum of a cilia-induced plain Couette flow $u_C(z)$ and a back-pressure-induced plain Poiseuille flow $u_P(z)$, where the subscripts 'C' and 'P' indicate Couette and Poiseuille flow, respectively The spacing between the channel floor to the measurement plane of maximum positive net flow is hence on referred to as the reference plane with position $z = h$. The measured velocity profile is approximated to:

$$u(z) = \frac{U}{H-h}(H-z) + \frac{1}{2\mu}\frac{dp}{dx}\left(z^2 - Hz\right) \quad \text{for } h < z < H \tag{11.3a}$$

$$u(z) = \frac{U}{h}z + \frac{1}{2\mu}\frac{dp}{dx}\left(z^2 - Hz\right) \quad \text{for } h > z > H \tag{11.3b}$$

where H is the channel height and μ is the dynamic viscosity of the fluid. The pressure gradient dp/dx and the cilia-induced velocity $u_c(z=0) = U$ are unknown parameters derived by nonlinear fitting of eqn (11.3a) to the measured velocity profiles in Figure 11.6a and b. Due to insufficient number of measurement points for $0 < z < h$ the volume flow rate in this region is estimated by inserting U and dp/dx from eqn (11.3a) into eqn (11.3b). The results are summarized in Table 11.1.

Note that friction effects of the side walls are neglected in these solutions. As indicated in Table 11.1, a cilia-induced maximum volume flow rate of

Table 11.1 Experimentally derived volume flow rates based on phase-averaged velocity profiles as shown in Figure 11.6a and b.

h (μm)	U (μm s^{-1})	dp/dx (Pa m^{-1})	dV_C/dt (μL min^{-1})	dV_P/dt (μL min^{-1})	dV_{net}/dt (μL min^{-1})
$d\alpha/dt > 0$					
40 as in Figure 11.6a	203 ± 10	6.3 ± 0.3	15.2 ± 0.7	−17.0 ± 0.7	−1.7 ± 1.0
60	202 ± 9	6.4 ± 0.3	15.2 ± 0.7	−17.5 ± 0.7	−2.3 ± 1.0
$d\alpha/dt < 0$					
60 as in Figure 11.6b	−93 ± 5	−3.5 ± 0.1	−6.9 ± 0.4	9.6 ± 0.4	2.7 ± 0.6
80	−92 ± 5	−3.6 ± 0.1	−6.9 ± 0.4	9.9 ± 0.4	2.9 ± 0.6

Results derived from the phase-averaged velocity profiles shown in Figure 6a and b by least square error fitting. From left to right: h: reference plane; U: reference velocity at $z=0$; dp/dx: back pressure gradient; dV_C/dt: cilia-induced maximum volume flow rate that is reached in the absence of back-pressure; dV_P/dt: maximum pressure driven volume flow rate derived from dp/dx; dV_{net}/dt: difference in volume flow rate $dV_P/dt - dV_C/dt$. $d\alpha/dt$ denotes the rotation sense of the magnetic actuation field.

$dV_c/dt = 15.2 \pm 0.7\,\mu L\,min^{-1}$ was determined from the results shown in Figure 11.6a. The choice of h defines the number of measurement points that are included in the fitting procedure of eqn (11.3a). As shown in Table 11.1, the effect of varying the reference plane height h for 20 μm was found to be smaller than the accuracy at which the volume flow rates could be determined. Largest deviations of the fitting curve $u(z)$ occur in the near proximity of the cilia where the assumption of a plain shear flow does not hold any more due to the occurrence of out of plane velocities.

As shown in Figure 11.6b a reversal of the rotation sense of the magnetic actuation field leads to a reversed but approximately 50% reduced cilia-driven volume flow rate of $dV_c/dt = -6.9 \pm 0.4\,\mu L\,min^{-1}$ compared to Figure 11.6a. Simulations by Khaderi *et al.*[23] indicate that the flow reversal is induced by an altered configurationally asymmetric cilia beat cycle and so is the difference in velocity magnitude.

The difference between cilia-induced volume flow rate dV_c/dt and pressure-driven volume flow rate dV_p/dt allows a qualitative comparison of the cilia performance in and closely around the chosen field of view (which covers only 0.24% of the channel bottom) to the average performance of the total cilia array. When the cilia performance in the chosen field of view is representative for the whole cilia array, the net volume flow rate dV_{net}/dt will be zero. From Table 11.1 it is evident that the cilia in the field of view induce a slightly under-average flow rate leading to an overall pressure dominated mass flux through the measurement volume.

11.4.3 Comparison with Existing Micro-fluidic Systems

Figure 11.7a and b compares basic characteristics of existing micro-pump systems (blue circles) reported by Iverson and Garimella[24] and Laser and

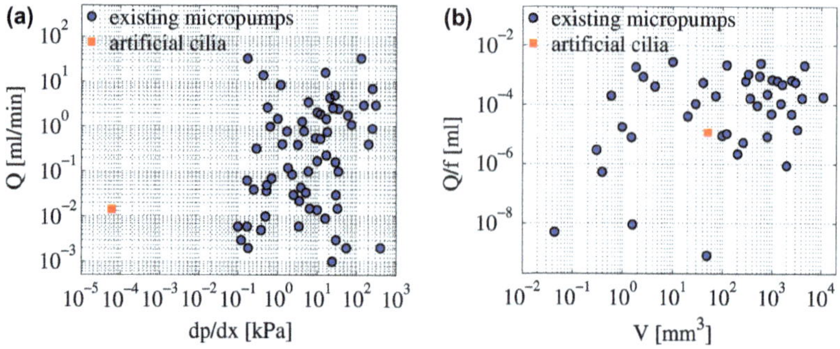

Figure 11.7 Performance of displacement and dynamic micro-pumps. The graphs are based on data given by Iverson and Garimella[24] and Laser and Santiago.[25] Circled and squared data points indicate displacement micro-pumps and dynamic micro-pumps, respectively. The star-shaped data points indicate the pumping device with artificial cilia described in this chapter. (a) Maximum volume flow rate as function of maximum back pressure. (b) Maximum volume flow rates per actuation cycle as function of the micro-pump size.

Santiago[25] with the present artificial cilia system (red square). The maximum volume flow rate is shown as function of the maximum back pressure on a log–log scale. Most micro-fluidic pumping systems operate in the order of 10^{-1}–10^{2} kPa or 10^{-3}–10^{1} mL min^{-1}. As shown in Figure 11.7b micro-pumps of sizes between 1 and 10^4 mm^3 typically transport volumes in the order of 10^{-6}–10^{-2} mL during each beat cycle. The square-shaped data points in Figure 11.7a and b indicate the artificial cilia device that was studied in this chapter. While the size and the volume flow rates generated by the artificial cilia pump lie well within the range of existing micro-pumping systems, its generated back pressure is more than three orders of magnitudes lower.

Our experimental findings are in good agreement with two-dimensional numerical simulations by Khaderi *et al.* who predicted volume flow rates and pressure heads of the same order of magnitude.[23] The low pressure heads of the present cilia system are the result of a pressure-induced backflow above the cilia array. We anticipate that higher pressure heads could be reached for higher cilia length to channel height ratios.

11.5 Summary and Discussion

In this chapter we report on the fluid manipulation of magnetically actuated rectangular artificial cilia performing a plane beat cycle in a closed micro-channel. Phase-locked μPIV measurements were performed to gain insight in the time-resolved and time-averaged velocity field which was compared to the in-plane cilia beat cycle.

The cilia exhibit two beat cycles during one actuation cycle. The nearly synchronous cilia motion induces an oscillating flow that changes direction twice during one actuation cycle reaching instantaneous velocities of up to the order of $10^3\,\mu m\,s^{-1}$ at 10 Hz actuation frequency. The corresponding time-averaged flow, however, only reached local velocities of approximately $130\,\mu m\,s^{-1}$ which is more than one order of magnitude lower than measured instantaneous velocities. Velocity fluctuations are dominating the flow since the cilia predominantly move synchronously forward–backward while exhibiting only a weakly asymmetric beat cycle. Despite the non-optimal beat cycle phase-averaged velocities induced by the studied cilia device are one order of magnitude higher up to date than reported in other experimental studies.[7,8]

At low actuation frequencies ($f < 6$ Hz) the net flow was found to be linearly proportional to the beat frequency while it appeared to increase exponentially at higher actuation frequencies, approaching a saturation velocity of maximum fluid transport. Since the magnetic energy input per magnet revolution is the same for all actuation frequencies, the total work done by the cilia filaments per beat cycle also stays constant. At low actuation frequencies the cilia move in a quasi-static manner solely due to the interplay of magnetic and elastic bending forces. With increasing actuation frequencies an increasing part of the work done by the cilia is dissipated by the fluid. It means that the effective kinetic energy transferred to the fluid per cilia beat cycle decreases as indicated in Figure 11.5d.

The cilia-induced maximum pressure drop and volume flow rate at 10 Hz actuation frequency were derived from the measured velocity profile to $dp/dx = 6.3 \pm 0.3\ \mathrm{Pa\,m^{-1}}$ and $dV_C/dt = 15.2 \pm 0.7\ \mu\mathrm{L\,min^{-1}}$. A volume flow rate in opposite direction is induced when the sense of rotation of the magnetic actuation field is reversed. In this case a pressure drop and maximum volume flow rate of only $dp/dx = -3.5 \pm 0.1\ \mathrm{Pa\,m^{-1}}$ and $dV_C/dt = -6.9 \pm 0.4\ \mu\mathrm{L\,min^{-1}}$ are created. We anticipate that the difference in net flow originates from an altered configurational asymmetry of the cilia beat cycle. The measurement results are in good agreement with numerical predictions of Khaderi *et al.*[23] These authors predicted a volume flow rate and normalized pressure head of the same order of magnitude with a two-dimensional discrete numerical model of synchronously beating cilia that are comparable in size, stream-wise spacing and channel height.

Large-scale field gradients in the actuation field used in the experiments induce a beat cycle phase-shift between the very distant cilia rows of approximately 60°. Numerical studies showed that besides configurational asymmetry the slight out-of-phase beating of cilia can also create a net flow.[4] However, the qualitative agreement between experimental and numerical results suggests that the measured net flow originates from a configurationally asymmetric cilia motion rather than the slightly out-of-phase beating of the cilia.

11.6 Outlook

For the design of future artificial cilia devices an increasing number of metachronal wave trains along the streamwise dimension of the cilia array would increase the fluid transport while likewise reducing fluid oscillations in the system.[26] Recent numerical studies give support to the suggestion that at low Reynolds numbers the net flow produced by a large number of metachronal wave trains may even dominate over the flow created by a configurationally asymmetric cilia beat.[27] Furthermore, fluid transport by metachronal coordination was found to be insensitive to details of the cilia deformation during the beat cycle.[4] We therefore think that the out-of-phase beating of cilia filaments offers a robust alternative transport mechanism that circumvents the difficulty of creating a well-controlled plane cilia beat cycle of significant configurationally asymmetry as it is required for synchronously beating artificial cilia.

Acknowledgement

The research was carried out at the Laboratory for Aero and Hydrodynamics, Delft University of Technology, The Netherlands.

References

1. E. M. Purcell, *Am. J. Phys.*, 1977, **45**, 3–11.
2. S. Gueron and K. Levit-Gurevich, *Biophys. J.*, 1998, **74**, 1658–1676.
3. B. Guirao and J.-F. Joanny, *Biophys. J.*, 2007, **92**, 1900–1917.
4. J. Hussong, W.-P. Breugem and J. Westerweel, *J. Fluid Mech.*, 2011, **684**, 137–162.
5. F. Fahrni, M. W. J. Prins and L. J. van Ijzendoorn, *Lab Chip*, 2009, **9**, 3413–3421.
6. J. Hussong, N. Schorr, J. Belardi, O. Prucker, J. Rühe and J. Westerweel, *Lab Chip*, 2011, **11**, 2017–2022.
7. A. R. Shields, B. L. Fiser, B. A. Evans, M. R. Falvo, S. Washburn and R. Superfine, *Proc. Natl. Acad. Sci. U.S.A.*, 2010, **107**, 15670–15675.
8. M. Vilfan, A. Potonik, B. Kavi, N. Osterman, I. Poberaj, A. Vilfan and D. Babi, *Proc. Natl. Acad. Sci. U.S.A.*, 2010, **107**, 1844–1847.
9. N. Coq, A. Bricard, F.-D. Delapierre, L. Malaquin, O. du Roure, M. Fermigier and D. Bartolo, *Phys. Rev. Lett.*, 2011, **107**, 014501.
10. J. den Toonder, F. Bos, D. Broer, L. Filippini, M. Gillies, J. de Goede, T. Mol, M. Reijme, W. Talen, H. Wilderbeek, V. Khatavkar and P. Anderson, *Lab Chip*, 2008, **8**, 533–541.
11. A. Alexeev, J. M. Yeomans and A. C. Balazs, *Langmuir*, 2008, **24**, 12102–12106.
12. S. N. Khaderi, M. G. H. M. Baltussen, P. D. Anderson, J. M. J. den Toonder and P. R. Onck, *Phys. Rev. E*, 2010, **82**, 027302.
13. A. Babataheri, M. Roper, M. Fermigier and O. Du Roure, *J. Fluid Mech.*, 2011, **678**, 5–13.
14. J. Belardi, N. Schorr, O. Prucker and J. Rühe, *Adv. Funct. Mater.*, 2011, **21**, 3314–3320.
15. Reproduced by permission of The Royal Society of Chemistry. http://pubs.rsc.org/en/content/articlepdf/2011/lc/c0lc00722f.

16. R. J. Adrian and J. Westerweel, *Particle Image Velocimetry*. Cambridge University Press, Cambridge, 2010.
17. R. Lindken, M. Rossi, S. Grosse and J. Westerweel, *Lab Chip*, 2009, **9**, 2551–2567.
18. S. T. Wereley and C. D. Meinhart, *Annu. Rev. Fluid Mech.*, 2010, **42**, 557–576.
19. M. Rossi, R. Lindken, B. P. Hierck and J. Westerweel, *Lab Chip*, 2009, **9**, 1403–1411.
20. R. D. Keane and R. J. Adrian, *Appl. Sci. Res.*, 1992, **49**, 191–215.
21. E. Gauger, M. Downton and H. Stark, *Eur. Phys. J. E Soft Matter Biol. Phys.*, 2009, **28**, 231–242.
22. A. Kloosterman, C. Poelma and J. Westerweel, *Exp. Fluids*, 2011, **50**, 1587–1599.
23. S. N. Khaderi, C. B. Craus, J. Hussong, N. Schorr, J. Belardi, J. Westerweel, O. Prucker, J. Rühe, J. M. J. den Toonder and P. R. Onck, *Lab Chip*, 2011, **11**, 2002–2010.
24. B. Iverson and S. Garimella, *Microfluid. Nanofluid.*, 2008, **5**, 145–174.
25. D. J. Laser and J. G. Santiago, *J. Micromech. Microeng.*, 2004, **14**, R35.
26. M. Li and J. G. Brasseur, *J. Fluid Mech.*, 1993, **248**, 129–151.
27. S. N. Khaderi, J. M. J. den Toonder and P. R. Onck, *J. Fluid Mech.*, 2011, **688**, 44–65.

CHAPTER 12

Measurement of Fluid Flow Generated by Artificial Cilia

A. VILFAN,*[a,b] D. BABIČ,[b] B. KAVČIČ,[c] G. KOKOT,[a] N. OSTERMAN,[a,c] I. POBERAJ[b] AND M. VILFAN[a]

[a] J. Stefan Institute, Jamova 39, 1000 Ljubljana, Slovenia; [b] Faculty of Mathematics and Physics, University of Ljubljana, Jadranska 19, 1000 Ljubljana, Slovenia; [c] LPKF Laser & Elektronika d.o.o., Polica 33, 4202 Naklo, Slovenia
*Email: andrej.vilfan@ijs.si

12.1 Introduction

Efficient pumping and mixing of fluids in microscopic channels is paramount in micro-fluidic applications, especially when circulation along a closed circuit is desired.[1] Several mechanisms have been developed to pump a fluid on a chip, including pneumatically driven peristaltic pumps,[2] acoustic streaming,[3] various techniques based on capillary forces (for droplet manipulation), electrokinetic effects (electrophoresis and electro-osmosis), rotary pumps[4] *etc.*

The efficiency of the ciliary pumping mechanism leads to the idea of using the same principle for designing artificial cilia that act as micro-scale pumps and mixers. The first attempt to fabricate artificial cilia was based on nanorod arrays manufactured from magnetic–polymeric composite materials.[5] By using a moving permanent magnet, it was demonstrated that the nanorods can perform a simple periodic motion. Furthermore, den Toonder *et al.* have proven that electrostatically driven cilia made of metal-coated polymer films can act as mixers and—outside the low Reynolds number regime—even as

RSC Nanoscience & Nanotechnology No. 30
Artificial Cilia
Edited by Jaap M J den Toonder and Patrick R Onck

Published by the Royal Society of Chemistry, www.rsc.org

pumps.[6–8] An alternative approach is based on light driven azo-doped liquid crystal elastomers,[9] although the speed of actuation that can be achieved with such a mechanism is presently still too slow for fluid pumping. Several theoretical studies have also addressed magnetic actuation of cilia.[10–12]

We recently succeeded in manufacturing self-assembled artificial cilia and proved that their asymmetric beating generated a directed fluid flow.[13] The artificial cilia were formed as flexible yet stable chains of super-paramagnetic colloidal particles. They were held together by an external magnetic field, which gave them structural stability and at the same time allowed the manipulation of their direction. We showed that a simple non-reciprocal beating pattern induced by conical rotation of the magnetic field resulted in nearly uniform fluid motion above the ciliated surface. At the same time, another group applied a similar principle to create chains of magnetic particles that were not anchored, but rather tumbled along the surface, which also has the potential of fluid pumping.[14] Magnetically driven cilia can also be actuated in a way that centimetre-scale metachronal waves are formed and even if the individual cilia beat in a reciprocal fashion they can pump the fluid at remarkable speeds.[15–17]

In this chapter we present a combined theoretical and experimental study of a system of self-assembled artificial cilia (Figure 12.1). The biomimetic cilia are constructed as long chains of spherical super-paramagnetic particles, which

Figure 12.1 An artist's view of an array of magnetically driven artificial cilia. Cilia are formed as chains of super-paramagnetic particles (black) that align themselves in the direction of the applied magnetic field. Their bottom ends are anchored to the surface by nickel dots (gold). Yellow dots represent tracer particles that were added to measure the fluid velocity.

self-assemble parallel to the external magnetic field. We rotate the magnetic field along the shape of a tilted cone, thus actuating the cilia in a simple non-reciprocal manner. By adding non-magnetic tracer particles we measure the time-averaged fluid velocity as a function of the height above the surface. To compare the measured flow with a model prediction we simulate the system by taking into account magnetic, gravitational and repulsive contact forces and describing the hydrodynamics in the Rotne–Prager approximation. A comparison of the predicted velocity profile with the experimental result shows a very good agreement.

12.2 Experimental Set-up

The experimental set-up (Figure 12.2) was built around an inverted optical microscope (Zeiss, Axiovert 200M, Achroplan 63/0.9W objective) equipped with optical tweezers, as described in several publications.[13,18–21] Three orthogonal pairs of water-cooled coils (the pair generating vertical field in Helmholtz configuration, the other two pairs fitted closely around it) were used to generate a homogeneous magnetic field within the sample. Specifically, all coils had a mean radius of 2.1 cm. The vertical pair with $n = 216$ turns was positioned 1.05 cm above and below the sample, and the horizontal pairs containing $n = 243$ turns at distance 3.65 cm away from the centre of the sample. The magnetic field per unit current had a density of 10 mT A^{-1} in vertical direction and 1.72 mT A^{-1} in both horizontal directions. The currents through the coils were individually regulated by a six-channel current source, which allowed us to generate a nearly homogeneous magnetic field of arbitrary direction and varying magnitude. The typical magnetic field density used in the experiments was 2–7 mT.

We assembled the cilia from monodisperse super-paramagnetic beads[22] (Dynabeads Epoxy M-450, Dynal Biotech, diameter 4.4 μm). The beads were coated with bovine serum albumin (Sigma–Aldrich, 10 mg mL^{-1} in ultra-pure water) in experiments with individual and rows of cilia, or with sodium dodecyl sulfate (5 mg mL^{-1}, 5 h in an ultrasonic bath) for 3×3 arrays. The

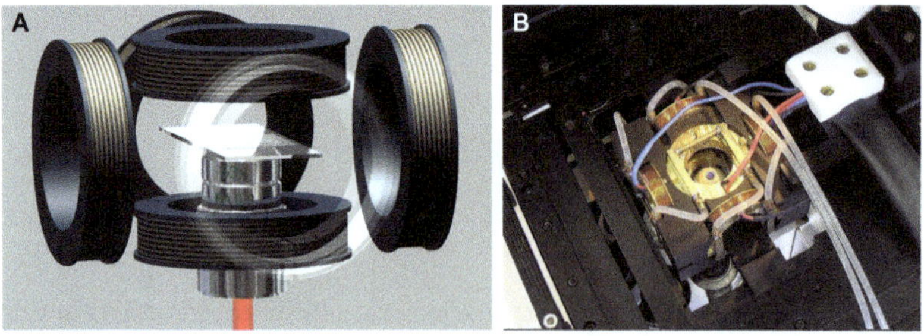

Figure 12.2 Magnetic tweezers set-up, built around an inverted microscope. The magnetic field was generated by three pairs of coils (each pair for one direction), shown schematically (A) and in top view (B).

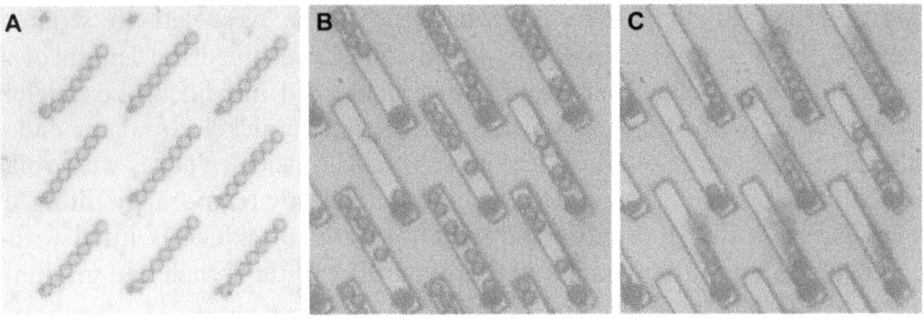

Figure 12.3 Two ways to assemble artificial cilia. (A) Super-paramagnetic beads were individually trapped by optical tweezers and assembled into chains, held together by an external magnetic field. They were anchored to the surface via nickel anchoring sites. (B and C) Guided self-assembly process: the beads were left to sediment into trenches in a photo-resist layer, which ensured that all chains had approximately the same length and width. The trenches measured 45 μm × 5 μm and had a depth of 5 μm. The bead diameter was $2a = 4.4\,\mu\text{m}$. Panel (B) shows the filled trenches after the removal of excess beads. In panel (C) the magnetic field was turned on and used to lift cilia from the trenches. Panel (A) is reprinted from Vilfan *et al.*[13]

measurements shown here were performed on uniform cilia which were assembled one by one with optical tweezers (Figure 12.3a). We also produced self-assembled cilia that were created by sedimentation of beads into trenches (Figure 12.3b and c). In order to anchor the cilia to the surface we created an array of ferromagnetic nickel dots. We used standard microscope glass slides each of which was coated with a 200–500 nm thick nickel layer produced by a conventional evaporation technique and etched to create nickel dots that were 5 μm in diameter.

The tracer particles were non-magnetic fluorescently labelled polystyrene spheres (Dragon Green, Bangs Laboratories, diameter 1 μm, used in Figure 12.7 and Figure 12.8) or silica spheres (Bangs Laboratories, diameter 1 μm, Figure 12.9 and Figure 12.10). Their movement was recorded with an Electron Multiplying CCD camera (Hamamatsu Photonics, C9100-13, Figure 12.7 and Figure 12.8) or a CMOS camera (PixeLINK, PL-A741, Figure 12.9 and Figure 12.10) and analysed from video recordings using a custom-written particle tracking software to obtain their trajectories.

Trenches that served the self-assembly of artificial cilia were fabricated in a 5 μm thick layer of photo-resist SU-8 2025 (Microchem, standard coating procedure, adhesion promoter TI Prime, Microchemicals GmbH). Direct illumination of the structures on the photo-resist was done with a UV laser (Omicron-Laserage GmbH, Bluephoton LDM375.20.CWA.L, 375 nm). The laser beam was steered by acousto-optic deflectors (A.A. Opto-electronic, DTSXY-400-405) controlled by the controller (Aresis, BSC-160) and focused by Zeiss LD Plan-neofluar × 10/0.4 Korr objective. The trenches were 5 μm wide and deep, *i.e.* slightly larger than the diameter of the beads, so that a single chain was formed in each trench when the beads were left to sediment (Figure 12.3b). The trench that was 45 μm long determined the length of an individual cilium, although some of the trenches did not fill up completely. The trenches were positioned such that there was one nickel dot at the end of each trench. After sedimentation, the excess beads outside trenches were rolled away with a rotating magnetic field, or alternatively by temporarily tilting the sample. The cilia that formed in the trenches were subsequently lifted with a magnetic field that initially had a direction parallel to the trench and gradually changed towards vertical (Figure 12.3c).

Glass plates with structured photo-resist were coated with bovine serum albumin (Sigma–Aldrich, 10 mg mL^{-1} in ultra-pure water) to prevent adhesion of the spheres. We used 200 μm spacers to ensure uniform separation between the glass slide and a coverslip, and the cell was sealed to prevent the evaporation of the fluid. At the same time the cell was thick enough so that the backflow did not alter the velocity field significantly.

12.3 Theory and Simulations

12.3.1 Methods of Flow Description

Artificial cilia, like their natural counterparts, operate at very low Reynolds numbers. If the typical velocity is of the order $v = 100\,\mu m\,s^{-1}$, the size of the objects $d = 10\,\mu m$ and the water viscosity $\eta = 0.001\,N\,s\,m^{-2}$, this gives a Reynolds number of $Re = \rho v d/\eta = 0.001$. This means that the fluid inertia on this length scale is completely negligible and the fluid motion responds immediately to the forces acting on it. Mathematically, the fluid is described by the Stokes equation:

$$\Delta \boldsymbol{v} = \nabla \boldsymbol{p} \tag{12.1}$$

It is complemented by the incompressibility condition:

$$\nabla \cdot \boldsymbol{v} = 0 \tag{12.2}$$

and the no-slip boundary condition on all surfaces. Cilia larger or faster than discussed here can actually reach Reynolds numbers at which inertia becomes important.[23] Depending on the complexity of the problem there are many different methods of solving the Stokes equation. One possibility is the method of finite elements, where one solves the Stokes equation on a mesh inside the fluid volume. Alternatively, boundary element methods rely on analytical solutions of the Stokes equation (usually the Oseen tensor) and determine the force density on all boundaries in a self-consistent manner. There are, however, many special cases where neither of the computationally intense methods is necessary. Instead, we can use an analytical or numerical expression for the mobility matrix $\boldsymbol{\mu}$ that connects the velocities $\boldsymbol{v}_i$ and angular velocities $\boldsymbol{\omega}_i$ of all particles with the forces $\boldsymbol{F}_i$ and torques $\boldsymbol{T}_i$ acting on them:

$$\begin{pmatrix} \boldsymbol{v}_1 \\ \vdots \\ \boldsymbol{v}_n \\ \omega_1 \\ \vdots \\ \omega_n \end{pmatrix} = \boldsymbol{\mu} \begin{pmatrix} \boldsymbol{F}_1 \\ \vdots \\ \boldsymbol{F}_n \\ \boldsymbol{T}_1 \\ \vdots \\ \boldsymbol{T}_n \end{pmatrix} \tag{12.3}$$

A prominent example is a collection of moving spheres in the presence of a planar boundary, a very good approximation for our system of artificial cilia. Theoretically, a multi-pole expansion allows the calculation of the mobility tensor to arbitrary precision.[24] Other examples include particles inside a spherical cavity[25] or particles between two planar walls.[26]

12.3.2 Rotne–Prager Approximation

In our case we deal with relatively small particles and small angular velocities, which means that we can neglect the rotational terms in the mobility matrix and approximate the translational terms up to the quadratic order in particle size.[27] Then the elements of the mobility matrix are given as:

$$\boldsymbol{\mu}_{i,j} = \left(1 + \frac{a^2}{6}\nabla^2_{x_i}\right)\left(1 + \frac{a^2}{6}\nabla^2_{x_j}\right)\boldsymbol{\mu}^{\text{Blake}}_{i,j} \tag{12.4}$$

with $\boldsymbol{\mu}^{\text{Blake}}$ denoting Blake's tensor,[28] which describes the flow of a point force in the presence of a planar boundary. Blake substituted the no-slip boundary with three images: an image-Stokeslet, a source doublet (D) and a Stokeslet-doublet (SD). The resulting mobility tensor can be written as:

$$\boldsymbol{\mu}^{\text{Blake}}_{i,j} = \frac{1}{8\pi\eta}\left(G^S(\boldsymbol{x}_i - \boldsymbol{x}_j) - G^S(\boldsymbol{x}_i - \bar{\boldsymbol{x}}_j) + 2z_j^2 G^D(\boldsymbol{x}_i - \bar{\boldsymbol{x}}_j) - 2z_j G^{SD}(\boldsymbol{x}_i - \bar{\boldsymbol{x}}_j)\right) \tag{12.5}$$

with:

$$G^S_{\alpha\beta}(\boldsymbol{r}) = \frac{\delta_{\alpha\beta}}{r} + \frac{r_\alpha r_\beta}{r^3} \tag{12.6}$$

$$G^D_{\alpha\beta}(\boldsymbol{r}) = (1 - 2\delta_{\beta z})\frac{\partial}{\partial r_\beta}\left(\frac{r_\alpha}{r^3}\right) \tag{12.7}$$

and

$$G^{SD}_{\alpha\beta}(\boldsymbol{r}) = (1 - 2\delta_{\beta z})\frac{\partial}{\partial r_\beta}G^S_{\alpha z}(\boldsymbol{r}) \tag{12.8}$$

Here $\bar{\mathbf{x}}_j = (x_j, \ y_j, -z_j)$ denotes the image position. The diagonal terms, describing the self-mobility of a particle, have a different form:

$$\boldsymbol{\mu}_{ii} = \frac{1}{6\pi\eta a}\begin{pmatrix} \beta_i^{tt} & 0 & 0 \\ 0 & \beta_i^{tt} & 0 \\ 0 & 0 & \alpha_i^{tt} \end{pmatrix} \tag{12.9}$$

with:

$$\alpha_i^{tt} = 1 - \frac{9}{8}\frac{a}{z_i} + \frac{1}{2}\left(\frac{a}{z_i}\right)^3 - \frac{1}{8}\left(\frac{a}{z_i}\right)^5 \quad \text{and} \quad \beta_i^{tt} = 1 - \frac{9}{16}\frac{a}{z_i} + \frac{1}{8}\left(\frac{a}{z_i}\right)^3 - \frac{1}{16}\left(\frac{a}{z_i}\right)^5 \tag{12.10}$$

An explicit form of the mobility matrix elements is given in the Supporting Information of Vilfan *et al.*[13]

12.3.3 Equations of Motion

The mobility matrix allows us to calculate the instantaneous velocities of all particles if we know the forces acting on them. The forces consist of magnetic dipole–dipole interactions:

$$\boldsymbol{F}_i^{\text{magnetic}} = \frac{\mu_0}{4\pi}\sum_{j\neq i}\left(3\frac{\boldsymbol{r}_{ij}(\boldsymbol{m}_i\cdot\boldsymbol{m}_j) + \boldsymbol{m}_i(\boldsymbol{r}_{ij}\cdot\boldsymbol{m}_j) + \boldsymbol{m}_j(\boldsymbol{r}_{ij}\cdot\boldsymbol{m}_i)}{r_{ij}^5} - 15\frac{\boldsymbol{r}_{ij}(\boldsymbol{r}_{ij}\cdot\boldsymbol{m}_i)(\boldsymbol{r}_{ij}\cdot\boldsymbol{m}_j)}{r_{ij}^7}\right) \tag{12.11}$$

with $\boldsymbol{m}_i$ denoting the magnetic moment of particle i, the gravity force:

$$\boldsymbol{F}_i^{\text{gravity}} = -mg\hat{e}_z \tag{12.12}$$

where m is the buoyant mass of a particle:

$$m = \frac{4\pi a^3}{3}(\rho - \rho_{H_2O}) \tag{12.13}$$

and the contact forces between touching beads and between a bead and the surface. We determine the latter in a self-consistent manner. The bottom bead of each cilium, which is anchored to the nickel dot, is also subject to a constraint force, ensuring that its position remains fixed. Then the velocity of bead i is determined as:

$$\boldsymbol{v}_i = \boldsymbol{\mu}_{ij}\boldsymbol{F}_j^{\text{total}} \tag{12.14}$$

We solve the equations of motion numerically using the Euler method.

12.3.4 Far Field of the Flow

Sufficiently far away from a cilium ($r \gg z_i$) the flow obtains a generic form that corresponds to the leading term in a multi-pole expansion:[29]

$$\boldsymbol{v}(\boldsymbol{r}) = \frac{3}{2\pi\eta} \sum z_i \left(F_{i,x}x + F_{i,y}y\right) \frac{z}{r^5} \boldsymbol{r} \tag{12.15}$$

where $\boldsymbol{r} = (x,y,z)$ is the position of the point where we observe the flow and (x_i, y_i, z_i) the position of i-th particle forming the cilium. ($F_{i,x}$, $F_{i,y}$, $F_{i,z}$) is the force this particle exerts on the fluid. The intensity of the flow is best characterized by the volume flow rate, which measures the volume of the fluid that crossed any half plane orthogonal to the direction of pumping per unit time (Figure 12.4a). If the pumping direction is along the x axis, the volume flow rate is:

$$Q = \frac{1}{\pi\eta} \sum z_i F_{i,x} \tag{12.16}$$

The volume flow rate generated by a single cilium is closely related to the velocity above a densely ciliated surface (Figure 12.4b). With ρ denoting the surface density of the cilia (number per unit area), the velocity in the region where the flow is homogeneous is:

$$v_c = \pi\rho Q \tag{12.17}$$

However, in our experimental set-up the ciliary array on which we measured the velocity distribution only counted up to 3×3 cilia, which is too small for the above expression to be valid.

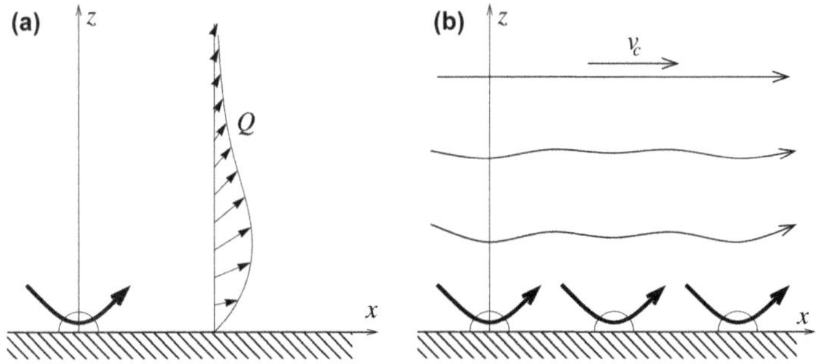

Figure 12.4 (a) The volume flow rate Q measures the fluid flux through a vertical cross section, orthogonal to the direction of pumping. The thick arrow symbolizes the complex near field flow past the cilium and the thin arrows the velocity at the cross-section. (b) Above a densely ciliated surface, the velocity field becomes homogeneous at a height sufficiently larger than the length of the cilia. From Vilfan,[29] with kind permission of the *European Physical Journal*.

12.4 Results

12.4.1 Flow Around a Single Artificial Cilium

The main broken symmetry in a beating cilium is that between the working stroke when it stretches out to move as much fluid as possible, and the recovery stroke, during which the cilium bends and sweeps along the surface so that it simultaneously minimizes the backflow and the viscous drag.[30] In that sense one can mimic the ciliary beat by rotating the magnetic field along the mantle of a tilted cone (Figure 12.5), although we are not able to bend the cilium during the recovery stroke. The time dependence of the magnetic field is the following:

$$\boldsymbol{B} = B_0 \begin{pmatrix} \cos\theta \cos\psi \cos\omega t + \sin\theta \sin\psi \\ \sin\psi \sin\omega t \\ -\sin\theta \sin\psi \cos\omega t + \cos\theta \cos\psi \end{pmatrix} \tag{12.18}$$

The time-averaged flow caused by this motion can be approximated (up to the order r^{-3}) as:[21]

$$\bar{\boldsymbol{v}} = \frac{C_N L^3}{2T\eta} \sin^2\psi \left(-\sin\theta \, D_y + \frac{3}{4} L \cos^2\theta \cos\psi \, Q_1 + \frac{3}{4} \sin\theta \cos\theta \cos\psi \, Q_4 \right) \tag{12.19}$$

Here:

$$D_y = \frac{yz}{r^5} \boldsymbol{r} \tag{12.20}$$

represents a pumping term (analog eqn (12.15)),

$$Q_1 = \frac{z}{r^5} \begin{pmatrix} -y \\ x \\ 0 \end{pmatrix} \tag{12.21}$$

a rotlet around the vertical axis, caused by the counter-clockwise rotation of the cilium, and

$$Q_4 = \frac{z^2}{r^7} \begin{pmatrix} 5xy \\ 5y^2 - r^2 \\ 5yz \end{pmatrix} \tag{12.22}$$

a rotlet around the horizontal axis. In these expressions we assumed that the centre of weight of the cone is located above the coordinate origin. We also replaced the chain of beads by a slender rod with a transverse drag coefficient[31] $C_N \sim 1.3\pi\eta$ and calculated the forces with the resistive force theory, rather than solving eqn (12.14). The angular dependence of these three terms is shown in Figure 12.6. The far field is generally dominated by the pumping term D_y, unless there is no tilt ($\theta = 0$) and the cilium produces just a whirl (Q_1). An example of a field generated by a single cilium is shown in Figure 12.7. It shows the full trace of a single particle, a collection of traces and the result of a simulation.

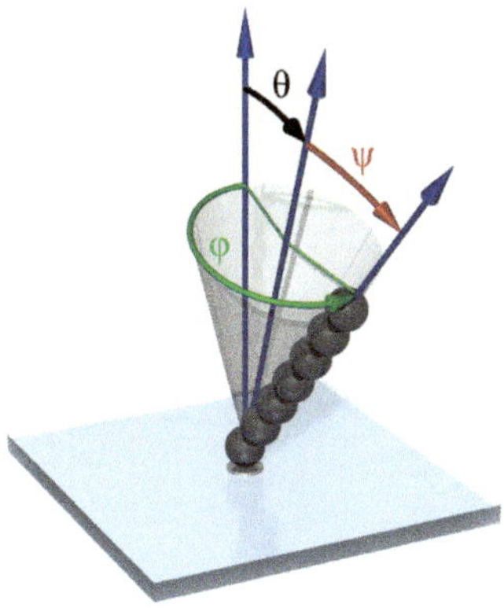

Figure 12.5 To mimic the asymmetry between the working and the recovery stroke of a cilium, we rotated the magnetic field such that it followed the mantle of a tilted cone. Due to the dipole–dipole interaction, the chain of beads largely followed the orientation of the field. From Vilfan *et al.*[21]

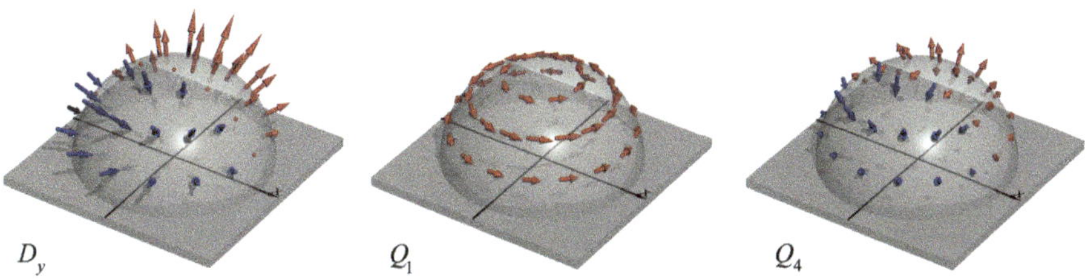

Figure 12.6 Angular dependence of the fundamental modes which describe the flow around a beating cilium. From Vilfan,[29] with kind permission of the *European Physical Journal.*

12.4.2 Flow Induced by a Row of Cilia

There are many cases where cilia appear aligned in one row, orthogonal to the direction of pumping. This arrangement is particularly common with cilia involved in feeding, for example in the sessile ciliate *Stentor*.[32] For an infinite row of cilia aligned at distances d along the x-axis and pumping the fluid in the y-direction the far-field expansion of the flow reads:

$$\bar{v} = \frac{C_N L^3}{2T\eta d}\sin^2\psi\left(-\sin\theta\frac{4yz}{3(y^2+z^2)^2}\begin{pmatrix}0\\y\\z\end{pmatrix} + L\cos^2\theta\cos\psi\frac{yz}{(y^2+z^2)^2}\begin{pmatrix}1\\0\\0\end{pmatrix}\right.$$
$$\left. + L\sin\theta\cos\theta\cos\psi\frac{z^2}{(y^2+z^2)^3}\begin{pmatrix}0\\3y^2-z^2\\4yz\end{pmatrix}\right) \quad (12.23)$$

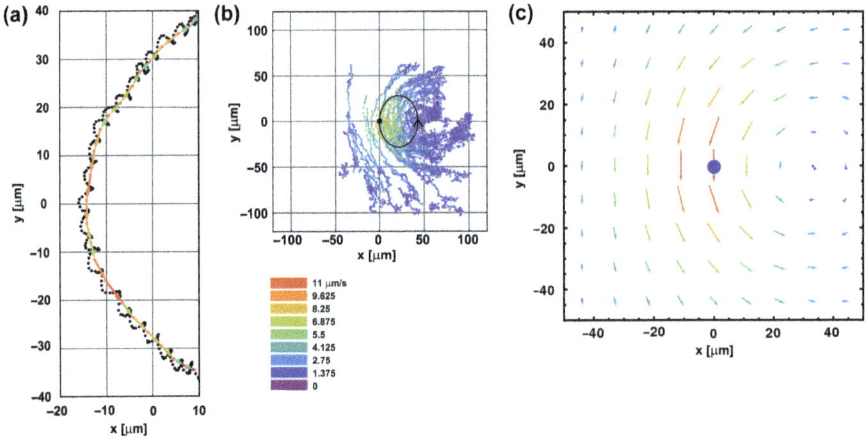

Figure 12.7 (a) Trajectory of one tracer particle above a single artificial cilium, consisting of seven beads. The cilium is anchored at the origin and following a field that rotates counter-clockwise along a cone tilted by $\theta = 40°$ with semi-cone angle of equally $\psi = 40°$. (b) Traces averaged over one period reveal the net motion of tracer particles. (c) Time averaged flow as calculated in a simulation based on the Rotne–Prager approximation. All velocities are measured or calculated at height $z = 55$ µm. Panels (a) and (b) reprinted from Vilfan *et al.*[21]

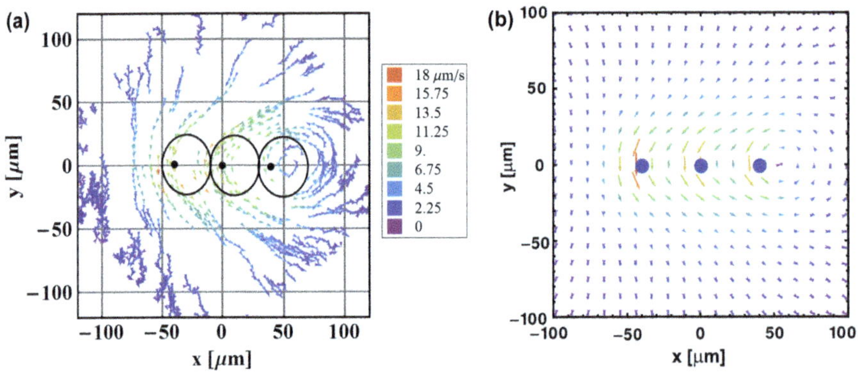

Figure 12.8 Time-averaged flow above a row of three cilia. The parameters are $f = 2$ Hz, $\psi = 50°$, $\theta = 30°$. The flow was measured at height $z = 40$ µm above the surface. (a) Experimental result. (b) Model calculation based on the Rotne–Prager approximation. Panel (a) reprinted, with permission, from Kokot *et al.*[20] Copyright 2011, American Institute of Physics.

At large distances the velocity decays as $1/r$ and points towards or away from the ciliary row. Directly above the row the flow decays with $1/r^2$. There is additionally some sideways motion caused by the rotation of cilia. A more detailed form of the flow above a row of three cilia is shown in Figure 12.8, along with an experimental result.

12.4.3 Flow Above a Ciliary Array

The situation that is closest to the majority of ciliated epithelia or microorganisms is a densely ciliated plane. Again, if the cilia are arranged on a quadratic lattice with lattice constant d, the resistive force theory yields a homogeneous velocity field for the flow sufficiently high above the plane, as compared with the length of the cilia. The flow velocity then reads:

$$\bar{v} = \frac{-\pi C_N L^3}{3T\eta d^2} \sin^2\psi \sin\theta \begin{pmatrix} 0 \\ 1 \\ 0 \end{pmatrix} \tag{12.24}$$

Our procedure for building self-assembled artificial cilia by means of sedimentation in trenches allows us to create large fields of cilia although they are not yet fully uniform in length. We therefore carried out the experiments on a

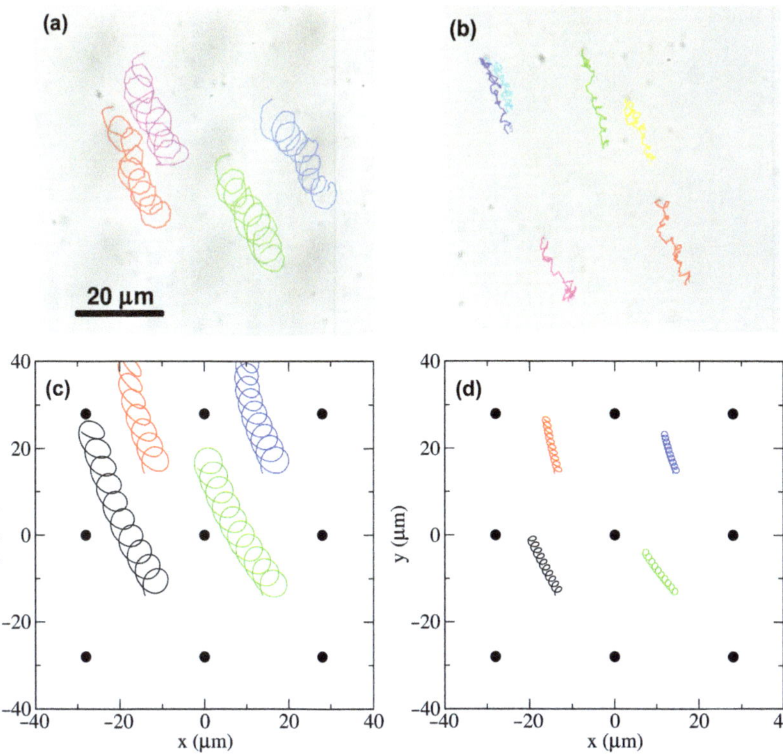

Figure 12.9 Flow above a field of 3 × 3 cilia, placed on a quadratic lattice with a spacing of 28 µm. The semicone and the tilt angles are both $\psi = 30°$, $\theta = 30°$ and the field is rotated with a frequency $f = 1$ Hz. Further, the cones are tilted in a direction 30° from the x-axis towards the y-axis. Traces of particles at a height of 40 µm (a) and 80 µm (b) above the surface. Simulation results under identical conditions are shown in (c) and (d). Panel (a) reprinted from Vilfan *et al.*[13]

field of 3 × 3 artificial cilia, assembled manually with optical tweezers. Although not sufficient to produce a homogenous flow, this field still captures the properties of a larger ciliated surface in an area directly above the cilia. Figure 12.9 shows an example of measured and simulated particle traces above such a field. Calculated time-averaged velocities are shown in Figure 12.10. Figure 12.10c also shows the comparison between the full Rotne–Prager calculation and the approximation based on the resistive force theory and far-field expansion (eqn (12.19)). Although the latter roughly reproduces the flow profile, its height dependence is not accurate at the relevant heights (up to three ciliary lengths). The Rotne–Prager approximation, on the other hand, shows a good agreement with the measured velocities within the range of statistical error.

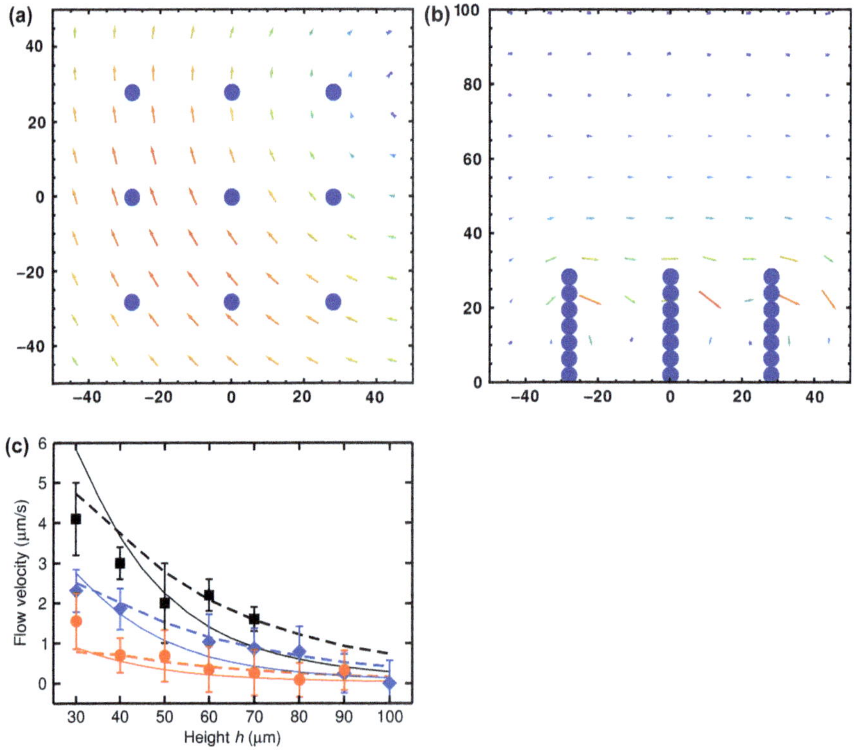

Figure 12.10 Calculated flow profiles above a field of 3 × 3 cilia. All parameters are identical to those in Figure 12.9. (a) The time-averaged flow profile at height $z = 28\,\mu m$ above the surface, calculated in the Rotne–Prager approximation. (b) Flow profile in the half plane defined by $x = 0$. (c) Measured (symbols) and calculated (Rotne–Prager; dashed; analytical approximation, continuous) flow velocities, averaged over the ciliated field, as a function of height. Different lines show data for $\psi = \theta = 20°$ (red), $\psi = \theta = 30°$ (blue) and $\psi = \theta = 40°$ (black).

12.5 Conclusions

We have demonstrated that the self-assembled magnetically driven cilia are capable of producing fluid flows with velocities of several micrometres per second. Due to the simple geometry of the set-up, which can be approximated as cilia immersed in a fluid with a single boundary, the velocity can be efficiently calculated by applying the Rotne–Prager approximation for bounded fluid. Alternatively, the resistive force theory gives analytical expressions that allow us to classify the flows by their multi-pole components and shows qualitative agreement with the numerical results.

The advantage of our system is that it allows for quick assembly and disassembly of cilia. The fact that the beads that form the cilium are not chemically bonded also allows for greater flexibility of the cilium; we can tilt it up to about 80° from the vertical position. Polymeric cilia[33] do not share this flexibility. However, bonded cilia allow significantly higher frequencies and densities, whereas self-assembled cilia become prone to disintegration.

Acknowledgement

We thank Jurij Kotar for his contribution to the magnetic tweezers, Anton Potočnik for the data on 3 × 3 ciliary arrays, Peter Panjan for nickel-coated plates and Holger Stark for discussions on the Rotne–Prager approximation.

References

1. T. M. Squires and S. R. Quake, *Rev. Mod. Phys.*, 2005, **77**, 977.
2. M. A. Unger, H. P. Chou, T. Thorsen, A. Scherer and S. R. Quake, *Science*, 2000, **288**, 113.
3. P. Marmottant and S. Hilgenfeldt, *Nature*, 2003, **423**, 153.
4. B. Kavčič, D. Babič, N. Osterman, B. Podobnik and I. Poberaj, *Appl. Phys. Lett.*, 2009, **95**, 023504.
5. A. Evans, A. R. Shields, R. L. Carroll, S. Washburn, M. R. Falvo and R. Superfine, *Nano Lett.*, 2007, **7**, 1428.
6. V. Khatavkar, P. D. Anderson, J. M. J. den Toonder and H. E. H. Meijer, *Phys. Fluids*, 2007, **19**, 083605.
7. J. den Toonder, F. Bos, D. Broer, L. Filippini, M. Gillies, J. de Goede, T. Mol, M. Reijme, W. Talen, H. Wilderbeek, V. Khatavkar and P. Anderson, *Lab Chip*, 2008, **8**, 533.
8. M. Baltussen, P. Anderson, F. Bos and J. den Toonder, *Lab Chip*, 2009, **9**, 2326.
9. C. L. van Oosten, C. W. M. Bastiaansen and D. J. Broer, *Nat. Mater.*, 2009, **8**, 677.
10. S. N. Khaderi, M. G. Baltussen, P. D. Anderson, D. Ioan, J. M. J. den Toonder and P. R. Onck, *Phys. Rev. E*, 2009, **79**, 046304.
11. E. M. Gauger, M. T. Downton and H. Stark, *Eur. Phys. J. E*, 2009, **28**, 231.
12. M. T. Downton and H. Stark, *Europhys. Lett.*, 2009, **85**, 44002.

13. M. Vilfan, A. Potočnik, B. Kavčič, N. Osterman, I. Poberaj, A. Vilfan and D. Babič, *Proc. Natl. Acad. Sci. U.S.A.*, 2010, **107**, 1844.
14. C. E. Sing, L. Schmid, M. F. Schneider, T. Franke and A. Alexander-Katz, *Proc. Natl. Acad. Sci. U.S.A.*, 2010, **107**, 535.
15. J. Hussong, N. Schorr, J. Belardi, O. Prucker, J. Rühe and J. Westerweel, *Lab Chip*, 2011, **11**, 2017.
16. J. Hussong, W.-P. Breugem and J. Westerweel, *J. Fluid Mech.*, 2011, **684**, 137162.
17. S. N. Khaderi, C. B. Craus, J. Hussong, N. Schorr, J. Belardi, J. Westerweel, O. Prucker, J. Rühe, J. M. J. den Toonder and P. R. Onck, *Lab Chip*, 2011, **11**, 2002.
18. J. Kotar, M. Vilfan, N. Osterman, D. Babič, M. Čopič and I. Poberaj, *Phys. Rev. Lett.*, 2006, **96**, 207801.
19. M. Vilfan, N. Osterman, M. Čopič, M. Ravnik, S. Žumer, J. Kotar, D. Babič and I. Poberaj, *Phys. Rev. Lett.*, 2008, **101**, 237801.
20. G. Kokot, M. Vilfan, N. Osterman, A. Vilfan, B. Kavčič, I. Poberaj and D. Babič, *Biomicrofluidics*, 2011, **5**, 034103.
21. M. Vilfan, G. Kokot, A. Vilfan, N. Osterman, B. Kavčič, I. Poberaj and D. Babič, *Beilstein J. Nanotechnol.*, 2012, **3**, 163.
22. G. Fonnum, C. Johansson, A. Molteberg, S. Mørup and E. Aksnes, *J. Magn. Magn. Mater.*, 2005, **293**, 41.
23. S. N. Khaderi, J. M. J. den Toonder and P. R. Onck, *Langmuir*, 2012, **28**, 7921.
24. B. Cichocki, R. B. Jones, R. Kutteh and E. Wajnryb, *J. Chem. Phys.*, 2000, **112**, 2548.
25. C. Maul and S. Kim, *J. Eng. Math.*, 1996, **30**, 119.
26. N. Liron and S. Mochon, *J. Eng. Math.*, 1976, **10**, 287.
27. J. Rotne and S. Prager, *J. Chem. Phys.*, 1969, **50**, 4831.
28. J. R. Blake, *Proc. Camb. Phil. Soc.*, 1971, **70**, 303.
29. A. Vilfan, *Eur. Phys. J. E*, 2012, **35**, 72.
30. N. Osterman and A. Vilfan, *Proc. Natl. Acad. Sci. U.S.A.*, 2011, **108**, 15727.
31. D. J. Smith, J. R. Blake and E. A. Gaffney, *J. R. Soc. Interface*, 2008, **5**, 567.
32. M. A. Sleigh, *Protozoa and Other Protists*, Press Syndicate of the University of Cambridge, Cambridge, 1991.
33. A. R. Shields, B. L. Fiser, B. A. Evans, M. R. Falvo, S. Washburn and R. Superfine, *Proc. Natl. Acad. Sci. U.S.A.*, 2010, **107**, 15670.

Subject Index

Note: Page numbers in *italics* refer to figures.